WHAT REVIEWEI

'As unscholarly as my reaction to this book may sound, it can best be summed up with, I love it. I know of no source that gives the kind of reliable and cogent data, relevant context and reasonable interpretation that I find in *Smelling Land*. Professor Scott's facts are straight and his conclusions are logically based on those facts—something very rare in most of today's energy/ environmental publications. His unique ability to "connect the dots" helps clarify and illustrate how each issue relates to the wider system.

Anyone with an interest in energy and environmental issues needs to read *Smelling Land*. Similarly, every university, corporation, and governmental body should have a copy in their library.'

—BEN BALL
Adjunct Professor of Management and Engineering (retired), MIT
Vice President, Strategic Planning (retired), Gulf Oil Corporation

'*Smelling Land* presents a passionate and compelling rationale for hydrogen—the "gold standard" of energy currencies. David Sanborn Scott eloquently makes the point that the urgent global challenge posed by energy use and climate change is not just about energy sources and applications. The challenge hinges on the energy carriers (or using the useful Scott-coined word, energy "currencies") that move energy from its sources to its services.

Smelling Land's anecdotal style coupled with uncompromising science will appeal to scientist and engineering specialists, yet it is easily accessible to general readers. This is a book that will both delight and surprise. But be warned—it will change your view of the future of planet Earth.'

—DR. PHILIP COCKSHUTT
Director, Energy Research,
National Research Council of Canada, 1976-1986
Executive Director, Energy Council of Canada, 1991-1997

'David Scott's new book, *Smelling Land*, takes the reader beyond the current series of books that describe the science of climate change and the need for reducing greenhouse gas emissions into a discussion of the economy and technologies needed for a sustainable future. Its critical virtue is that it provides not just suggestions or targets for incremental changes, but a practical blueprint for something fundamentally different: a solution to problems rooted in our current industrial and community practices. For those who take the climate challenge seriously, *Smelling Land* is a book that must be read.'

—THE HON. DAVID ANDERSON
Director, Institute for the Environment, University of Guelph
Minister of Environment for Canada, 1999-2004

'Scott takes us on a personal journey of discovery, his odyssey, that led him to conclude hydrogen and electricity will one day become the twin currencies that drive civilization. He carries us along through each stage of that journey as a mentor, so we understand how he arrived at his conclusions without imposing those conclusions on us.

Scott is a prophet in the old sense, but he's not a soothsayer. Instead, he exhorts us to be clear in our language and thinking as we confront today's energy and environmental challenges. This is his greatest gift: teaching us how to look at challenges from all sides, inside out and backside-to-front, and teaching us how to think. We can each draw our own conclusions about the potential for hydrogen, but Scott gives us the tools to consider, to imagine and to assess, so that we can set sail on our own odysseys.'

—DR. MARK PETRI
Program Manager, Emerging Energy Technologies
Argonne National Laboratory, Chicago

'In my opinion, this book will be unanimously judged a compelling and seminal exposition of our energy future, and a formidable guideline for civilization's handling of the crucial climate disruption threat now upon us. . . This book is unique among its peers in addressing the problem of climate disruption while naming hydrogen as the sole remedial solution.'

—WILLIAM J.D. ESCHER
From his review in the International Journal of Hydrogen Energy

'David Scott's perspectives and writing style are unique, and from my reviewing process I can say that **Smelling Land** is as technically sound a book as it's possible to have. Speaking as one who has spent the last two decades working on the issue of how to provide sufficient yet sustainable energy supply, I find **Smelling Land** important, thought-provoking and timely.'

—DR. ALISTAIR MILLER
Senior Scientist Emeritus
Atomic Energy of Canada Limited

'David has done what he does best: "use language as the machinery of thought" to detail the future of energy systems. A must read for an insight into the Hydrogen Age by the author who created much of the nomenclature. The chapter on "metastability" strikes me as being of special significance.'

—DR. GEOFFREY BALLARD
Co-Founder, Ballard Power Systems
Member, Order of Canada

'*Smelling Land* is a wonderful example of the "energy-of-change": hydrogen energy. For years, David Scott has thought about hydrogen systems, their technologies and infrastructures. Hydrogen provides storability and transportability for renewables; it helps decarbonize fossil fuels; and, through fuel cells, hydrogen can power vehicles and generate electricity in homes. My hope—no, my conviction—is that *Smelling Land* will reach readers all over the world, not just scientists and engineers. . . Scott knows his physics and his wealth of language offers many exciting hours of reading. Truly, Scott is an *homme de lettres.*'

—DR. CARL-JOCHEN WINTER
Vice-President for Europe, the International
Association for Hydrogen Energy
Vice-President, German Aerospace Centre (DLR), 1972-1992

'*Smelling Land* needs to be read by everyone working in the energy area—hydrogen, solar, nuclear, fossil. The book ties them all together. I also recommend it for all students entering or thinking about entering engineering. This is not a text book, it is a technological philosophy book that can steer career choices.

To me, Part Four, "Earth's Energy System," is the heart and soul of *Smelling Land*. It explains the concepts of energy, entropy and exergy in a clear, simple, even lyrical way. A gut-level appreciation of these concepts is essential if we are to think seriously about Earth's energy systems, upon which humankind's survival depends.'

—DR. KENNETH SCHULTZ
Operations Director, Energy Group
General Atomics, San Diego, CA

'In an engaging, highly readable but never trivial manner, **Smelling Land** sets out how our energy system really works. David Scott's credentials for doing so are impeccable. He has served at the highest levels of the international hydrogen community. The book can be read and enjoyed at several levels by science and non-science audiences. David Scott is known for his almost clairvoyant ability to see the forest (energy systems) above its trees (individual technologies)—and then to explain how the trees are evolving.

Smelling Land is timely and urgent. To my mind, it gives the single comprehensive strategy that can attack climate change. I know of no other book that does this so well—and none which so well equips the reader to judge new and stimulating ideas.'

—**A.K. (SANDY) STUART**
Chair, Export Development Canada, 1993—1997
Member, Order of Canada

SMELLING LAND

The Hydrogen Defense
Against Climate Catastrophe

David Sanborn Scott

ENHANCED EDITION

Library and Archives Canada Cataloguing in Publication

Scott, David S. (David Sanborn), 1935
    Smelling Land : the hydrogen defense against climate
catastrophe / David Sanborn Scott.

Includes bibliographical references and index.

ISBN 978-0-9809674-0-1

    1. Hydrogen as fuel. 2. Energy development--Environmental aspects.
3. Power resources--Environmental aspects. I. Title.

TP359.H8S382 2008   333.79'4   C2008-901616-5

Published by the Canadian Hydrogen Association www.h2.ca

Interior design: The Queen's Printer of British Columbia

Printed and bound in Canada on 'Silva Enviro Edition Natural,'
which is composed of 100% post-consumer fiber.

All truth passes through three stages.
First, it is ridiculed.
Second, it is violently opposed.
Third, it is accepted as being self-evident.

ARTHUR SCHOPENHAUER

To the memory of my parents,
Alberta Sanborn Scott and Gilbert Beresford Scott

PREFACE TO THE ENHANCED EDITION

The enhanced edition includes a new chapter, 'If Shocked Awake, How Fast Could We Get to the Hydrogen Age?'

An additional appendix, 'Bits and Pieces,' covers topics that came alive during media interviews following the first edition, issues that should have come to prominence after the December 2007 United Nations Climate Change Conference held in Bali, and a few topics that I simply thought were fun—like the tie between the invention of steam engines and the abolition of slavery. I've also borrowed an idea from home-movie DVDs, to identify, at appropriate points in the text, how readers may access 'bonus chapters' that expand on various concepts. These were chapters left on the cutting room floor.

Of course, like any author, I also welcomed the opportunity to correct typos and to have another shot at getting-the-words-right.

DAVID SANBORN SCOTT
VICTORIA, CANADA
FEBRUARY 2008

XIV

Preface to First Edition

Smelling Land: *The Hydrogen Defense Against Climate Catastrophe* is about civilization's energy system and how it impacts climate, and how that impact can be greatly reduced.

Smelling Land also touches on the energy systems of life and of Earth. It's intended for poets, historians, legislators and industrialists every bit as much as for engineers and physicists. It's a story that explains why using hydrogen is the only way we can harvest clean, sustainable sources so their energy can be used to fly airplanes and power such free-range vehicles as cars, trucks and ships— all without releasing a drop of carbon dioxide into the atmosphere. And it's a story that explains why, by using hydrogen, we'll be following the trail Nature blazed when she brought Earth alive.

During the turmoil of wondering how to write a book for both general and technical readers, I happened to be reading Richard Dawkins' *Climbing Mount Improbable*. In his introduction, Dawkins thanked Charles Simonyi for his counsel on writing for a broad audience: 'Try to inspire. . .with the poetry of science [and engineering] and make your explanations as easy as honesty allows but at the same time do not neglect the difficult. Put extra effort into explaining to those readers prepared to put a matching effort into understanding.'

It is to this that I have aspired.

David Sanborn Scott
Victoria, Canada
March 2007

CONTENTS

🌶 Hot peppers indicate somewhat more demanding chapters

Part One

SMELLING LAND

1. A CABIN BOY'S LEGACY

In which we begin to see things the other way around.

I no longer stroll a tidal shore without thinking of the cabin boy. Without wondering about his fate and why that fate was to be swinging by his neck, from a yardarm, dead, when the British fleet struck the Isles of Scilly.

That cabin boy has stuck in my head since 1974 when my young son Doug and I were part of a small crew bringing *Samphire*, a 39-foot sailboat, from Cork to Southampton. We had sailed south from Ireland to the Isles of Scilly that lie off the western tip of England and had turned east in a gentle breeze, but with a gale in the forecast. Day's end in a small boat at sea with uncertain weather brings a curious mixture of tranquility and foreboding.* Astern, the Sun and the Scillys slid below the horizon. Looking aft, our skipper, Sammy, mused, 'You know, during a thick fog a famous British admiral drove his fleet hard aground on those islands.' Then Sammy spoke of a footnote to that history.

It is told that when the fleet was still some distance off the islands, a cabin boy came to the admiral's stateroom to say, 'Sir, I smell land. I think we should heave-to until the fog clears.' Advice from a cabin boy to an admiral of the Royal Navy was neither expected nor welcome—especially not in 1707, and especially not to Admiral Cloudesley Shovell. The boy was reprimanded and sent away. Yet he must have gone on deck for another whiff, because soon he was back at the admiral's door, no doubt apprehensive, but not enough to stop him repeating his warning, 'Sir, I smell land. I think we should put about.' And that is why he was swinging from the yardarm when the fleet crunched ashore on the land he had been smelling.

It's an interesting tale of the era. But for me the intriguing part of this story is the question: What was the cabin boy smelling when he said he was smelling land? He was smelling *life*. Life in the intertidal zone, that marvelous jumble of seaweed and kelp, crabs and mollusks—all of whom make their home

---

* For those interested in yachting history, a Force 9 storm did develop that September night and led to the sinking, with loss of life, of British Prime Minister Edward Heath's yacht *Morning Cloud III*.

between the high and low watermark of tidal coasts. The cabin boy lived his life on the sea—in those days cabin boys were seldom allowed ashore even when in harbor—so a whiff of damp seaweed was a whiff of land. Yet what he was smelling, of course, is exactly what we smell when we say we are smelling the sea.

The cabin boy was correct, but his was unconventional whiffery.

So for me, 'smelling land' became a metaphor for seeing things the other way around. Sometimes smelling land is just raw fun. Sometimes it brings the exquisite and lonely joy of stumbling upon a new insight. Sometimes smelling land is helpful. And some-times it is essential if a problem is to be resolved or an opportunity captured.

Taken as a parable, smelling land also warns of immediate dangers to the smellers, at least if they dare speak of what they have smelled, or worse, of the implications. Fortunately in modern times, at least in most places, the penalties have been reduced. We no longer swing people from yardarms. People may lose jobs, or promotions or the warm feeling of being taken seriously. Academics certainly lose research grants. Smellers-of-land on the corporate ladder are shunted off to staff positions—thought safe on the flanks, but dangerous to corporate stability when in positions of line responsibility.

The parable also warns of longer-term dangers to ships of state, corporations, or large institutions should they discourage the whiffers or trivialize what they say. And if we want to understand how civilization's energy system is evolving, if we want to avoid dangers and capture opportunities—because the system and its future are rich in both—smelling land is essential.

2. CHARTING THE COURSE: TOWARD A CLEANER, RICHER HYDROGEN AGE

In which we review the purpose of our voyage and begin to learn the ropes.

Two specters threaten civilization.

First is the prospect of environmental and economic devastation caused by unprecedented climate volatility and change. Until the 20th century, for more than 600,000 years the carbon dioxide (CO_2) content of Earth's atmosphere had *never* gone above 300 parts per million by volume (ppmv). Throughout this enormous stretch of time, carbon dioxide and Earth's temperature have tracked each other like two dogs on a leash. As CO_2 levels rose or fell, so have global temperatures, through both ice ages and times when we would have been basking at the poles. Now human activity has pushed atmospheric CO_2 to more than 380 ppmv. Moreover, no matter what we do, it is sure to reach at least 450 ppmv, which will be 45% higher than at any time during those 600,000 years. We are in trouble, the severity of which is impossible to overstate.

The second threat is a panorama of collapsing nations, caused by the local depletion and global maldistribution of quality fossil fuels, especially oil. Because at any moment, exploding geopolitics could shatter international oil trade and strangle industrialized nations.

If we remain tethered to fossil fuels, the first peril *will* become reality and the second might. So for Earth's geophysiology and the civilization it sustains, climate volatility is much more critical than depletion. Indeed, long before we could burn all our fossil resources in the ground, the resulting climate disruption will have destroyed our way of life. So it's the global environment, not depletion, which slams an absolute cap on fossil-fuel use.

Either climate catastrophe or oil strangulation will bring vicious worldwide wars as nations, cultures and peoples fight for survival.

It might already be too late. But if we have a chance, that chance is built upon a single straightforward strategy. We must rapidly adopt technologies that use only the twin, carbon-free energy *currencies*, hydrogen and electricity, and these currencies must be manufactured from sustainable energy *sources* that do not emit carbon dioxide. Together, hydrogen and electricity can provide the full menu of civilization's energy services—from flying airplanes and fueling cars, to powering computers and heart pacers, and to any service that may come along in the future. Electricity will continue to dominate communication. Hydrogen must come to dominate transportation.

Transportation is the real problem. We seem to have lulled ourselves into believing we can resolve the transportation issue by *reducing* fossil fuelling. But that's not enough. We must eliminate all fossil fuelling.

There is another misconception. Some folks think our problems can be solved if we use carbon-free energy sources like wind, solar and nuclear power. Unfortunately, far too few people then go on to ask: How can these carbon-free sources power free-range vehicles like airplanes? The answer is: By using these sustainable sources to harvest hydrogen from water—and then using the hydrogen to fly those airplanes—and push any other vehicle now pushed by gasoline. Remarkably, there is no other option.*

Today, we can look back on the wonderful things bequeathed by the first hundred years of electricity. And we can look ahead to an equally wonderful first hundred years of hydrogen. Hydrogen will not push aside electricity because these two energy currencies are synergistic. By the time we've reached the 22nd century, electricity and hydrogen (hydricity)† will have become our staple energy currencies. From today's vantage—a time when many people think our energy future has so many uncertainties—it will be surprising to learn that once hydricity is established, it will be with us as long as civilization survives.

---

* You might think I've forgotten about 'electric' cars, by which we mean vehicles whose motive energy is carried in batteries. I haven't. Battery vehicles can work for limited ranges and if we have enough re-charging time. But they can't replicate the performance of today's vehicles. And certainly won't work for ships or airplanes. You'll find fundamental performance comparisons of different drive trains set out in Chapter 38 'Fuelcells: Chip of the Future.'

† See Appendix A for the origins of the word 'hydricity.'

Together, hydrogen and electricity can take us to a brighter tomorrow—at least for those of us living in developed nations.

About the prospects for developing nations, I'm less confident. Still, what is certain is this: If we in the industrialized nations do not get our energy-environmental house in order, the rest of the world has no chance at all. Absent our house in order, we will neither be able to help less fortunate people nor serve as role models, as we seem so fond of claiming. Without appropriate energy systems in rich nations, the poorer nations will, at best, go through development paths that mimic those we've followed—consuming immense quantities of increasingly low-grade fossil sources, spewing carbon dioxide into the global commons, doing their bit to bring on climate instabilities that will ultimately destroy the economies, environments and quality of life for us all.

If you think that last sentence an overstatement, try to imagine the effect of the Gulf Stream shutting down, wildly distorting the climate of northern Europe and eastern North America and, because the Atlantic oceanic conveyor powers the circulation in all oceans, changing the climates of the entire world. Or try to imagine rising ocean levels that ultimately force the evacuation of the Netherlands, Florida, Bangladesh and more than half the world's largest cities. And while you're at it, try to imagine the military conflicts that would result.

These last few paragraphs are filled with declarative statements. *Smelling Land* will set out the rationale and build the understanding that, I believe, will sustain these statements. Still, my dominant purpose is to encourage rationality. It's rationality that's important.

Thinking is better than suffering, so it astonishes that so many highly motivated, well intentioned people believe personal deprivation is a prerequisite for making the world a better place. Adjusting lifestyles to serve society makes sense. Unthinking deprivation doesn't.* In fact we'll find the many themes of *Smelling Land* will coalesce into one: Cleaner is richer.

But why hydrogen and electricity? Unique among energy currencies, both hydrogen and electricity can be manufactured from *any* energy source, which

---

* Chapter 3 'Turning out the Lights' gives but one illustration.

of course includes all non-fossil sources. When used, neither currency emits carbon dioxide. And they're interchangeable. Hydrogen can be converted into electricity, electricity into hydrogen. This interchangeability will provide a flexibility and robustness that are largely absent from today's energy system.

Today, climate scientists and the media's reporting on climate change speak almost solely to the phenomena causing the problem, to the changes we might expect, and to the coping strategies we might adopt. Sometimes people speak to important but wholly insufficient, actions—like using fuel-efficient vehicles or windmills. Almost no one speaks of the single overarching strategy, hydrogen systems, that can give us a chance to escape climate catastrophe.

Although conceptually straightforward, implementing a hydricity strategy will require a major evolution in our energy system infrastructures. It will also require time—up to a decade for the first components, more than half a century for others. In contrast, we can quickly introduce helpful but insufficient starter tactics (such as hybrid cars and windmills). But while deploying starter tactics, it's critical that we simultaneously begin to implement longer-term sustainable hydrogen systems. It can be done. And that is what *Smelling Land* is all about.

Those readers new to thinking of hydrogen as the essential response to the climate threat may believe hydrogen systems are only for a distant future. So I should briefly remark on how hydrogen is used today. Since the 1920s, pipelines have crisscrossed Germany's Ruhr Valley carrying hydrogen from one industrial plant to another. Today, they also crisscross France, much of the US Gulf states and many other regions of the world. Most of us know the hydrogen-oxygen pair powered the flights to the Moon—and today send up satellite after satellite. Immense quantities of hydrogen are also used to manufacture ammonia for fertilizer and for upgrading and refining oils. All industrialized nations have laboratories, industries and universities developing H_2-fueled technologies, from fuelcell flashlights to hydrogen storage tanks. And each year more and more cities introduce hydrogen-fueled buses. Still more hydrogen (H_2) is used for manufacturing float glass and microelectronic chips, for hydrogenating foodstuffs and so on.*

If you haven't been thinking about using hydrogen to fuel transportation, don't be dismayed. Few people have. Yet it's now poised to be the fuel that reshapes the 21st century.

Readers coming aboard *Smelling Land* are, in many ways, similar to a crew climbing aboard a sailing ship during the 17th century. Some will be seasoned sailors—in our case, scientists and engineers. But for many of our crew— those without formal technical training but with a lust for adventure and understanding—this will be their first ocean passage. Landsmen will want to quickly learn the ropes—or at least the most essential ropes. Most seasoned sailors will want to strengthen what they know, so they can better enjoy the new landfalls we'll find along the way.

To prepare for our journey, I'll take a bit of time to review some of the things we already know, a few ideas we once knew but have misplaced in the attic of our minds, and a few concepts to which we've not yet been exposed. In short, I'll use the rest of this chapter for shipboard orientation.

What is hydrogen? Hydrogen is Nature's simplest element. An atom of hydrogen has one electron circling one proton. It can't get simpler than that. Hydrogen is therefore the lightest element in the universe—which gives it great advantages as a fuel for airplanes and why it was used to fly us to the Moon. At normal pressures and temperatures hydrogen is a low-density gas. So carrying hydrogen on board vehicles is, perhaps, the major challenge to using it as a transportation fuel. When hydrogen is carried aboard road vehicles today, it is usually compressed and stored in high-pressure tanks. But when hydrogen fuels rockets, it is first liquefied by chilling it to very low temperatures, about 20 K (-253°C

---

* In excess of 41 billion kilograms of hydrogen is produced annually throughout the world. This represents a value in excess of US$700 million. This data may be found in *Chemical Economics Handbook*, SRI Consulting, 2003 and *Canadian Hydrogen, Current Status and Future Prospects*, DALCOR Consultants Ltd. 2004.

or -424°F). Later I'll explain why I expect liquid hydrogen (LH$_2$) will ultimately become the staple fuel of transportation.

Hydrogen is invisible, odorless and non-toxic. It contains more energy per unit mass than any other chemical fuel—almost three times as much as gasoline. And unlike the waste product from fossil-fueled technologies, the only waste product of hydrogen-fueled technologies is pure water.

Unlike fossil sources, pure* hydrogen (H$_2$) is not found naturally on Earth—so we must *mine* it from materials like water (H$_2$O) or methane (CH$_4$). Mining hydrogen requires an energy source. When, for example, the hydrogen is mined from water, we need energy to do the mining. So how we 'get' hydrogen is important. If both the energy for mining, and the material from which it's mined, are clean, then using the hydrogen will leave a much cleaner world. But when we use dirty sources of energy or material—as we do for most hydrogen production today— then, although using this hydrogen might help, it's not environmentally benign.

As with any new fuel, people should ask about safety. Reporters will surely drag out the *Hindenburg* airship accident to show, they will say, that hydrogen is dangerous. Yet hydrogen had nothing to do with the *Hindenburg's* fall from the sky. Indeed, hydrogen may be the safest of fuels, as will be explained in Chapter 33 'But Doesn't Hydrogen Explode?'

I must also say what hydrogen isn't—at least the hydrogen we discuss in this book. A *very* rare form (isotope) of hydrogen is tritium. Tritium does not occur in Nature and is both difficult to manufacture and distinct from normal hydrogen. Yet because tritium is one ingredient in thermonuclear weapons, some people have nicknamed these devices 'hydrogen bombs.' Nothing in this book has anything to do with thermonuclear weapons. Rather, we'll be talking about how ordinary, garden-variety hydrogen can fuel cars and airplanes, skidoos

---

* 'Pure' means hydrogen not tied to another chemical element, such as oxygen (O) in water (H$_2$O).

and tractors, even heart pacers and cell phones—and how it can also be used as a feedstock for many commodities. Nothing to do with bombs.

The Hydrogen Age versus the hydrogen economy: When I use the phrase 'Hydrogen Age,' it describes a time when hydrogen will be the dominant energy currency. By then, hydrogen systems will have visibly and sharply reshaped the delivery of energy services—by replacing an old, dirty, inefficient and systemically brittle Fossil Age with a clean, efficient and flexible Hydrogen Age.

I'm less attracted to the term 'hydrogen economy' because our economy will always be composed of many facets. We sometimes describe the second half of the 20th century as the Electricity Age, because electricity-powered technologies were the dominant 'new thing' reshaping culture. Yet during these years, many aspects of our economy had nothing to do with electricity.

What is a fuelcell: In spite of what you often hear, a fuelcell is not an energy source. It is a technology that converts a fuel's energy, most often hydrogen, into direct-current electricity. Like any fuel, hydrogen gives up its energy by combining with oxygen, usually drawn from air. Like a microelectronic chip, which needs electricity to function, a fuelcell needs a fuel to function. (The remarkable analogies between fuelcells and microelectronic chips will be discussed in Chapter 38 'Fuelcells: Chip of the Future.')

The technology I've called a fuelcell is, today, usually called a 'fuel cell.' I prefer the composite word for several reasons. When a noun constructed from two nouns comes into general use, we often first hyphenate, then amalgamate—such as home work, home-work and homework. Moreover, if the double-noun name is particularly awkward in usage—as in 'a hydrogen-fueled fuel cell'—it seems there are good reasons to accelerate the coming word fusion.

What is electricity? At one level, electricity is simply a group of charged particles 'on a roll' through electrically-conducting material. In our energy system the charged particles are commonly electrons, each with a single negative charge. We should understand why the charged particles are rolling: Particles with opposite charges attract each other. This means negatively charged electrons try to move to where there is a larger population of positive charges. Electrons can only travel though electrically-conducting material, typically metals. They can't pass through non-conducting material, such as rubber. So we use electrically-conducting materials to give pathways to the electrons, and insulating materials to set boundaries on their travels.

A few more features of electricity, including how it delivers power and the difference between alternating current and direct current are discussed in Appendix B-1. But we won't need these ideas until much later in our voyage.

What is natural gas? Natural gas is another fossil fuel. It was given the adjective 'natural' because it came naturally from the ground. About a century ago, natural gas began to replace 'coal gas,' which was then in common use. Coal gas contains a mixture of carbon monoxide (CO) and hydrogen. And because carbon monoxide is a poison, coal gas was poisonous. Natural gas doesn't contain poisons.

Natural gas comes in several grades and usually contains different impurities, such as sulphur compounds. Methane (CH_4) is its dominant ingredient. As its chemical formula tells us, each methane molecule contains one atom of carbon and four of hydrogen and so methane has the highest ratio of hydrogen to carbon of any hydrocarbon fuel. Therefore, when burned, it emits less carbon dioxide (per unit energy delivered) than any other fossil fuel. Like hydrogen, methane is odorless. So to ensure your nose warns you when natural gas leaks, odorants (trace nose-tweakers such as mercaptons) are put into the gas shipped into your home. When you get a whiff of natural gas you're smelling the added odorants, not methane.

What are greenhouse gases? The dominant greenhouse gases are water (H_2O), methane (CH_4) and carbon dioxide (CO_2). For now, all we need to know is that they consist of molecules that usually contain three or more atoms. Per molecule, greenhouse gases trap more outbound thermal radiation departing Earth than do the gases that overwhelmingly make up Earth's atmosphere—the diatomic molecules oxygen (O2) and nitrogen (N$_2$).

Energy and exergy: During the early part of this book, I'll only use the word 'energy.' Energy is a familiar word, so we all think we know what it means. But when we get to Part Six, 'Earth's Energy System,' we'll find that energy doesn't behave the way we often expect it to. The result is that much of our common understanding of how the energy system works is, in fact, a misunderstanding. Fortunately, during the first part of our voyage, the way we use the word 'energy' shouldn't mislead.

Part Six explains what exergy is and how it differs from energy. This will allow deeper and sometimes critical insights into how energy systems work.

Laws of Nature and laws of people: When joining a ship's crew, old-time mariners quickly learned the difference between shipboard laws and laws enforced ashore by the local constabulary. Similarly, before we set out smelling land, we should be clear on what distinguishes the laws of Nature from the laws of people.

A law of Nature is a truth humankind has uncovered about how Nature works. If we got the law right, it is never broken. In contrast, laws of people are developed precisely because they can be broken. We don't need to legislate that energy, mass or momentum must be conserved, because whether or not it's legislated, our universe always behaves that way. In contrast, human laws are imposed to describe things that *can* be done but which we *don't want* done, like murder. What we've uncovered to be Nature's laws are descriptive. What we've chosen to be civilization's laws are prescriptive.

This sharp, but frequently forgotten difference too often confounds jurisprudence. Let's be sure it doesn't confound us.

Every odyssey has unpredictabilities. Yet it makes for a happier ship if everyone knows the general sailing plan, even if we can't anticipate all the adventures we'll have along the way. In the voyage that is *Smelling Land* (enhanced edition), we'll start with a new way to enjoy watching how energy systems work, and then tighten up what we mean by environmental intrusion and sustainability. Next, we'll consider Earth's energy system to discover what keeps our planet, and us, alive. All the while, we'll be building a platform that will allow us to better appreciate why and how our world can be on its way to a brighter future. Astonishingly, we'll also discover that for the last five billion years, Nature has been preparing our world for a cleaner, richer Hydrogen Age.

Part Two

Civilization's Energy System

3. TURNING OUT THE LIGHTS

In which a parable reminds us that words shape action, and points to the dangers of ignoring the system and forgetting the services.

Smack in the middle of news-rich October 1973, the Organization of Petroleum Exporting Countries (OPEC) skewered the world's economies with its famous oil embargo.* It was not OPEC's first oil disruption. It was just the first we remember, because it worked. Which meant it had to be *named*. In the autumn of 1973 there were many things to be named—'Watergate,' 'The Saturday Night Massacre' and the 'Yom Kippur War.'

To catch the mind, a name must be simple and memorable. Finding a straightforward, pay-attention name for the oil embargo seemed easy. News dispensers must have thought more people would understand the importance of energy than of oil, so the adjective was changed from 'oil' to 'energy.' Then the noun was changed from 'embargo' to 'crisis,' because crisis jumps with pay-attention. *Voilà*, we had an 'energy crisis.'

I doubt chicanery drove the renaming of the oil embargo. Instead, I expect it was our desire for simplicity and zing—which are not bad intentions if what we rename remains clear. But this time our renaming caused confusion. Everyone decided that having an energy crisis meant we were running out of energy. And if we were running out of energy, it became socially responsible to 'save' energy—any kind of energy, as long as we saved it.†

Everyone likes to be thought socially responsible, especially those for whom survival depends on public approval. Large institutions and governments, the

---

* In their book *Energy Aftermath*, Tom Lee, Ben Ball and Richard Tabors point out that the supply disruption never achieved embargo status.[1] Rather, these were short-term supply discontinuities caused by modest productivity reductions aimed primarily at the United States and the Netherlands, but aggravated by confusing, inefficient and ineffectual regulations for marketing and distribution within these countries and made worse by panicked political leadership and excitable media.

† The pervasive, troubling misunderstanding of 'energy conservation' is the subject of Chapter 21 'Conservation, Confusion and Language.'

great adopters of conventional wisdom*, began developing policies appropriate for an energy crisis rather than for an oil embargo. Policies whose echoes reverberate today. But for now, let's stay with the 1970s.

During those years I was living in Toronto, Canada's largest city and the capital of the province of Ontario. Ontario was blessed with one of the world's great electric utilities, in those days called Ontario Hydro—a name born of times when most electricity was generated hydraulically from Niagara Falls, and a name retained long after other sources of electricity became more important. In the 1970s, Ontario Hydro was the second-largest electric utility in North America, and served a land area about ten times that of England and more than twice that of Texas. Three energy sources provided its electricity. More than one-third came from hydraulic sources, about another third from uranium and a little less than one-third from coal. (Today more than 50% comes from uranium.)

Usually several years pass before large institutions catch up to sudden shifts in conventional wisdom. Annual reports tell much about a corporation's emphasis-of-the-day. By 1978, five years after the oil supply disruption, Ontario Hydro had caught up. Its annual report emphasized the utility's responsiveness to the energy crisis. Bound within covers displaying a kindergarten poster admonishing us all to save energy, the report's centerfold showed two photographs of downtown Toronto. Both photos were taken on the same day in early February. The first, during late afternoon while the city was disgorging people, is a magnificent city scape of brilliantly lit skyscrapers set against the cold pink sky of a mid-winter sunset. The second photo was taken later that evening, after dinner, when folks were home relaxing. The city is black. The sky is black. All is black because Ontario Hydro had encouraged its corporate customers—the banks, the large legal and accounting firms, the mining company head offices—to join them in the business of saving electricity by turning out the lights.

---

* In 1956, John Kenneth Galbraith introduced and defined the marvelous phrase 'conventional wisdom' in his book, *The Affluent Society* [2].

But Toronto is cold in February. Sometimes very cold. Lights help warm the offices. In fact, while a 100-watt light bulb may give off only about 5 watts in light, *all* 100 watts delivered as electricity to the light bulb eventually becomes heat. A 100-watt light bulb provides the same heat to your living room as a 100-watt heater.

So, when we turn off the lights we must turn on something else. Usually it is the furnace. And when we turn out the lights, we turn down those energy sources that generate the electricity to feed the lights. In Ontario, that means turning down hydraulic, uranium and coal, but not oil. In 1978, less than 2% of Ontario's electricity was generated from oil. In contrast, oil overwhelmingly fed the furnaces that heated buildings and homes. So turning out the lights backed out renewable hydraulic along with indigenous uranium and coal, and pulled in more foreign crude. Turning out the lights required importing tens of millions of dollars *more* foreign crude.

Yet it gets even sillier.

During late afternoon, Toronto's electricity-powered subways are running fast, jammed with people on their way home for dinner, each person and each train on the hindquarters of the person and train ahead. In a million homes, electric ovens are cooking dinner. The city gulps electricity. Later on, in the quiet of evening, the ovens are off, the trains run less frequently with fewer cars and smaller loads, and the metropolitan area picks at electricity like a child picks at spinach. Later yet, in the stillness of the night when few photographers are taking pictures, even more lights are off. But it's the coldest time of day and the furnaces rumble.

Something else is relevant. Two things dominate the cost of producing electricity: Servicing the capital cost of the generating station and the ongoing cost of fuel to run the generators. The management and maintenance costs are usually small compared with capital and fuel costs. A utility must have sufficient installed capacity to meet peak loads, so any increase in peak load demand is, systemically, very expensive. Once installed, the capital cost remains constant whether or not electricity is produced. So the *marginal* cost of producing more electricity during off-peak times is just the cost of supplying energy to the generators. In Ontario, this is the cost of the uranium or coal or falling water. For coal-fired generation, the coal costs about 50% of the cost of the electricity produced, for nuclear it is about 10%, for hydraulic it is almost nothing. When the electricity grid is drawing part loads, any well-managed utility will use its low marginal-cost sources and shut down generators where the cost of input energy is high.

This means that during times of low electricity demand, the cost of producing a bit more electricity is trivial compared with the cost of delivering extra power at peak load.

Yet it was during off-peak times, times when Ontario Hydro was already shutting down generating capacity because stoves and subways grew quieter, that it encouraged corporate Toronto to turn out their lights so it could shut down even more capacity. Just when it would have been especially easy and less expensive to use renewable hydraulic and other indigenous energy sources to back out foreign crude, out went the lights and on went the furnaces to suck in more foreign crude.

Of course, everyone agreed that Hydro responded gallantly to the 'energy crisis.' What could be more selfless than encouraging people to buy less of your product? Even the kindergarten cohort cheered. But in Canada, turning out the lights in February made the disruptive effects of the oil embargo even more disruptive. Turning out the lights made Ontario more dependent on imported crude, not less.

Throughout North America and most of the industrialized world, other corporations and governments were finding their own ways to do foolish counterproductive things. It is not that Ontario Hydro in the 1970s should be singled out. Rather it is the careless choice of words—the seed of sloppy thinking and the stuff of flawed conventional wisdom—that stands accused*.

---

* Since the first edition of Smelling Land, people often ask me, 'Is this kind of thing still happening?' My answer is, 'Every day!' Recently, several provincial governments in Canada have toyed with legislating a ban on incandescent lightbulbs. The federal government is considering making the ban nation-wide.

Three engineers, headed by Prof. Bryan Karney, Chair of the Division of Environmental Engineering at the University of Toronto, and two of his colleagues have recently completed an analysis showing the different impact of such a ban on three provinces, Ontario, Quebec and

So for me, turning out the lights became a parable that warns: *Words shape actions.* It warns that if we can't describe a problem, we are unlikely to solve it. We might even make it worse. Flipping things over, if we can't describe an opportunity, we are unlikely to capture it, and we might twist it into a problem. Energy system history is shot through with billion-dollar examples.

That is why, watching our evolving energy system—trying to make sense out of what has happened, what could or should happen, sniffing out hidden dangers and opportunities—I've become resolute about the need for care with words. Not so much in applying Miss Grundy's rules about split infinitives or modifiers caught dangling. Rather, care when choosing simple words to keep ideas straight. Sometimes they are difficult ideas. More often they are just unfamiliar or confused ideas. Yet it is exactly those ideas that must be clearly understood if we have any hope for rational public policies or sound business strategies.

The parable of turning out the lights leaves us with a few more notions.

It reminds us that if we want to use less imported crude, or reduce environmental impact, or create jobs, or establish innovation-intensive industries or promote regional development, we must consider the *system*. We cannot optimize anything from strategies born of myopic attention to one piece of the system.

The parable also reminds us that *services* drive the system. Few of us would consider a light bulb a cogeneration technology. But in the winter, in any northern city, it is. It cogenerates two services: Light and heat. If people had thought about the second service and how it fit into the system, they might not have so quickly turned out their lights.

Finally, we are brought face to face with the fact that good intentions alone are not good enough. Good intentions alone can be dangerous. They bring self-satisfaction that can anaesthetize thinking and sideline rationality.

Thinking *is* better than suffering.

Alberta. Because Quebec's electricity is hydraulic or nuclear generated, banning incandescent bulbs will cause a substantial increase in annual CO_2 emissions. In Ontario, with its mix of hydraulic, nuclear and some coal, it will be a close-run thing. In Alberta—where all the electricity is fossil-fuel generated—we should bring on the compact fluorescents[3].

4. A CONCEPT: ENERGY CURRENCIES

In which we are introduced to energy currencies—the invisible part of our energy system, yet the key to understanding where hydrogen fits.

Words shape action. And the story of turning out the lights shows that familiar words, carelessly used, often confound clear thinking.

On the other hand familiar words, carefully used, *can* encapsulate a new idea—if the new way we use the old word is clear from its context. This often happens when new technologies come along. We use, but do not confuse, the old words 'crash' and 'virus' when talking about our computers. 'Software' is a new word that grew out of the old word 'hardware'—previously used to describe hammers and chisels—as both 'hardware' and 'software' took on new meanings when computers became part of our daily lives. Our living language does similar things when hunting for ways to describe emerging cultural phenomena. What would we do without 'spin?'

To understand our energy system and how it works, we need a new concept—the concept of 'energy currencies.' We will need to be clear about how energy currencies differ from energy sources to differentiate between the *role* of such things as gasoline or electricity, which are currencies, and the *role* of such things as crude oil or uranium, which are sources. One difference is that currencies are harvested *from* sources: Gasoline is harvested from crude oil and electricity can be harvested from uranium. But there are many other ways to distinguish between currencies and sources.

If you are in the energy business, about now you might say, 'this fellow is doing a lot of sputtering about what we know are "energy carriers" or "secondary energy sources." What's the big deal?' At first glance there is no big deal. But a closer look will show that mainstream vocabulary camouflages nuances. To miss a nuance can be to miss the point.

In 1980, I realized that terms such as 'energy carrier' were causing us to miss a nuance and, in turn, to miss a critical point. So I began searching for a name that would at least suggest the nuances. That's when the analogy with monetary currencies led me to 'energy currencies.' Monetary currencies—dollars, yen or euros, for example—enable financial transactions, but are not themselves sources of wealth. An energy currency enables energy transactions but is not, itself, a source of energy. That part of the analogy is straightforward. Other parts, the nuances, are more abstract but often more valuable because they lead to insights.

No single monetary currency can be used for every kind of financial transaction. Each currency, be they dollar bills or telecurrencies (electrons running about in microelectronic chips and photons flying to and from satellites), can only be used for certain transactions. For example, a euro coin might be used to buy a chocolate bar. A telecurrency can move a billion dollars from New York to Tokyo in a wink. But the currencies for these two transactions are not easily interchangeable. We would not expect to fly a cargo plane of euro coins to Tokyo to capture a weekend spike in interest rates. And today it would be unusual to use a telecurrency to buy a chocolate bar—although this is not quite as preposterous as it might seem. We are beginning to find chocolate-bar vending machines activated by debit or credit cards.

Following our analogy with monetary currencies, each energy currency can only be used for a *defined range of transactions*. Examples? Well for one, it is difficult to use electricity to fly an airplane and difficult to run a computer on Jet A. Turns out there is another truth embedded in this analogy: The limitations energy currencies impose on energy transactions are usually much more restrictive than the limitations monetary currencies impose on financial transactions. It's more difficult to fly an airplane on electricity than to buy a chocolate bar with a telecurrency.

Next let's examine how the analogy carries over to *convertibility*. Some monetary currencies are easily convertible. It is fairly easy to change dollar bills to telecurrencies, or telecurrencies to pesos. Other monetary conversions can be a bit more difficult, like changing Bangladeshi taka into dollars, especially in Magnolia, Texas. But with time, even in Magnolia, it might be done. Moreover, it's often easier to convert one way than the other. In Magnolia it will be easier to convert US dollars to taka, than taka to dollars.

Energy currencies are also convertible, but the restrictions on convertibility are often more severe. We can easily convert gasoline to electricity—for instance, by using the alternator in your car or a motor-generator at your cottage. But the reverse, directly converting electricity to gasoline, is *not* possible. One-way convertibility is typical of exchanges between many of today's energy currencies. These one-way gates severely restrict our ability to design efficient and, most importantly, robust energy systems.

There are also analogies regarding currency-to-currency *exchange costs*. The cost of monetary currency exchange is the amount we pay for the exchange transaction—say, to cover costs of having an exchange booth at the airport, the salary of the attendant and profits to the owners.

By analogy, we always put more energy into an 'energy conversion booth' than comes out. The difference is the energy price we pay, because the efficiency of any energy conversion device is always less than 100%.* We're willing to pay this price because the *next* currency is what we need for the *next* step in the system chain. Consider the alternator in your car. More energy goes in by the alternator belt than comes out along the wires. But to power the lights and radio we need electricity. Radios don't run on gasoline, so the price we pay for converting gasoline energy to electrical energy is worth it.

When we talk about energy conversion, we speak of efficiency. When we talk about monetary conversion, we speak of exchange rates. Yet by whatever name, the price we pay for energy conversion is usually a higher percentage of the incoming energy than is the percentage we're charged for money exchange.

So for both energy and money there are patterns:

- Individual energy and financial currencies are restricted to certain transactions—and for energy currencies the restrictions are usually more severe.

---

* We can get tricked on this. A heat pump salesperson may tell us a heat pump puts more heat energy into our home than it used in electric energy to run the heat pump. The salesperson is correct. But the heat pump also took energy from the environment. The total input energy coming from electricity *plus* the energy from the environment is always more than the heat delivered to our home. And, of course, we *paid* only for the electricity. That's really what the salesperson is saying. The magic of heat pumps is explained in Chapter 24 'Exergy Takes Us beyond the Lamppost.'

- The currency-to-currency exchange for both energy and money is usually more difficult in one direction than the other—and for energy, it's sometimes only possible in one direction.
- There are always costs for currency-to-currency exchange—and for energy, the cost is usually a greater fraction of the input.

So for any energy conversion process there is always an energy cost. Yet what strikes me is how many times I've encountered well-meaning people, often reporters, who will say something along the lines of, 'I've heard we must use more energy to *make* hydrogen than we get *out* of hydrogen! If that's true, then hydrogen must be impractical.' What nonsense. Yet it shows that many people don't realize that all along the system chain, each time we convert energy from one form to another, we pay a conversion price. These conversions are done to make the system more efficient, not less.

Let's illustrate this concept by following the trek energy takes from a nuclear reactor in a power plant to your TV set. Heat from the nuclear reactor's fission processes is converted into energy in the steam that leaves the reactor. Then a turbine converts the energy of the steam into energy carried by a rotating shaft. Then a generator converts the rotating shaft energy into medium-voltage electricity which a transformer converts into high-voltage electricity (for efficient transmission to cities). When it gets to the city another transformer converts the high-voltage electricity into low-voltage electricity (making it safe enough to enter your home) and then, finally, your TV set converts the electricity into sound and light (albeit not necessarily enlightenment). Each step is from one intermediate currency to the next. And at each step we pay an energy conversion price. But it's worth it, because the *system* benefits, because we can't jam nuclear reactors into TV sets.

I have always been troubled by the public's obsession with finding new energy sources. Sources are seldom the problem. Nature has given us plenty of sources—solar, wind, tidal, geothermal, fossil fuels and uranium. The trick is to harvest these sources at competitive costs, with low environmental intrusion, and in ways that deliver currencies suitable for feeding the technologies that deliver the services we need, or want. Currencies are the *gatekeepers*, allowing (or preventing) sources to power different services.

Proponents of new energy sources, each hoping theirs will help save the world, seldom think about the currencies their source will make. Yet the lack of a suitable currency will always be a systemic showstopper—often blocking the use of new sources *where they are really needed*. How can wind, solar or nuclear sources be used to fly airplanes without first manufacturing a currency suitable for airplanes? They can't. But because this is seldom considered, a massive conceptual disconnect flaws public thinking and the quality of public policy.

In March 1989, an example of this disconnect crashed in upon the news. A new energy source had been found, 'cold fusion'—but there was no thought of the currencies it might make.

Before we talk about this cold fusion experiment, it's good to remember that both hot and (if it exists) cold fusion are nuclear processes, which means mass is changed into energy according to the famous Einstein equation $E = mc^2$. During a fusion process, energy is released when two or more light elements fuse into one heavier element.* The total mass of the heavier element is slightly less than the combined mass of the original fusing elements. The difference is the amount of mass converted into energy—the 'm' in the equation. To calculate the energy released, we simply multiply this difference in mass (m), by the velocity of light squared, c^2. Seems easy, doesn't it?

We know hot fusion works. That is how the Sun works and how thermonuclear bombs work. But bombs release energy with a bang. If we want to use fusion peacefully, we need a peaceful, slow way to release the energy. We must avoid bangs. For many years engineers and scientists have tried to use extraordinarily

---

* Fusion is one of two ways to change mass into energy. The other is fission. In fission, atoms split apart rather than fuse together. In fission, the total mass of the fission fragments is less than the original mass of the unsplintered element. Fission is the process used for 'conventional' atomic bombs, for propelling submarines, and for generating electricity. The erroneous idea of a close relationship between nuclear electricity generation and nuclear bombs—which exists in the minds of many people—is one of the reasons the peaceful use of nuclear power has been so restricted around the world, especially in North America. We'll return to this in Chapter 31 'You've Got to Be Carefully Taught—Know Nukes.'

exotic technologies, employing such things as very high magnetic fields, very high temperatures and pressures—and even higher piles of money—to control fusion so it can deliver its energy peacefully. But so far commercial success for hot fusion remains a long way off.*

Then in 1989, Stanley Pons and Martin Fleischmann reported they had achieved cold fusion, what some called 'fusion in a test tube'—fusion without the need for high temperatures, pressures, magnetic fields and dollars. They had found a new way to harvest fusion as a *source* of energy. The media yelped. Here was our salvation!

Yet during those first days of cold fusion excitement, how many commentators from the ABC, the BBC or the CBC, or journalists from *The Times* of London, New York or Los Angeles asked: If cold fusion works, what currency will it make—and, most important, what services will that currency energize? None asked because they lacked the concept of currencies. It's tough to ask about something when there isn't a name for that something.

To get our world in order we must find a currency that can be used for all our transportation needs and can be manufactured from any non-carbon source. Yet because the concept of energy currencies is so little known, billions of monetary currencies are spent on energy sources while little is spent on developing energy currencies that will allow these sources to be used where they are most needed. Electricity is fine for many things. But as we said, it won't fly airplanes and, in spite of some people's hopes, electric batteries are really very poor at propelling cars.

We don't have an energy shortage crises, but we do have an energy currency crises.

So the question becomes: Is there a currency that can allow energy harvested from wind and uranium to fly airplanes? Happily, the answer is: Yes! It is a currency that can be made from any energy source. It is entirely renewable.

---

* I believe it will be a long way off for a long time. But from having met many of the engineers and scientists trying to develop this dream, I am impressed with their talent, dedication and commitment, and so I hope I'm wrong.

Using it can be environmentally sweet. And it's much the best fuel for airplanes. It goes by the name hydrogen.

5. ENERGY CURRENCIES: THE KEY TO SYSTEMIC LIBERTY

In which we realize that material has been the smokescreen hiding the role of currencies, and that we win systemic liberty by breaking the material bonds between currencies and sources.

What camouflaged energy currencies? Was it a conspiracy of multinational oil and automobile companies, cooking up misinformation, sending us down false paths? When something important is hidden, many of us wonder why. I doubt it was a conspiracy. My hunch is it was nothing more sinister than having little need to distinguish sources from currencies—at least not until recently.

The need to differentiate between these two systemic roles has crept up on us, crept up as the system became more sophisticated and interconnected. Moreover, while language—the machinery of thought—is quick to invent words for new *things*, it is slower to invent words for new *roles*. Roles are more abstract, tougher to keep in the head, difficult to make crisp. Perhaps we haven't yet had time to see the role of currencies taking shape upon the stage.

Material may have helped keep things opaque. In the past, the material of an energy currency was usually the same as its source. It is tough to see the difference between a currency and its source when the stuff of one is also the stuff of the other. More recently, the stuff of one is less likely to be the stuff of the other. Let's look at how this transition took place.

Consider a homesteading family. Trees were the homesteaders' main source of energy. When cutting down a tree for firewood, they'd think it silly to distinguish wood-in-the stove from wood-in-the tree, naming one a currency and the other its source. Yet wood-in-the stove plays the systemic role of a currency, which is spent to warm the cabin or cook food. But in homesteading times this linguistic distinction would have been the puffiest of word puffery.

Then things began to change. The stuff of currencies began to be distinguishable from the stuff of sources. Coal shoveled into the boiler of a steam locomotive belching and hissing its way into a prairie town differed, at least in some ways, from the coal in the ground. Before it got to the locomotive, it had been scrubbed, sorted and broken into shovel-sized chunks. Later, when diesel locomotives displaced steam locomotives, the stuff of the next fuel, diesel, differed even more from the stuff of its source: Crude oil. Still later, as automobiles became our favorite way of getting about, the stuff of gasoline became even more distinguishable from crude oil than diesel. Compared with crude, gasoline is less viscous, more translucent and so on. Gasoline and diesel may both be children of a common crude-oil parent, but of the two, gasoline is the one you'd least recognize as a child. Yet all these currencies still have material ties to their sources. Electricity completely broke the line.

Today, electricity is the currency that feeds our telephones and microwaves, and that dominates almost every aspect of our life except for transportation and keeping warm. The material of electricity (electrons on a roll) bears no similarity to the material of the source used by electricity-generating stations—no relationship at all to the stuff of coal, natural gas, water power or uranium. So a pattern has developed: The progressive decoupling of the *stuff* of currencies from the *stuff* of their sources. This pattern has been unfolding over the past two centuries, becoming ever clearer during the 20th century and now in the early 21st.

The decoupling of a currency from its source material brought systemic liberation. Severing material bonds that had glued source and currencies together for thousands of years allowed us to:

- multiply the *services* that can be provided,
- expand *where* these services can be delivered, and
- increase the number of *sources* that can be used.

Together, all three have brought extraordinary systemic flexibility and business opportunities.

To witness one example of this evolution, let us travel back in time to Niagara Falls—one of many places to watch the dawn of the Electricity Age.

In the early 1850s, the energy from Niagara Falls was used to rotate shafts. These shafts delivered the energy of the falls to any technology that could be attached to the end of the shaft. The technology might be a saw to cut logs, or a millstone to grind wheat. The currency, in this case the power transmitted along the rotating shaft, had been harvested from the energy of thundering water. Of course, rotating shafts can only do their rotating over short distances. Drawings of early industrialized towns show rotating shafts, slapping belts and humming gears that, as the mechanical era reached its zenith, could deliver energy over several miles.

Entrepreneurs of the day first looked for locations to install water wheels that turned shafts. They would then lease the rotating shafts to other entrepreneurs who would attach saws, millstones, lathes, or looms for their own businesses. The lessor (owner of the rotating shaft) filled a role resembling today's gas or electric utilities. The lessee filled the role of manufacturer. But it all had to be done within a few miles of the falls.

Then came the idea of sticking an electric generator on the end of the shaft to make a different currency, electricity. *Voila*, now the energy of Niagara Falls could be shipped all the way to Buffalo or Toronto by transforming mechanical energy (falling water and rotating shafts) into the new energy currency, electricity. Now the energy from Niagara could be delivered several hundred kilometers away. What's more, after it got to these distant places, people could use the electricity to feed electric motors that transformed the energy back into rotating shaft power—like the axle of a streetcar. Electricity not only increased the distance over which Niagara's energy could be used, it expanded the services Niagara Falls could deliver.

If in those days there had been a Society of Streetcar Horses that met to discuss the quality of their hay and other matters of common interest, the society might have sent a deputation to the Streetcar Authority. Fighting to protect their jobs, the deputation would have pled unfair trade practices by electric motors, because motors were using this newfangled currency. Hay was the only currency horses could use, and the only energy source for hay was sunlight. Motors, using electricity, could tap *any* energy source, which, the horse deputation could have argued, gave motors an unfair advantage.

Snapping material bonds meant any source could compete for the business of pulling streetcars. If it were happening today, we might say that electricity allowed the 'deregulation' of streetcar energy sources. A free market among sources had been created. People were free to use the best source they could find—best for cost, availability and convenience, or for the environment. Hay no longer had a monopoly. Streets had fewer horse buns.

However, taking a streetcar to work was not a new service. All electricity did was introduce a new way to power the streetcars. As electricity became an everyday thing, people began thinking of new ways to use it, and eventually that brought a cornucopia of new services. At first these were derivatives of existing ones, although delivered more reliably and safely. One of these services, arriving at about the same time as electric streetcars, was illumination, as people began switching from coal gas and oil lamps to electric light bulbs.

Eventually, electrical technologies began offering services that earlier generations could not have imagined. One way or another, the first of these services involved communication. It started with the telegraph and moved quickly to wireless Morse-code. Yet sometimes in its early years—before we really understood its value—the wireless was turned off. That was the case aboard the *California*, when she was sitting quietly, hearing nothing, doing nothing, while a few miles away the *Titanic* was sinking.*

Years later, the microelectronic chip made electricity-powered communication wildly contagious—leaping continents and oceans, burrowing into automobile dashboards, traveling down mine shafts, flying out into deep space.

One reason information technologies so quickly penetrated every corner of our planet is that electricity can be the child of any energy source. Any country or region can provide electricity, using any available source. A fax machine is blind to the source that made its electricity. In contrast, the Coal and Iron Age of the 19th century was trapped in countries that had both the specific energy

---

* I'm disappointed that, in a movie that cost so much and paid so much attention to creating accurate scenes of now-departed technologies, the directors of the 1997 movie *Titanic* neglected this heart-wrenching true story—to rather focus on a comparatively vacuous, highly improbable love affair, and preposterous scenes like being handcuffed in a sinking cabin and, later, paddling about in 2°C water.

source, coal, *and* iron ore to be smelted by the coal. This allowed Great Britain immense global power for more than a century.

But no single energy source has a lock on the Information Age. This allows the Information Age's benefits to flow easily among all countries, no matter what their energy mix, or even whether these countries are rich or poor in energy sources. Japan and Korea's energy source poverty has not prevented them from exploiting all aspects of the Information Age.

These are the patterns of the past and present. I believe the future will continue snapping the shackles of source-currency material bondage, continue snapping the links between the stuff of sources and the stuff of currencies.

This pattern will persist because more and more services will be provided by electricity. But there is a deeper reason. The next currency that will gain widespread use—and profoundly impact culture—will be hydrogen. In a world shot-through with uncertainty, one thing about which I am most certain is that electricity and hydrogen together, will become the staple currencies of civilization although hydricity* is likely to take the best part of a century to reach full bloom†.

Like electricity, hydrogen contains no material trace of its energy source.

Will source-currency decoupling continue the gift of systemic liberation? You bet. Without hydrogen, most new sources will never be able to power transportation except in trivial ways. So we have another pattern. Today, by manufacturing electricity, any source can supply all parts of the communications market. Tomorrow, by manufacturing hydrogen, any source will be able to supply all parts of the transportation market. Of course, using hydrogen has another advantage. The exhaust is water—often water vapor. A bit of water vapor and nothing else coming out the tailpipe, will move us toward a cleaner world.

---

* See Appendix A for source of the word hydricity.

† Unless the world wakes up to the severity of our climate threat and follows the path outlined in Chapter 45.

As we've observed, when electricity was first introduced, it provided services people already knew they wanted—such as public transportation and illumination. Much later it brought blessings undreamed of by homesteaders. So it will be with hydrogen. Today it is easy to think about, and plan for, hydrogen buses and airplanes.* But we can't divine the many services that second-and third-generation hydrogen technologies will bestow.

We'll talk more about hydrogen throughout this book. For now, the good news is that decoupling the stuff of currencies from the stuff of sources will continue into the future. We've seen it characterize the evolution of the energy system throughout the 20th century. With the introduction of hydrogen, it will continue throughout the 21st century and far beyond. Indeed, while progressive source-currency decoupling may have characterized energy systems during the 20th century, it will be a *leitmotif* of energy systems in the 21st:

- Allowing non-fossil sources to be used where they are most needed, especially in transportation;

- Bringing environmental benefits, economic strength and cultural growth in ways we cannot yet fully imagine; and,

- Spreading all these good things, quickly, around the planet.

These are the patterns that can illuminate opportunity for investors, business strategists, national planners, and for young people choosing careers.

I'll conclude this chapter by reinforcing the idea that to discriminate between sources and currencies is to differentiate between *roles*, not between *things*. The roles things play depend on what part, or how much, of the system we are considering.

To illustrate, if Earth is the system and sunlight its primary energy source, coal may be considered a currency with the special task of storing the sunlight's energy throughout the eons. But if the system is an electricity-generating station, coal reverts to its more familiar role as an energy source. Or think of hydrogen and sunlight. On Earth, hydrogen will be an important currency. But

---

* Chapter 40 'Contrails Against an Azure Sky,' talks about those wonderful birds, LH_2-fueled aircraft.

in the bowels of the Sun, hydrogen is the energy source of nuclear fusion that powers the inner turmoil of our most important star.

Standing on Earth, we think of sunlight as our planet's dominant energy source.* But if we're drifting past Jupiter, gazing wistfully back at our Earth from afar, sunlight assumes the role of a currency, carrying energy from the Sun on an eight-minute trip to Earth. This sunlight, the solar system's energy currency, has a monopoly on powering photosynthesis—upon which all life depends. Sunlight has the unique role of bringing Earth alive.†

---

* Compared to the Sun, extremely minor energy sources include geothermal energy, which is driven by the nuclear decay process deep within Earth's interior, and the tides which are driven by the decay in the Earth-Moon relative motion. (Some have responded to this last statement by saying, 'I thought gravity pulls the tides.' They're correct. Gravity does pull the tides, but it's the decay in the Earth-Moon energy that provides the energy to do the pulling—energy that is dissipated as the tides move over the ocean bottoms and rush through tidal estuaries.)

† When we arrive at Chapter 22 'Entropy and Living Planets' we'll find a much deeper and I think beautiful explanation of how sunlight brings Earth alive.

6. THE ENERGY SYSTEM AND ITS FIVE-LINK ARCHITECTURE

In which we sit back, watching the energy system, marveling at the serenity of its architectural imperturbability, when all the while its innards are stirred by innovation, time and culture.

The strange thing about the energy system is that we almost never think about it as a system. We think about its bits and pieces—about electricity generation or oil cartels, airplanes or fax machines—but not about the overall system. Repeatedly, bits-and-pieces thinking has caused us to miss opportunity and to blunder into misadventure. Funny, because we understand the importance of systemic thinking in so many other aspects of modern life.

Take our financial system, where each of us has an opinion about how almost everything sends ripples through the economy. Some folks are very well paid to explain and anticipate the ripples, like the economists we see on the nightly news telling us how combinations of a cold winter in Europe, El Niño off Chile, and the price of pork bellies in Chicago—or fears of terrorists and hurricanes will influence pension fund performance and, therefore, shape our retirement. And then they tell us how the retirement trends will affect holiday resorts and the livelihood of Polynesian waiters. Economists are burrowed so deeply into the system, they become part of the system.

Most of us have taken a malfunctioning car into a service station because, say, we thought the battery had a problem—only to have a good mechanic tell us it's the alternator, or the voltage regulator, or, more embarrassing, a blown fuse. To find the problem with your automobile, good mechanics draw upon their knowledge of the architecture of your car's electrical system.* Similarly, a good physician will use her knowledge of your architecture to tell you the pain in your thigh results from a pinched nerve in your back. For car maintenance or

---

* When writing this I'm reminded of Robert Pirsig's wonderful, metaphysical trip celebrated in his book *Zen and the Art of Motorcycle Maintenance*[4].

health care, it is no surprise that to search out troubles we need a good concept of automobile or human architecture.

Connections are the essence of how things work—living, technological or cultural things. It's also fun to watch how systems sustain their neighboring systems, connecting and disconnecting, feeding and being fed—a symphony of synergies.

But for energy, the full system—from the sources through to the services it provides—is almost never considered. This makes it tough to anticipate the ripples. Instead, oblivious to the system, we go around turning out the lights.

To understand how our energy system works, we need in front of us, stuck in the mind, a picture of its architecture. We need a vision of how the bits and pieces fit together—what they do for each other, not just what they do themselves. Yet when we begin to conceptualize the energy system's architecture, we run smack up against the question: How much detail do we need? Einstein is reported to have said, 'Everything should be made as simple as possible, but not simpler.' That is a profoundly important instruction.[5]

So we must find an energy system architecture that is as simple as possible, but not simpler. It must be an architecture that can help us hunt for opportunities, solve problems and avoid misadventure—all without tangling our dendrites.

To get started, let's review how people commonly think of the energy system. Take the oil business. People view it as beginning with exploration and production, then refining, and then the marketing and delivery of fuels and petroleum products. Or people might think of the electricity system as starting with an energy source, say coal or uranium, which is fed to a generating station. The station then delivers electricity through distribution and delivery networks right up to the meter at your home. These are everyday views of our energy system, and they can be depicted by the first three links in the chain of Figure 6.1.

Figure 6.1   A three-link systemic architecture running from sources to currencies.

In this chain, the sources might be crude oil, coal or uranium. The harvesting technologies might be oil refineries, generating stations and their distribution networks. The currencies are petroleum products like gasoline and diesel, and electricity.

But these three links don't go far enough. After the electricity gets through the meter and into someone's home, it travels through a toaster, a refrigerator or a TV. These technologies sit in what I've called a 'service technologies' box. In turn, the services they deliver—toasting bread, preserving food, delivering entertainment and information—sit in the 'services' box. If we were following the oil chain, the end service might be transportation. We have now arrived at what people want: Services.

Energy is a means to an objective. Services are the objective.

For me, this obvious, simple idea—that services, not sources, drive the system—took a long time to sink in, a long time to marinate in my head before it became the *first* thing I thought, not an afterthought I tried to remember. I tended to plod through the system, from sources to services, both because that's how we've been taught to think *and* because, as an engineer, I know that's how energy flows. But energy flow flows counter to the way the system works. The system's purpose is to supply services, and all else follows from that purpose.

Yet to think from services to sources required that I think from right-to-left, which I found difficult. So, in Figure 6.2, I flipped the diagram to begin with services on the left. It is easier to redraw diagrams than to retrain a brain which, in the West, has been taught to look at pages left-to-right since kindergarten. Flipping the diagram makes it easier—at least for me—to remember how things work. And seeing how things work is essential for smelling land.

Figure 6.2 A five-link systemic architecture running from services to sources.

During a graduate course I introduced this idea of services driving the system. On the final exam I sometimes included an unusual question such as 'Make up your own question and answer it' (students find this much more difficult than you might expect). Once, to one of these odd-ball questions, a student, Wendy McDonald, gave me a new way to think of the system that I hadn't thought of before. She viewed the energy system as a demand-supply system. People demand a service, the service demands a service technology, the service technology demands a currency, and so on back to energy sources. Why hadn't I thought of that? It's so obvious.

Now let's return, for a moment, to the more common view of energy systems. Perhaps the reason people think of a three-link chain is because, as shown in Figure 6.3, that's what the financial pages of our newspapers call the 'energy sector.' But the financial pages amputate the service technologies and services. Once the utility has delivered the electricity, or the oil company has delivered the gasoline, that's all there is to report. So the financial headlines bleat, 'Electricity Deregulation' or 'Prices at the Pump Rise on Fears of. . .' But as we've noted, people want energy services.

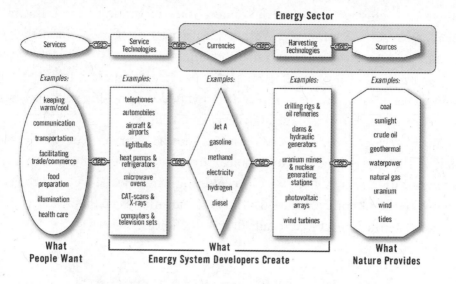

Figure 6.3 The systemic architecture with examples for each link.

We've observed that, although the system is driven from services to sources, the *energy flow* is from sources to services. So there are flows both up and down the chain: Energy right-to-left, demand left-to-right. So in Figure 6.3, I've removed the suggestive arrows to emphasize that direction depends on what you want to consider. In addition, because examples always help visualize the abstract, I've added a few typical ones under each link of the chain. The examples are arbitrary; there are no implied horizontal connections between them.

Figure 6.3 includes some of the services people want and many of the sources Nature provides. As I've said, the three right links are what financial analysts call the energy sector. The three middle links identify some of the technologies people have created. It's evolution *within* these central links that shapes and reshapes culture.

The chain is always driven by the services people want, such as food, comfort, transportation and health care. And it is always supplied from the unchanging menu of sources that Nature provides, including sunlight, water power, coal, oil and uranium. In some regions, different items on Nature's menu of sources may sometimes be in short supply or of lower quality. Other regions may have an abundance of one source and a paucity of another. But overall, the sources

Nature sets before us today are the same as they have always been. But mere existence is not the full story. We must learn how to harvest these sources, economically, and with minimal environmental intrusion.

I believe this five-link architecture is true to Einstein's dictum to keep things as simple as possible but not simpler. Mind you, deeper within each link, sub-architectures nestle beneath the high-level architecture.

Now let's watch the machinery work by choosing a service from the left and tracking the consequences of our choice through to sources on the right. Pick communication—after all, we are the chattering-class apes. If we want to get in touch with grandmother, Figure 6.4 can help our mind's eye follow our choices and the systemic impact that results from our choices.

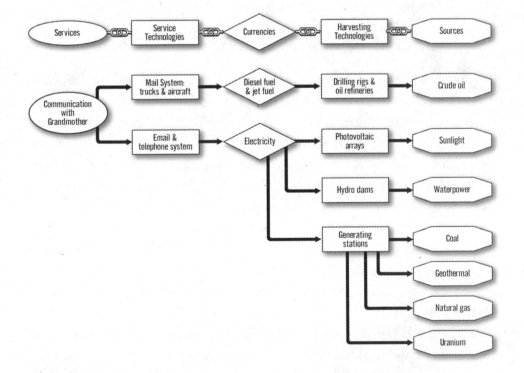

Figure 6.4 Two systemic pathways to deliver your message to grandmother.

If we want to send a message, we have a choice of several service technologies: The telephone, a letter or an email. If grandmother lives across town, we'll likely choose the telephone. But if she lives across an ocean, we might choose the postal service especially if it's a birthday greeting. If a letter is our choice, we've automatically selected postal-service technologies: Trucks, airplanes, sorting machines and so on. Because they use so little energy, let's neglect sorting machines and concentrate on the vehicles that carry the mail. Today, trucks and airplanes use conventional liquid fuels—gasoline, diesel or Jet A—and, moving the next step to the right, we find these fuels in the currency link. Farther right are the harvesting technologies—oil refineries, and their ancillary distribution, delivery and control systems. Finally, we get to the far right, to the sources, where we find crude oil sitting smugly, secure in the knowledge that no other source can push it aside. We might not know the grade of crude, whether it's light Arabian crude or syncrude from the Athabasca oil sands. Yet we can be sure the source won't be hydraulic power, uranium or wind. Once we decide to use the postal system, the chain running back to the energy source becomes frozen.

Services drive the system. They pull the chain. Sources just sit there, waiting to be called.

Next, let's pick a different way to reach grandmother. Today's prototype grandmother might have a computer beside her sewing machine. Then we could send her birthday message by email. Computers are just another service technology stuck on the end of our electricity supply system. As we move from the service to service technologies, and then to currencies, the chain remains frozen. We are without options. But stepping farther to the right after the currency box, we encounter a great thaw. We now have many options among harvesting technologies and therefore among sources.

Electricity can be generated from any energy source. Mostly, generating stations use coal, water power, uranium, natural gas or oil. But in Denmark, northern Germany and on the Golan Heights, they might use wind. In Iceland, they sometimes harvest geothermal, and Saudi Arabia is experimenting with sunlight. When electricity is the currency, we have the liberty to choose from the full menu of Nature's sources. Figure 6.4 shows which pathways point to choices, and which do not.

Our five-link chain can be applied to the energy system since the time of cave-dwellers. If we choose the 19th century, Figure 6.5 shows horses and hay, rotating shafts and waterwheels all under their appropriate links. The five-link architecture is appropriate for yesteryear, just as it is today.

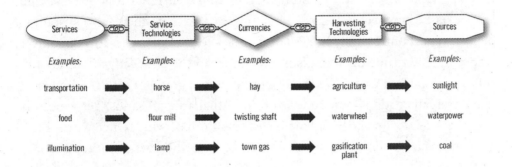

Figure 6.5 Examples of service delivery chains from yesteryear.

The reason to have the systemic architecture stuck in the mind is to use it as a platform for anticipating the future. For that, it helps to know what things change and what things do not. If we reexamine Figures 6.3 and 6.5 we see:

- The five-link architecture never changes;
- The menu of Nature's sources never changes;
- The *categories* of service people want never change, *but*
- The things that occupy the three middle links do change, and,
- The sources we *select* from Nature's menu do change.

Just after the telephone had been invented, some British Rail managers harrumphed, 'It might be needed in America, but we have messenger boys.' In contrast, the starry-eyed mayor of an American city was heard to exclaim, 'Why! I can imagine a day when every city will have one.' In those days, how many of our ancestors traveled back and forth across continents and oceans to visit their scattered families or to enjoy a winter interlude in the Sun? And do you think you would have found, in the pantry of a Canadian homesteader on the snow-driven prairies of Saskatchewan, oranges from Florida, Branston pickles from England and two kilos of French brie? The quantity and quality of services have expanded beyond our forebears' dreams—all the result of changes *within* those three central links. But nothing has altered the timeless architecture.

To sense the cadence of change for the overall system, let's pick a single service—land transportation. Figure 6.6 shows the most common features for almost 300 years—chopped into 70-year intervals, plus or minus a decade or so. The patterns are clear and they can help us imagine the picture for other services—such as keeping warm, communication and illumination.

During the first 250 years of Figure 6.6, once we select the transportation technology the available choices in the three links farther to the right are frozen. For both the 1770s and the 1840s, the no-other-choice source is sunlight. When we get to 1910, the no other-choice source has become coal. For the 1980s, it's oil. But when we arrive at 2050, the staple currency of transportation will be hydrogen—and that will open up a plethora of energy source options.

Earlier, we witnessed this broadening of energy source options when talking about electrical service technologies. Now we see that if the currency is either electricity or hydrogen we open the horn of plenty to Nature's full menu of sources. Figure 6.6 understates the breadth of this horn of plenty and it understates how many options there will be for harvesting technologies. But we can surely see that the 2050s will present many more choices for transportation energy sources than we have today. The 2050s will exhibit the source liberation we described in Chapter 5.

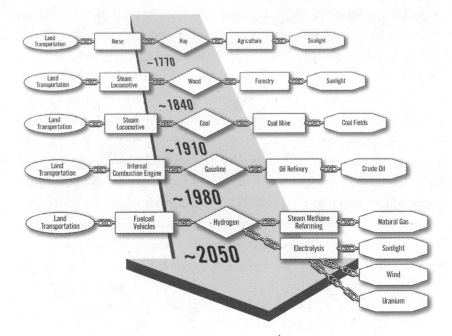

Figure 6.6 Evolution of the land-transportation chain over more than two centuries

Now let's think about architectures within architectures.

Our five-link architecture is the highest-level picture of the energy system—the one that is as simple as possible but not simpler. Yet sometimes we must drop down to the next level of architectural detail. If we decide to seek opportunities in one of the service technologies, such as urban transportation, we will need to know something about the architecture of urban transit systems. Then, perhaps, we will need to know the architecture of a bus, then of its powertrain, then its fuel-delivery system.

A *hierarchy* of architectures may be needed to find which bolt to tighten. Or, at the other end of things, to know how to develop a computer model that shows the best way to reduce greenhouse gas emissions, or the best way to set corporate strategy. But it's always valuable to climb back up through these hierarchies to see that five-link chain sitting there unperturbed, helping you get a fix on your business.

I often think it would be fun to introduce this five-link architecture to a class of six-year-olds. To engage the little people in history we could ask them to make up five-link chains—cut from bits of yellow, blue and orange paper and festooned with sparkles to demonstrate how they thought transportation, or communication, worked when their parents were children. They would probably laugh at what their parents did. With their imaginations stirred, they could then put together chains for when their grandparents were children, or the pioneer days, or the Roman Empire or the Ming Dynasty. And with the catalyst of a good teacher, it would give the kids a feel for the extraordinary intimacy between culture and technology. If some six-year-olds undertake such a project, I hope they will show me what they've done.

It would set my neurons a-flutter.

7. Whales and Whiskey Barrels

In which tensions between attractors and barriers, the tugs and snags of change, are seen guiding energy system evolution.

On a Sabbath afternoon during the waning summer of 1859, Pennsylvania blacksmith William A. ('Uncle Billy') Smith decided to use his day of rest for a stroll in the countryside. The stroll took him past a well he was drilling for Colonel Edwin L. Drake, a 'crazy fellow' raised in Vermont. Drake, using his self-bestowed military rank to impress the locals, had commissioned Uncle Billy to drill a well outside Titusville—but not for water. Uncle Billy was drilling for *oil*.

The colonel and his back-east financiers had decided to try drilling for illuminants, fuel for oil lamps. Everyone knew the scheme was preposterous— no one *drilled wells* for oil! Whale oil, kerosene (also called coal oil) and town gas were the illuminants of the day. Everyone knew whale oil was the best, but high cost and limited supplies restricted its use. Kerosene and town gas were also used because they were somewhat less expensive. Kerosene was made from the oily drippings oozing from underground coal seams. Town gas, sometimes known as coal gas, was manufactured from coal in the town's 'gas plants.' But kerosene supplies were limited, and town gas tended to blow up houses and poison people—in both reality and literature. The suicide that concludes Arthur Miller's *Death of a Salesman* was accomplished by town gas. (About half the heating value in town gas came from poisonous carbon monoxide. Today, natural gas contains no carbon monoxide.)

There were many barriers that blocked better lighting in the late 19th century. But the high cost and limited supply of illuminants were the most important.

That is why Uncle Billy leapt with joy when he saw a sticky dark liquid floating atop the water in his well. Titusville boomed. Land prices rocketed. Frenzied drillers smacked holes into the ground hunting for 'rock oil'—a name chosen to distinguish it from whale or coal oil. More and more rock oil wells gurgled. Soon supply exceeded demand and so, of course, rock oil prices plummeted.*

Brandishing the weapon of innovation, Uncle Billy and his friends had smashed through the barrier of high-cost illuminants. Price, once a barrier, flipped to become an attractor pulling energy system evolution into an era of brighter living rooms. Innovation unleashed this systemic evolution. Innovation is always the trigger—this time it was the idea of driving holes into the ground rather than harpoons into whales.

But as the price barrier was smashed a new barrier jumped up in its place, like another duck in a shooting gallery. The next barrier was a shortage of whiskey barrels in which to store and transport rock oil. For a short while, the price of barrels rose to be several times the price of the oil they contained. Some oil-well drillers, trapped within a locally saturated market and unable to get the barrels they needed, began losing money. Suppliers of whiskey barrels got ready for early retirement. Yet the momentum for change was there. And soon, so would be the barrels.

Thinking about these events, watching the actors change while the theme repeats, my mind jumps to our five-link energy system chain, to watch it slip-sliding, snagged and pulled, along the arrow of time. Figure 7.1 paints this slip-sliding image on a two-dimensional page. Still, for me, our slip-sliding chain is best imagined in the color and dynamics conjured in the mind's eye.

Sometimes the chain sweeps along quickly, as it did during the decades following Uncle Billy's Sunday afternoon stroll. During those decades, the innovation of drilling for oil provided much more than oil; it provided a platform for a surfeit of related innovations. Today, we might call them spin-off innovations. Sometimes the chain gets snagged, often by just one link, or a corner of one link. If it's truly snagged, system development just about stops and the chain sits there, trembling, tensions building, as the links which are free to advance try to get on with their destiny, try to take things into the future—while the snag becomes more and more frustrating.

---

* This history is well recounted by Daniel Yergin in *The Prize*.[6]

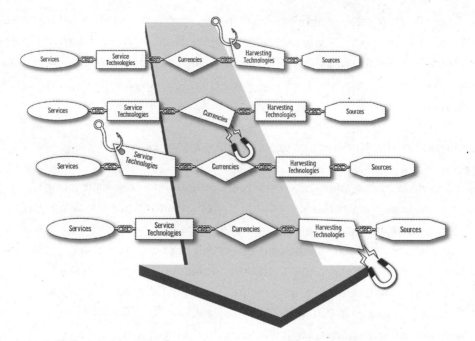

Figure 7.1 Schematic of hooks (barriers) and tugs (attractors) on energy system evolution.

I've glibly used such expressions as 'into the future,' expecting the direction to be self-evident. For example, I expect few will question that, for transcontinental travel, the evolution from horses to aircraft, was 'into the future,' not the inverse. But for smelling land, it's important to have a general principle on which we can ground our judgment about whether a proposed innovation is truly towards the future, or might, in fact, be a step backwards. For now, I'll suggest that energy system evolution always seems to be toward improving the *quality* and *convenience* of the services, while simultaneously improving *economic efficiency, energy efficiency* and *environmental gentility*.

Because it's so important, we'll frequently return to these five points that guide systemic evolution. Our first opportunity occurs in the next chapter, 'The Gods of Energy Planning Foolishness,' when we'll put more meat on these ideas. Later we'll return to the criteria in Chapter 15 'Template for Sustainability,' as we try to tighten up what we mean by sustainable energy systems.

In spite of our rhetoric about change, today is pretty much a time of stuckness for our energy system.* Most stuckness comes from a poor understanding of how the system works and which parts are most important—and why. And what are the real barriers to appropriate systemic change and what are merely well intentioned but misguided slogans.

Barriers and attractors may be grouped into categories:

Technical barriers and attractors: A contemporary technical barrier is the difficulty of storing low-carbon fuels like natural gas and hydrogen on board cars and airplanes. On the other hand, an emerging technical attractor is the fuelcell. Fuelcells operate at high energy-conversion efficiencies. Moreover, due to their modular construction—a stack of electrode plates that can, in principle, be cut to different shapes and stacked together in different sizes—fuelcells will become more suitable for mass-production than are today's internal combustion engines.

Economic barriers and attractors: A contemporary economic barrier is the high cost of liquefying natural gas and hydrogen, which is the most severe impediment to using these cleaner fuels in transportation applications. Often economic attractors and barriers come in the shape of fiscal and monetary policies, such as tax incentives and interest rates. Together, these policies influence such things as the availability of capital and the time required to recoup investments in innovation. Monetary and fiscal policies can be either barriers or attractors. Policies allowing long timelines to recoup investment encourage innovation. Policies demanding short timelines discourage innovation.

Cultural barriers and attractors: The perception that we were running out of natural gas, prevalent in the late 1970s and early 1980s, is an example of

---

* What I mean by 'stuckness' should be clear from context. For those wanting a beautiful discussion of stuckness, read Robert M. Pirsig's *Zen and the Art of Motorcycle Maintenance*[4].

a cultural barrier. Today's perception that nuclear energy is unacceptably dangerous—or that there is no way to deal with the waste—is another cultural barrier, slowing the use of probably the greenest of energy sources. Sometimes a cultural barrier is conjured, like the oil shortages claimed after hurricane Katrina. Cultural attractors include the public's desire for clean air, and its growing awareness that atmospheric emissions will bring catastrophic climatic instabilities. Of course, initiatives responding to these concerns will be challenged by vested interests. And that, in turn, will present a cultural barrier to compete with the original cultural attractor.

Legislative barriers and attractors: Past legislation that prevented natural gas from being used in electricity generation was a barrier blocking the use of this comparatively cleaner, low-carbon fuel. Today's automobile emission laws constitute a legislative attractor, encouraging the use of these same low-carbon fuels.

Supply barriers and attractors: These can be real or imaginary, temporary or sustained. The whiskey barrel shortage was a real, but temporary, supply barrier.

Barriers and attractors are usually linked. Supply barriers, for example, induce economic barriers. And there are many more complex linkages. The difficulty of storing low-carbon fuels on board vehicles (our example of a technological barrier) could be beaten if we reduced the high cost of liquefaction (our example of an economic barrier), which could, in turn, be achieved by technical breakthroughs in cryogenic refrigeration (which would then be a technological attractor). Then, of course, a cultural barrier might arise in public fear of cryogenic fuels, in which case we might find ourselves facing a legislative barrier forbidding cryofuels—bringing us back to square one, blocked again from using low-carbon fuels in transportation.

Examples of barriers and attractors shaping systemic evolution are linked and fluid. For us, it is the *concept* that is important, not a few examples. The concept can be used as a methodology, as a way-of-thinking—by investors hunting for business opportunities, by professors chasing research grants, or by legislators trying to glue together legislation that will lead to a better world rather than to simply mollify special interests. Most importantly, barrier-attractor analysis can be a powerful technique for sniffing out opportunities and dangers. And it's damn effective for smelling land.

For nations and corporations, the larger the barrier the larger the opportunity, and the more vigorous the energy system evolution when the barrier is breached. Barriers hold back evolution. So removing a snag releases tensions that have developed and grown while the barrier was in place. The power from the release of a barrier is the power that speeds change. Opportunities always lie within the turmoil of rapid change. People who smash barriers have an inside track at getting rich, but it's no guarantee. Pushed aside by the big guys, oil pioneer Colonel Drake died an impoverished, discouraged man.

For the small entrepreneur, inventor or investor, the mix of opportunity and danger comes in different proportions than for large corporations. It can be briefly exhilarating for small innovators when they break the largest barriers. But soon the unleashed torrent whirls them this way and that, from one danger to another. Having triggered the next wave of systemic evolution, small entrepreneurs often don't have the resources to avoid drowning in the maelstrom. If we were offering prudent advice, we might warn them away from the largest barriers or attractors, counsel them to seek out the small ones hidden deep within the system. Lesser barriers and attractors can yield colossal opportunities for little folks, with correspondingly lower risk from the big players.

But that is not the way it works. Usually, innovators are people seeing no bounds to their abilities or opportunities, willing to try anything no matter how audacious. And the truth is we need them, using their ideas, bashing away at barriers until the barriers begin to leak, even if our sense of fair play is troubled when they pay the cabin boy's price.

Introducing an attractor also accelerates systemic evolution, but seldom with the vigor of removing a barrier. That's because the system usually develops greater tensions when trying to break past a snag than it does when responding to the tug. Anticipating a tug is a tenuous kind of thing. Snags are more concrete. Easier to see.

Anticipating the tug of an attractor is especially difficult for *technical* attractors. In the energy system, few businesses properly anticipate the effect of a technical attractor. When they try, most often they get it wrong, being either too starry-eyed or too dismissive. Getting it right requires understanding what the technical attractor does, being able to judge how what it does will impact systemic evolution, and mixing all this with entrepreneurial energy, management skill and a measure of luck. A tall order.

Industry is better at anticipating the effect of *legislative* attractors (or barriers). The captains of industry know how new legislation will affect their business. Usually they have been trying to accelerate, or slow, proposed legislation for years. That is what they pay most of their lawyers and all their lobbyists to do.

Now let's continue the saga that began with Uncle Billy's famous stroll. Soon Titusville had lots of whiskey barrels. Illuminants became plentiful and comparatively cheap. Other innovators rode the coat tails of abundant illuminants by inventing, manufacturing and distributing better lamps. Yet as cities and homes became brighter, lamps still produced soot, grime and fires. So the barrier to better illumination had shifted from high prices, to a shortage of whiskey barrels, to environmental degradation and risk to libraries and lives. Indeed, before electric lighting, the risk of fire caused many libraries to refuse 'artificial illumination'—that is, anything but sunlight. Gore Hall at Harvard was one.

So a receptive world was waiting when in 1882, Thomas Edison, his banker by his side, pushed a switch starting a generating station that fed electricity to his light bulb. Innovation had destroyed barriers, this time danger and dirt, and had simultaneously introduced attractors, this time convenience and brighter rooms.

A pair of innovations did the job—a new service technology, light bulbs, and a new energy currency to feed light bulbs, electricity.

Most people saw blessings in clean, safe, brighter illumination. A few saw dangers, among them the folks who owned kerosene and coal oil companies.

Then, in the 1890s, a horseless carriage came around the corner, its internal combustion engine burping, banging and needing gasoline. The horseless carriage pulled the energy system into the future. A future in which oil lost its illumination market, retained its lubrication market, and gained an entirely new market powering transportation. Several decades later, oil found yet another market as a feedstock for 20th century materials, as in most of our sailboats, much of our cars, and well, look about you, almost everything from toothpaste tubes to garden chairs. Dangers for some are always opportunities for others. Yet when it's not sheer good luck that separates winners from losers, it is usually a sense of where systemic evolution is taking us.

Now we'll venture into the deeper future. *Déjà vu* will characterize energy system evolution during the 21st century. Again, the evolution will center about a pair of innovations. The first will be a new service technology, this time fuelcells. And once again the second will be a new energy currency to feed the technology, this time hydrogen.

Like the innovation wave that began with electricity flowing through light bulbs, the innovation wave that starts with hydrogen coursing through fuelcells will just be the beginning. A myriad of service technologies will then enter to enrich our lives, all using hydrogen.

On March 2, 2006, the very day I was putting the final touches on this chapter, my friend Ged McLean, a former colleague from our Institute at the University of Victoria who had left to found Angstrom Power, phoned me to say, 'David, this will be a short call. But I want you to know that I'm calling you from a fuelcell-powered cell phone.' 'My God!' I thought. 'How wonderful.' I felt excitement that might have approached that of Thomas Watson, Alexander Graham Bell's collaborator, when in 1876 he first heard Bell's voice through their 'telephone.' Then, recalling the starry-eyed American mayor mentioned earlier, I thought, 'Why! I can imagine a day when every teenager will have one.'

Of course, it's easy to imagine a fuelcell telephone. But can we imagine the-as-yet unimagined? Of course not. Think back to the innovation wave spawned by electricity when new electric technologies spread outwards from their light bulb nucleation site. Can you imagine how difficult it would have been for people in the 1890s to imagine cellular telephones or email? Or could they have imagined eating breakfast when a neighbor (using an electricity-powered telephone) called to say, 'Turn on your TV'—and then sitting in their living room watching, live, on an electricity-powered screen as the second aircraft flew into the World Trade Center.

To imagine, in the early 1900s, all the services electricity-powered technologies would bring by the early 2000s would have been impossible. When the hydrogen innovation wave begins expanding out from its nucleation sites early in the 21st century, the results will mimic those of the earlier electricity wave—in patterns though not details. Hydrogen, too, will trigger a swarm of now unimaginable service technologies. So truly unimaginable we can't list them. We'll just have to wait.

8. THE GODS OF ENERGY PLANNING FOOLISHNESS

In which, after watching ill-found policies reverse energy system evolution, we look for ways to distinguish changes likely to live from those doomed to die.

We've been talking about barriers and attractors that will hold us back, or pull us toward, better energy systems. We've discussed, in some detail, technical barriers and attractors. Now we'll turn to *culture* and *legislation*. Since barriers are usually stronger than attractors, and legislation is always the child of culture, I'll pluck the next story from our rich history of legislative barriers.

In 1954, the United States Supreme Court introduced price regulation of natural gas transported across state lines. Prices were fixed below their market value, artificially encouraging consumption and discouraging the hunt for new supplies. It was a formula for trouble. Yet the trouble was postponed for almost two decades until the mid-1970s—by a massive surplus of natural gas. This surplus delayed the collision between increasing demand and declining supply, giving time to build the violence of the ultimate confrontation.*

High on their mythical mountaintops, mists swirling about them, frolicked the gods of Energy Planning Foolishness. Like the gods of love and war who are the stuff of Greek and Nordic legend, the gods of EPF get much of their fun from meddling in the affairs of mortals—stirring up mischief with a little pull here and a little push there. So during these times they were delighted at the prospect of the coming collision and set about to nudge mortal events towards the best time for EPF knavery. Why not, the gods asked themselves, arrange the collision to coincide with a time already disrupted by other energy troubles,

---

* Natural gas problems resulted from the limited infrastructure for *delivering* natural gas (interstate and local pipelines), which caused a lot of gas to be 'flared' at the wellhead. This flared natural gas is called 'associated gas,' because it is produced in 'association' with pumping oil. It would have been better to encourage investment in delivery infrastructures rather than employ the blunt tool of price regulation at the wellhead.

such as oil embargoes and exceptionally cold winters in eastern America? They chortled at their idea of waiting for cold winters in the east, because they knew natural gas was delivered from the west through dated, undersized pipelines— so their prank would also stir up east-west conflict. Not the east-west conflict of traditional geopolitics, but rather homegrown conflict within the good old United States.

It was all arranged for the mid-1970s. Eastern schools closed down for weeks for lack of natural gas to keep them warm. 'Freezing in the dark' entered our lingo and the cultural response was swift. Everyone decided we were running out of natural gas. What was worse, many people saw the big guys protected from shortages while ordinary folks suffered. 'Why,' people asked, 'should large industries and electric utilities be allowed to gulp natural gas when the evening news shows our school children shivering?'

To appear decisive—to be seen as *doing* something—legislators scurried about for proposals. This gave the Gods of Energy Planning Foolishness another sweet opportunity, because they knew the supply shortfall had nothing to do with fundamental shortages of natural gas in the ground. They knew the supply snag had been caused by an earlier generation of lawmakers, the 1950s generation, who introduced legislation that slowed exploration for new supplies and delayed the repair of old (and the construction of new) infrastructures to deliver natural gas. But the gods also guessed these realities would not influence the public's gut response. The gods were right. The 1970s legislators passed a Fuels Use Act that made it illegal to use natural gas for electricity generation.[7] Carried forward by the unquestioned belief that we were running out of natural gas, laws were written to send utility executives to jail if they insisted on burning the stuff. Other laws exhorted and cheered the users of coal. All for the good of society.

In the last chapter, we spoke of how technical, economic or supply barriers can stop, for a while, the orderly evolution of the energy system toward a brighter future. Now we see that foolish legislation can slam orderly evolution into reverse.

To legislate a preference for coal over natural gas was to blow against the winds of historical pattern. For more than a century the energy system had been evolving towards lower-carbon fuels like natural gas, and away from higher-carbon fuels like coal. Low-carbon fuels had long been capturing an increasing market share among energy sources. Trains ran better, further and cleaner on diesel than on coal. Home heating was better and cleaner with natural gas than with oil.

Natural gas-fired generating stations produce electricity much more efficiently and cleanly than coal-fired stations (coal-fired stations run at about 43% efficiency, combined-cycle natural gas stations run at about 63% or more). These comparative efficiencies made shutting out natural gas from electricity generation even more preposterous. Yet none of these realities stopped legislation that encouraged coal and forbade natural gas. Billboards and full-page magazine advertisements repeatedly told us that coal was 'the fuel of the future.' Some still do.

Thinking back to those times, I recall a late-1970s trip to Norfolk, Virginia where I couldn't resist a harbor tour. The tour boat took us past the decommissioned passenger liner *United States*, then past an impressive fleet of both commissioned and mothballed warships, and finally past Jacques Cousteau's small, forlorn *Calypso*, bobbing against a particularly rickety jetty, a rusty fragment of her TV image. What most sticks in my mind are not the ships, tugboats or even the shockingly small *Calypso*. What sticks in my memory is trying to keep ahead of the coal dust landing on my clothes, getting inside my shirt, itching under my belt—coal dust carried out to our tour boat by a fresh offshore breeze. I got an especially painful micro-lump stuck in my eye, just as the tour boat announcer, pointing to huge coal piles on the docks and bubbling with tour boat announcer exuberance, told us, 'This is only the beginning!' Norfolk, he said, was about to play a major role shipping 'the fuel of the future' around the globe to an energy-hungry world. As we floated within the cloud of coal dust coming across the water, he really did use those words. With tears pouring down my cheek from a small bit of the future lodged in my left eye, I had my doubts.

The Gods of Energy Planning Foolishness always have their best chance for trickery when they exploit our conventional wisdom. For as Karen Blixen says in her beautiful line from *Out of Africa*, 'When the gods want to punish you, they answer your prayers.'

With this warning, we should identify overarching principles that we can use to test the legitimacy of emerging conventional wishful thinking—which I'll call conventional wishdom—before it has us serving up sacrifices to the Gods of Energy Planning Foolishness. A good place to start is to return to our barrier-attractor imagery. But to use the imagery we need to know what it is that the system is evolving towards. We need a concept of the future, or at least of the patterns of the future. We need a sense of where the system is going and how it is getting there. Otherwise, how can we know which are barriers and which are attractors?

Probably the best compass to use when seeking the direction to the future is to tighten and refine the ideas we began setting out in the last chapter. The compass is another template: Energy system evolution always takes a path, however meandering, that improves the:

- *quality* of services,
- *convenience* of services,
- *economic efficiency* of the system chain delivering the services,
- *energy efficiency* of the system chain delivering the service, and
- *environmental gentility* of the system chain delivering the service.

'Environmental gentility' may seem a soft idea, well intended but fuzzy words signifying little. But to me it is a sharp idea. Sharp, because environmental gentility means reduced environmental intrusion—the amount by which any action, of people building a city or of beavers damming a stream, intrudes upon Nature's flows and equilibria.

The five parameters above are useful for judging whether an innovation is likely to live or to fizzle—or whether legislation is likely to help or hinder. Some might want to collapse these five into two—*improved quality* and *lowered costs*. But I prefer five. Five begins to give texture and substance that are too easily lost with the more simplistic notions of quality and cost. Innovations most likely to succeed usually satisfy all five. Some will satisfy most, but not all—although, in time, sustained changes usually satisfy all.

Worldly left-brainers may decide that economics drives it all. Right-brainers will have a hard time being definitive. Dedicated environmentalists may argue that environmental gentility is the only thing that matters. But here is the message: Business planners who dismiss *environmental gentility* from testing business plans are likely to lose a lot of money. Well intentioned environmentalists who neglect *quality, convenience* or *economic efficiency* are unlikely to move things towards a better world.

Let's return to a story we've already recounted. Let's test how our five parameters for success fit with the tale of electric streetcars displacing horse-drawn streetcars. If we view streetcar evolution from this new perspective, we see it's in harmony with our template for changes likely to take root.

Quality and *convenience* came from the ability to deliver more passengers at higher speeds.

Environmental gentility came, in part, from fewer horse buns. Once when explaining this, someone responded, 'I'd rather have a few horse buns than streets crowded with streetcars and traffic.' But the imagery of a few horse buns is embedded within the idea of only a few people using public transit. The mythical two-horse town comes to mind. But environmental intrusiveness must be judged in context of the service delivered—which must include the number of passengers transported and how far. Can you imagine the depth of horse buns if, today, all commuters traveling in and out of Manhattan, Tokyo, London, Buenos Aires, Paris, Toronto, or Boston, were pulled by horses?*

Switching from horses to electric motors also improved *economic efficiency*, although in some towns in the early days, cost advantages were not immediately clear. But in the end, going electric certainly was less expensive.

---

* The environmental benefits didn't all come at once. For example, electric streetcars initially brought the visual pollution of overhead wiring until some streetcars dove underground to become subways.

Figure 8.1 The five-link chain when horses pulled streetcars.

Energy efficiency probably leaped the furthest ahead. We can sense this by comparing the service-to-source chain for horse-drawn streetcars (Figure 8.1) with the service-to-source chain for electric streetcars (Figure 8.2). We must remember the very poor energy efficiencies of agriculture (the harvesting technology link) and horses (the service technology link), compared with the much better efficiencies of the electricity chain.

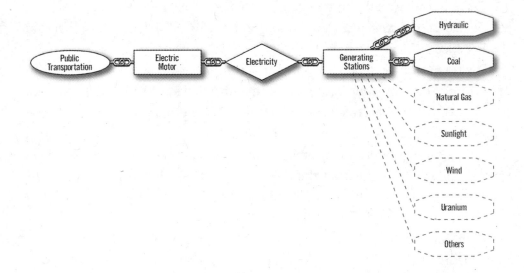

Figure 8.2 The five-link chain for the first electric streetcars.

The electric streetcar chain ends in an opening funnel that can draw upon many sources. Coal and hydraulic options were the most prevalent during the late 19th and early 20th centuries. But you can see five other sources, in dashes, included in Figure 8.2, ready to be called upon later in the 20th century.

No doubt our future will continue with a full complement of foolish innovations, investments and legislation—there are no obvious signs that we're getting wiser than our ancestors. Yet I hope some foolishness can be deflected by applying our five parameters as a 'go, no-go' test before proceeding. For a time horizon of a half-century or longer, you'll find these tests lead to the increasing use of energy currencies that can be manufactured from any energy source—and, in particular, from carbon-free sources. And therefore lead to the twin currencies, hydrogen and electricity.

Part Three

INTERMEZZO

9. Mozart, Metaphor and Math

In which, watching Salieri listen to Mozart's manuscripts, we wish it were as easy to hear the beauty of equations.

A metaphor is a shrunken parable, or so I sometimes think. Metaphors store truth in compressed speech, so they share with equations the ability to encapsulate ideas. But unlike equations, metaphors *shun* precision. For as Lawrence Joseph wrote in *Gaia: The Growth of an Idea*,[8] 'Metaphors are like myths, often certifiably false yet purportedly chock-full of veracity and wisdom.' I think that, sometimes, the value of a metaphor is its imprecision. There are times for untouched photographs and times for impressionists. Yet the softness, the imprecision, makes metaphors vulnerable to 'yabut' attacks—yabut it doesn't hold in this or that circumstance, yabut the story didn't really happen that way. We must acknowledge the yabuts, but not, necessarily, discredit the idea.

Even if the details are imprecise—for instance, even if there had never been a cabin boy or an Admiral Shovell, let alone a cabin boy who was hanged for his curious habit of sniffing the breeze for hints of land—parables often encapsulate truths. Or, considering a parable from a well-known book: Can we verify there was a Samaritan walking down the road just in time, having the time, and taking the time to aid a mugged Israelite lying in the ditch? Was Jesus ever asked to *prove* his story? Or do we think Jesus just made it up? Is the message weakened by whether or not it was a precise factual record? Or do we believe Christ simply knew that, at one time or other, there must have been a Samaritan that helped a bleeding Israelite?

Metaphors are useful because their encapsulation gives a toehold for understanding, which can lead to the next level of understanding as we peck away at the metaphor's fuzzy edges. The cliché, 'the exception proves the rule,' speaks to pecking at these edges. Metaphors are squishy but still useful, because they're a catalyst to the mind as it reaches for understanding.

In a wonderful scene from the movie *Amadeus*, Mozart's wife, Constanze, surreptitiously takes a pile of her husband's compositions to court composer Antonio Salieri. She hopes to demonstrate Wolfgang's skills and thereby persuade Salieri to appoint him tutor to the emperor's niece. She accomplishes no such thing. Instead, as Salieri reads Mozart's compositions the music fills his head—and jealousy fills his heart. Unlike Salieri, few of us can read music, so the film doesn't zoom to the manuscripts. Instead, a cinematic trick helps us hear the notes on the page. As the camera focuses on Salieri's face, a *music-over* envelops us within Mozart's genius—and helps us understand Salieri's pain.

I wish there were a cinematic trick to show the beauty in equations. To show, for instance, how the four equations of electromagnetic theory can be folded together to give birth to a single, magnificent, wave equation—which, in turn, can describe radio waves returning home from a Mars probe, or moonlight reflected from an undulating ocean, or a rainbow. I wish we knew how to film a *physics-over*.

In this book, we'll use a few equations to help make ideas crisp. All will be simple equations.* What's more, I have a big advantage over Mozart. If you compare the manuscript for Mozart's Mass in C Minor with the few equations we'll encounter, most folks will find reading the equations *much* easier.

As we set out to think about equations, etymology is a good place to start. The roots of the word equation are 'equate' and 'equal.' *The Oxford English Dictionary* says an equation is a 'statement of equality. . .that indicates a constant relationship between variables.' *Webster's Unabridged* says an equation is 'the act of equating or making equal' and 'equally balanced.' So the first step in appreciating equations is to know that the stuff on the left is equal to the stuff on the right.

The next question is: What kind of stuff? Equal in what way? The two sides of an equation can be equal in several ways. They might be equal in energy (if the equation embodies Nature's law of energy conservation), or equal in material (if it symbolizes mass conservation), or equal in momentum, or electrical charge,

---

* It turns out that the simple equations describe the most important engineering concepts.

or chemical elements, and so on. Sometimes they can be equal across more than one property at the same time.

There are two ways to write equations. Physicists and mechanical engineers commonly use an equal sign (=) between the left and right sides, often equating cause and effect. Chemists and chemical engineers often use an arrow (→), to indicate the direction of a process. But in all cases, the left and right sides are balanced.

For our next step in examining this equation business, we'll choose a process we learned in high school—splitting water into its molecular components. It's a process central to the coming Hydrogen Age. The equation that describes splitting two molecules of water $(2H_2O)^*$ into two molecules of hydrogen $(2H_2)$ and one of oxygen (O_2) is:

$$2H_2O \rightarrow 2H_2 + O_2$$

This water-splitting equation balances material. It tells us that the amount of material in the water entering equals the amount of material in the hydrogen and oxygen leaving. The left side of the equation is the mass that goes in, water (H_2O). The right hand side is the mass that comes out, hydrogen and oxygen $(H_2$ and $O_2)$. Yet the equation gives still more information. It also says the number of hydrogen atoms entering must equal the number leaving. And it requires a balance in oxygen atoms—two atoms go in (one in each of the two water molecules) and two come out (both in the single oxygen molecule).

Experienced equation readers will know there is a similar equation that can be used to balance energy as well as material. To remember the energy balance, we can rewrite the equation to include the input energy, knowing that the output energy is the hydrogen's potential to do work:

$$2H_2O + \text{net input energy} \rightarrow 2H_2 + O_2$$

---

* Those whose high school chemistry is locked in a far distant past, might recall that the Arabic numeral in front of the molecular group denotes the number of molecules (here the upper case 2 in $2H_2O$ means two molecules of water). Also each letter (H and O)—or sometimes a group of letters, when the first is a capital and the second is lower case, such as 'He' which designates helium—represents a chemical element. A subscript following an element's symbol (the subscript 2 in H_2) indicates the number of atoms of that element in the molecule.

This equation works for all water-splitting processes. But if we have a *specific* process—electrolysis, for example—we can then refine the equation to more completely describe what's happening.

For electrolysis, the process-specific energy-material equation can be written:

$$2H_2O + \text{electricity} \rightarrow 2H_2 + O_2 + \text{waste heat}$$

We must account for all energy inputs and outputs. So in addition to the input electricity and output waste heat, we must also remember any energy that enters or leaves via material. Obviously, the output hydrogen carries chemical energy. But the input water and output oxygen might also contain energy if, for example, they are at temperatures or pressures *above* environmental conditions. This form of energy is called 'thermomechanical.'*

The input energy is the sum of the electrical energy *plus* any thermo-mechanical energy carried by the water. The output energy is the sum of waste heat, *plus* the chemical energy carried out in the hydrogen, plus any thermo-mechanical energy that might be carried by the hydrogen and oxygen.

The above three equations all describe electrolysis. It's a matter of choice. The choice should be one of convenience, a matter of what special features of the process we want to consider.

For me, high school too often became an exercise in memorization, which may be one reason I did so poorly. Today, I never memorize equations. Rather, I try to understand the physics and then write a convenient mathematical visualization to describe the physics. From this perspective, equations can be considered a way of describing phenomena we already understand. Then we can use the precision and compaction of the equations to understand them better.

Of course we must appreciate that, just because we can write an equation that describes electrolysis doesn't mean that every time we put electrical energy into water the result will be hydrogen and oxygen. If we heat a pot of water with an electric emersion coil, we'll simply get a pot of hot water and probably

---

* Or depressed below the environmental conditions, as we'll learn in Chapter 23 'It's Exergy!'.

some steam. So this equation doesn't tell us what always happens when we add electrical energy to water. It tells us what could happen—and will happen if we choose a specific process.

While we're talking about the water-*splitting* (hydrogen making) equation

$$2H_2O + energy \rightarrow 2H_2 + O_2 \qquad\qquad 9.1$$

I can't resist writing the inverse, a water-*making* (hydrogen using) equation

$$2H_2 + O_2 \rightarrow 2H_2O + energy \qquad\qquad 9.2$$

In wonderfully compressed form, these two equations capture the essence of our coming Hydrogen Age. The water-splitting equation (9.1) says that (*if* we use an appropriate technology) we can take energy from any source and use it to split water into hydrogen and oxygen. The hydrogen product can be stored for later use, while the oxygen product can be used for industrial processes or thrown out to the atmosphere.

The water-making equation (9.2) says that, when we use the hydrogen, we get both the energy and the water back. The released energy can power cars or fly airplanes, while the water is given back to the environment from whence it came, or used for drinking (a slightly more circuitous route back to the environment).

This small package of two equations efficiently encapsulates the closed cycle of hydrogen systems—a closed cycle that lies at the root of hydrogen-age environmental gentility. Once we know how to read the equations, they can illuminate profound things about our future. But if we limit ourselves to words alone, it's impossible to gain these insights with the precision, brevity and numeracy that equations allow.

About now, remembering your high school chemistry class, you might ask: Instead of using two equations, why don't we make things even more efficient by using a single equation with two arrows—one aiming to the right, the other to the left? Like this:

$$2H_2O + energy \leftrightarrow 2H_2 + O_2$$

Chemists frequently use this dual arrow to emphasize that changing the external conditions can drive a process either forward or backward. But the reason I prefer two equations when describing the Hydrogen Age is that the water-splitting equation applies where the hydrogen is *manufactured*, while the water-making equation applies where the hydrogen is used—which may be high in the sky powering an airplane, or along a country road powering your car.

Marguerite Duras wrote, *'J'écris pour savoir ce que je pense'* ('I write to know what I think'). Physicists often set down equations to encapsulate what they know, or are beginning to guess, and then use them to learn more about what they know, or guess. So beyond their essential and obvious value when designing new technologies, equations are used to:

- Test the truth of what we've guessed,
- Understand more clearly what we know, and
- Use what's known to learn things as yet unknown.

We've already discussed metaphors as one way to compress understanding. Formulae are another.

A formula is a recipe. It's not an equation. An everyday formula might be used for mixing pigments in a paint store to get a special color for your living room, or for mixing flours and yeast to make bread. There is nothing elegant or graceful about formulae although they can still be useful. A formula may tell you how to get a certain color, but it won't give you a grand principle for *all* colors. Similarly a recipe for bread does not give a general, powerful insight about all foods, or even all breads, it just tells you how to make one kind of bread. And if the formula is used with a little inaccuracy, the result won't be much different than if the baker achieved perfect accuracy.

We know, of course, that the chemical formula for water is H_2O. A chemical formula is a symbolic description of a substance, like water, and therefore is a kind of amalgam—having the precision of an equation but without an equation's ability to predict.

A formula can be either a blunt, application-specific workhorse or a description of a material. It might give you a rule for how to mix white, red, canary and a touch of lamp black to achieve a rusty hue—but never an insight into Van Gogh's artistry. Equations float above formulae. Equations are in a different league. Because they *can* give wonderful insights into Nature's artistry, like her rainbows.

Part Four

UNDERSTANDING ENVIRONMENTAL INTRUSION

10. FROM OIL LAMPS TO LIGHT BULBS: PATHWAYS TO ENVIRONMENTAL GENTILITY

In which we realize there are always two ways to clean the place up—add collectors to catch pollutants, or change the process to stop making pollutants.

We're told we can't have it both ways. Conflicts between people intent on protecting our environment, and others intent on protecting their income, are media staples. Responsible faces above dark blue suits on the nightly news solemnly explain how we must balance the need for environmental protection against the need for economic growth—too much of one, they say, will bring too little of the other. They give clear examples. If we want to reduce dirty emissions from coal-fired electricity generating stations we can install scrubbers and bag houses. But then we must pay more for our electricity to offset the costs of the scrubbers and bag houses—*and* the energy to run them.

Yet something seems amiss, or perhaps incomplete. Arguments for the balanced approach always assume environmental protection requires imposing a constraint, such as adding a pollution collector on some established way of delivering a service. Installing scrubbers on coal-fired generating stations is just one example. Requiring catalytic converters on our automobile exhaust systems is another.

Perhaps we should try smelling land, try looking at things the other way around. If we sit back to reflect on *how* the delivery of energy services has evolved during the past two centuries, we find that imposing constraints is just one way to clean up messes. Let's pick up the threads of the systemic evolution, sniff out the patterns and look for messages.

People always want better energy services. In the mid to late 1800s, improved lighting was high on everyone's list of priorities. This desire for better lighting sent young boys half way around the world in dangerous, appallingly uncomfortable sailing ships to hunt for whale oil—the highest-quality illuminant of the day, affordable only by the wealthy. Later, the need for illuminants pulled other men into the interior of Pennsylvania to drill for rock oil that could be burned in the latest Vienna lamps. In some cities, town gas, made by gasifying coal, fueled the lamps. But because town gas was a mixture of carbon monoxide (CO) and hydrogen, sometimes carbon monoxide leaked into homes poisoning families if it didn't first kill them in an explosion.

As we entered the 20th century, we began to provide illumination in a new way, with electric light bulbs. Light bulbs didn't dump greasy soot into the parlor. Washing the curtains, walls and upholstery became a less frequent chore. No doubt the improved air quality in the parlor also brought health benefits— long before anyone used the phrase 'air quality' or concerned themselves with airborne carcinogens.

So people began switching from lamps to electric light bulbs. This gave lighting its great leap forward. Initially electric lighting wasn't cheaper—only the best homes in the best areas could afford it. Moreover, people didn't change from oil (or town gas) lamps to light bulbs because we were running out of oil or coal. Light bulbs won because they were *better*. Better is the correct word. Light bulbs brought better illumination—brighter, steadier light, and so on. They were also more convenient. Folks could now pull a string attached to the light bulb on the ceiling. Much easier than trimming the wick, refilling the oil bowl and lifting the chimney to fire up the lamp. Things got even more convenient when people could just flip a wall switch. Light bulbs delivered better service. *That* is why they knocked out oil lamps.

Beyond convenience and better illumination, two other things made light bulbs better. They were less dangerous to people and to the environment. Of course, safety and environmental protection are different sides of much the same coin—distinguished, perhaps, by how quickly the damage occurs. This may be the difference between acid rain or carcinogenic emissions that kill forests and people slowly but reliably, and a pipeline explosion or dam failure that kills them quickly and unexpectedly. Gradual damage that we know we're causing is environmental intrusion. If it is a surprise—or we claim it is a surprise—we can say it was an accident.

The primary motivation for changing to light bulbs was better lighting. Yet environmental gentility and safety came along as fringe benefits. Today, our primary motivation might be reversed—environmental gentility often drives change and we have become obsessed with safety. So let's dig even deeper into the principles that underlie what was happening when we changed from oil lamps to light bulbs—because the best principles are more powerful when generalized.

Light bulbs in the living room were a *process change*. Illumination came to the parlor in a different way, employing a different service technology, using a different process, requiring a different currency. We often think that to stop pollutants making messes we must add collectors to catch the pollutants. The change from oil lamps to light bulbs shows that we can also change the process and thereby stop *making* the pollutant.

The moral is that there are always two ways to clean the place up:

- add a collector to catch the pollutant, or
- change the process to stop making the pollutant.

Usually we first think of adding a collector. Yet changing the process is usually better—not only for environmental improvement, but because it often brings ancillary benefits such as greater convenience, higher efficiencies and better service.

What about cost? In the early 20th century, illuminating your home with electric light bulbs was much more expensive than with oil lamps. But soon it became cheaper. Sometimes immediately, but almost always in the end, using process change as a strategy for environmental protection brings economic benefits—strengthening corporate and national well-being.

In the end, cleaner is cheaper. Illumination is but one of many examples.

When I was four years old, I lived with my parents in a northern Quebec paper-making town called Kenogami.* Coal heated our home. I can't remember what the furnace looked like or where we stored the coal, but I do remember the black, airbrushed feathers above the heater outlets. I also remember the colossal cold-air return dominating the front hall floor. I remember crawling

---

* Kenogami was later swallowed by the communities of Chicoutimi and Jonquiere.

out over the middle of its grating and lying there, spread eagled, neither my hands nor feet reaching beyond the grate, imagining I was flying, my mind teasing the cold-air return, daring it to suck me down the miles into the bowels of the basement and on into the furnace. When I think back on the size of the ducting, I realize the hot air from the furnace was circulated by convection. No fans pushed hot air through our Kenogami home.

Later we moved to Belleville, Ontario. That furnace I can remember. It was a huge, seemingly alive, octopus-like monster in the basement. A different part of the basement was boarded off for the coal bin. I'd follow my Dad down into the basement to clean out the firebox and sift clinkers from the ash.* I think we put the clinkers out with the garbage. I know that we scattered the ash over the ice on the sidewalk in the winter so old folks wouldn't fall. But for us kids, the ashes spoiled the sledding.

When I became a young father with my own children, our first home in Oakville, Ontario, had an oil-furnace. Our oil-furnace house was much easier to keep clean than my childhood coal furnace houses. Yet it wasn't entirely clean. Moving a wall picture was tricky because you could see where the picture had previously hung. The basement still smelled of oil, in part because the oil-tank sat in the basement. The furnace made a lot of noise—made a whump when it ignited—and some bits of greasy soot hung around to remind us that we really did still have a furnace. As my kids grew older they wanted a recreation room. First we had to clean out the basement and gain some space by getting rid of the oil tank. We installed a natural gas furnace. The old oil tank went off to the scrap yard, the basement had more usable space, the kids had a recreation room and the fuel oil smell vanished.

Today I live on the southern tip of Vancouver Island, on the west coast of Canada. A heat pump warms our home, and that same heat pump also supplies

---

* I've found most people born after 1950 have never heard of 'clinkers.' But just about everyone born before has. That is a pretty good testament to when they stopped heating homes with coal. Clinkers are just like they sound, chunks of non-combustible oxides, typically silica, calcium and alumina. I think of clinkers as big cinders. You could get a cinder in your eye as a steam locomotive went by. But clinkers are too fat to fly, so they won't get in your eye. Grey ash is even finer than cinders. Of course, these are not technical definitions, just remembrances from times passed.

cooling on those few days when sea breezes don't cool us naturally. The heat pump sits outside and is fed electricity made by hydroelectric generators. Unlike a furnace, a heat pump can run forward and backward, pumping heat from outside into the living room when it needs warming, or pumping heat from the living room to outside when it needs cooling. Heat pumps deliver more thermal energy to your home than is in the electricity they use. And it's the cleanest home heating I've had. Looking back over my home-heating history, two thoughts pop up. First, with each change our homes became more comfortable. We got better service. Second, each time the home-heating process changed we also cleaned the place up—both inside and outside—but we *never* installed a collector on the chimney.

Changing the process made things cleaner for two reasons. First, each successive process had improved efficiency. Coal furnaces were comparatively inefficient beasts. About 40% of the coal's energy went to warming the house; the rest went out the chimney, either as unburned fuel or simply as heat to warm the great outdoors. Later, my oil furnace put about 60% of the oil's energy into warming the house, the rest up the chimney. The gas furnace, about 85%. A heat pump lifts heat from the environment and pumps it into your home. Using this trick, we can get more than 200% as much energy into our house as was in the electricity used to run the heat pump. (The magic of heat pumps will be explained in Chapter 24 'Exergy Takes Us beyond the Lamppost.') Each process change led to more heat from less energy. And if nothing else changes, the pollution from delivering an energy service is proportional to energy used.

But something else *did* change. And that brought the second reason for environmental benefits. Each new heating system required a different fuel—a different energy currency. And that was what enhanced the environmental benefits.

One way to measure an energy currency's intrusiveness is its carbon content. When carbon burns, it enters the atmosphere as carbon dioxide, the most significant of the anthropogenic greenhouse gases. These emissions from human activity will increasingly bring climatic instabilities during the first half of the 21st century and catastrophes during the second.

Curiously, the carbon content of fuels often provides an indication of the fuel's other environmental intrudants, like sulphur compounds, heavy metals and carcinogens. So carbon is a kind of canary*—because as carbon content rises, so too, most often, do many of these other unpleasantries. For this reason, I'll use carbon content as the proxy for overall environmental intrusion from fossil fuels, especially coal.

About 80% of coal's energy content comes from carbon; almost all the rest is from hydrogen.† About 65% of the energy in oil is carried by carbon, the rest by hydrogen. About 45% of the energy in natural gas is carried by carbon, the rest by hydrogen. None of the energy in electricity is carried by carbon—and if generated by hydraulic or nuclear power, no carbon is used, or released, during electricity production.

So the reduced pollution from my home heating is the product of *multiplying* the effect of improved efficiencies *times* the effect of cleaner currencies. If you want to calculate this yourself, you can *multiply* the reciprocal of efficiency (1/ efficiency) *times* the percentage of energy that comes from carbon. You will find that the coal furnace emits about twice as much CO_2 as the oil furnace, which emits about twice as much CO_2 as the natural gas furnace. The heat pump doesn't emit any—although the *system* delivering the electricity might, if the electricity is fossil-generated.

Changing the process rather than adding a collector is a largely unplanned environmental improvement strategy that is going on all around us. The major steel-making industries are now introducing direct reduction processes. These processes bypass the need for coke ovens and blast furnaces to bring in energy savings, lower emissions and better steel.

---

* Most readers know that caged canaries were taken down into coal mines to act as early warning systems for carbon monoxide buildup. Canaries (and mice) were noticeably more sensitive to carbon monoxide poisoning than people. So if the canary fell off its perch, the minors knew they must, very quickly, get their tails out of the mine. Indeed, for this job, that's why canaries won over mice. A dead mouse might be mistaken for a snoozing mouse.

† There can be other constituents of coal such as sulphur that also release heat when burned ($S + O_2 \rightarrow$ heat $+ SO_2$). Unfortunately, the resulting airborne SO_2 released from the stack typically turns into airborne sulphurous acid.

Computer modeling is used to improve product designs and industrial processes. Design options can be quickly optimized to reduce material and energy requirements. Early in 1994, the inaugural issue of a new technical journal, *Environmentally Conscious Manufacturing*, arrived on my desk. Manufacturing without messes. I was struck by a recurring theme: Authors repeatedly gave examples of new, environmentally responsible processing that also brought cost-savings and better products.

Naturally, it doesn't always immediately work this way. Sometimes it takes a little longer. Consider the search for new refrigerants to replace chlorofluorocarbons (CFCs) in refrigerators and air conditioners. Environmentally, CFCs are chemical moths chewing holes in the fabric of Earth's stratospheric ozone shield. To date, no replacement refrigerant gives as good performance as the old CFCs. But let's wait and see. Perhaps it's because the *service technology* (the refrigerator) hasn't much changed, just the refrigerant. In our home-heating and illumination examples, the service technology did change. We didn't merely substitute a new currency in an old technology. Before we could switch from whale oil and coal gas to electricity, we had to invent light bulbs.

Imagine yourself elected as a 'green' legislator in the late 19th century. Your constituents are complaining about soot and dirt from oil lamps. Some think the acrid smell is causing grandfather's headaches. Pesky people are demanding 'something' be done. They want you to push for legislation requiring catalytic converters on the chimneys of all future oil lamps—to catch those greasy emissions before they darken the parlor. Moreover, they want subsidies to help them install those catalytic converters on their existing lamps. Will you satisfy your constituents by introducing catalytic converter legislation? Or will you try to promote electric lighting, perhaps by encouraging community electrification?

Most of your constituents want catalytic converters; people like direct legislation for direct trouble. Many fear newfangled electric lighting, perhaps influenced by lobbyists who, to show the dangers of electricity, have been electrocuting a stray dog each day in Central Park, Manhattan.* Can you convince your constituents that cleaning the place up will be easier and safer than they expect? Can you convince them to change the process rather than add a collector?

Now, in the first decades of the 21st century, a little more than a century after people began switching from lamps to light bulbs, we'll increasingly face that same question. Specifics will differ, but the *issue* will be the same.†

Process changes will take us to a brighter, cleaner, richer future. Collectors won't.

---

* The lobbyists were explicitly protesting alternating current.

† Today, the much promulgated idea of carbon sequestration is a classic of the add-a-collector genre. Chapter 32 'Harvesting Hydrogen' and Chapter 17 'Hydrogen: The Case for Inevitability' discuss where sequestering might work and where it surely won't.

11. TIDAL FLATS AND AIRPORTS:
QUANTIFYING ENVIRONMENTAL INTRUSION

In which we begin developing a template to help us sort out what matters from what doesn't.

It's the January rainy season on Canada's west coast. I'm on a twenty-minute connector flight from Victoria to Vancouver. The Dash 7 flies over the Gulf Islands, then over the San Juan archipelago and finally over the Strait of Georgia. Passengers look down upon the islands and on the ships ending or beginning their voyages that link North America to the world. I always enjoy this flight.

Today, the takeoff is in rain showers but there are breaks in the clouds allowing the morning Sun to spotlight random islands along our route. The aircraft is landing in Vancouver from the west, droning in over the tidal mud flats of the Fraser delta. The glide path appears to be taking us into a rice paddy—the slick runway, a mere wavelet above the ocean. I'm overtaken by an image: The latest rain shower could be an advance paratroop battalion for a coming oceanic attack upon the runways.

I think of the many airports where the glide path's last few kilometers skim above water, many with runways projecting out over the sea. My mind jumps back to news of aircraft toppling off the ends of these runways and into Tokyo Bay, Boston Harbor, Hong Kong Harbor and, indeed, into harbors around the world. None of this should be surprising, since the airports of eight of the world's ten largest cities are only a few meters above sea level. Indeed, some whole countries—countries having among the world's highest population densities like the Netherlands and Bangladesh—sit just a few meters above, or below, ocean level.

Before the wheels touch the runway my thoughts turn to the prospect of rising ocean levels caused by climate disruption. Many are persuaded this will be *the* environmental issue of the 21st century. I'm among these. Most blame will be laid on emissions from the energy system. Additional blame will be ascribed to deforestation and other land use changes.

As our aircraft taxis toward the terminal, I check the morning newspaper. What will be the environmental story of the hour? Today there are two: 'Parents Fear Results Will Come Too Late,' which is about exposure to electromagnetic fields from electric transmission lines, and 'Residents Opposed to Spraying,' about a projected infestation of gypsy moths that a government agency hopes to prevent by poisoning the larvae. All these thoughts remind me of how difficult it is to sort out what matters and what doesn't, from the plethora of environmental issues that get the presses rolling and our juices roiling.

So whatever you might think about our climatic future, this morning's flight and newspaper reinforce my belief that we need some underlying general principles to help us weigh the significance of the many environmental threats that compete for our attention.

We've talked about why changing processes rather than adding collectors is almost always the best way to clean the place up. But what do we mean by a 'clean place?' What does environmental gentility mean? To me, it means *to intrude as little as possible on Nature's flows and equilibria.**

Let's introduce clear principles that can put meat on what may seem a comparatively abstract idea. One way or another, *material* is always involved when we alter Nature's flows or equilibria.† So we must consider, at each link in the five-link chain, the impact of the:

- material *taken from* the environment (for example, the effect of strip-mining coal to feed steel mills, or cutting down trees to make paper),

---

* Equilibrium is one of those things set by the beholder's eye. Thermodynamicists will say a system is in equilibrium if, when isolated from all material or energy exchanges with its surroundings, it does not change with time. We might think of this as a kind of homogeneous internal equilibrium. But I'm using equilibrium the way some ecologists and atmospheric scientists might use the phrase 'labile equilibrium' to account for reactive processes. Labile equilibrium is an approximate steady state over comparatively long periods, where the continuous flow of material and energy sets up a sustained pattern of material and energy distribution.

† This is true even when material is not the direct intrudant. The newspaper story about exposure to electromagnetic waves from electric transmission lines is one example. These electromagnetic fields are caused by the movement of electrons, guided by the transmission lines and driven by spinning generators back at the power station. By the way, I know of no credible evidence that electromagnetic waves from transmission lines are a health hazard.

- material *diverted within* the environment (for example, the effect of dams that reconfigure river flows to harness waterpower for electricity generation), and

- material *put into* the environment (for example, the effect of dumping more than 60 million metric tonnes of carbon dioxide and 360,000 metric tonnes of sulphur dioxide (SO_2) into the atmosphere, daily).

Yet it is not enough to consider just the *direct* impact of the material. We must also understand how this material modifies other aspects of Nature's equilibria.

Let's develop this idea using the example of the CO_2 we dump into the atmosphere when we burn either fossil or biomass fuels. There is no direct impact of CO_2 upon people. It's not toxic. It's not smelly. And except in very high concentrations—above 1% of the air we breathe—carbon dioxide causes no health problems.* Rather, it is carbon dioxide's *intervention* in the *outbound* flow of infrared radiation from Earth to the universe—restricting the pathways, jamming the pipes—that changes CO_2 from something that really doesn't matter into something that will make all previous anthropogenic disruptions seem trivial. That's because, if less energy leaves Earth (as radiation to the universe) but the same amount keeps coming in (as radiation from the Sun), then the difference is an energy buildup on (or within) Earth. This energy buildup will cause warming. More important, the resulting planetary energy imbalance will do many other things, lots of them unpredictable, like altering atmospheric circulation that changes the intensity and frequency of tropical storms or the patterns of rainfall and cloud cover. This energy build-up will also increasingly accelerate the melting of polar ice caps, thereby further raising ocean levels that have already begun rising due to the thermal expansion of ocean waters—which is what got me thinking about this business while on the glide path into Vancouver.

---

* CO_2 buildup in our lungs (and blood) triggers our need-to-breathe response. That is why SCUBA divers can return to the surface from great depths by continuously exhaling as the pressure drops in the surrounding water—without feeling the need to take another breath. Of course, they must exhale; otherwise they'll explode.

The chlorofluorocarbons (CFCs) that civilization has sent into the sky are another example of how we intrude upon Nature's equilibria and flows. As we've said, CFCs are the stuff of yesterday's refrigerators and aerosol cans. They alter Nature's stratospheric equilibria because, upon decomposition, the released chlorine acts as a catalyst for processes that destroy ozone. In turn, because ozone filters out ultraviolet (UV) radiation from inbound sunlight, the reduced ozone lets more UV reach Earth's surface. This additional UV is causing folks more severe sunburns and increased skin cancers. Moreover, some scientists think it might also be a reason for the disappearance of frogs, toads and other species particularly sensitive to UV radiation.

The CFC example demonstrates a three-step intrusion. First, the intruding material acts as a catalyst for reactions that alter Nature's chemical equilibria in the upper atmosphere. Second, this resulting lower ozone concentration allows more UV to reach Earth's surface. Finally, higher UV levels cause sunburns and, perhaps, frogicide. Then again, some will see benefits. Sunblock manufacturers might see their stock prices rise.

When speaking of dangerous amounts of material intrudants in the environment, the question becomes: How much is too much? Too much means *compared to the amount in the environment before people came along.* Engineers and scientists say, *compared to the background level.* To help us with the idea of too much relative to normal background levels, let's think about a bowl of chicken soup.

It should not be surprising that adding a teaspoon of arsenic to your soup—say, as arsenious oxide (As_4O_6)—will cause you a bigger tummy ache than adding a teaspoon of water. Although you added the same amount of arsenic as water, it was the arsenic that caused grief because the normal level of arsenic in a bowl of soup is about 1.2-millionths of a gram. Compared to that, a teaspoon of arsenious oxide contains about 15 grams of arsenic. So the arsenic-fortified soup contains about twelve million times as much arsenic as the unfortified soup. By contrast, an additional teaspoon of water represents an increase of a mere 2% in the amount of water already in your soup. Your bowl will contain just a bit more of slightly thinner gruel, but it's unlikely to give you excruciating pain throughout your gastrointestinal tract.

'Well!' you might say, 'that's because arsenic is a poison and water isn't.' But now imagine increasing the water in the original soup above its normal level by that same 12 million times. This would mean your original bowl of soup expanded into about 3.6 million liters. Your bowl of soup would fill about three Olympic swimming pools. So if you were forced to consume a bowl of soup enriched by *either* 12 million times the usual amount of arsenic or 12 million times the usual amount of water, in the normal bowl-of-soup-slurping time, you'd have a tummy ache. Indeed, in either case it would be your last bowl of soup.

I have provided no details of what caused this arsenic or water overload to bring about your untimely end—said nothing about how these excess quantities affected your physiology. But you don't need to know the details to know that increasing the normal level of any of your food components by such a large percentage will certainly bring trouble, even if you are spared the fine points of why you would croak. It's not that 12 million times the normal amount of something will always have equally bad effects as 12 million times of another substance. But there is surely a pattern here.

The background levels of arsenic and water in *our* world determine how *our* physiology will respond to soup fortified by these commodities. Life on Earth was born and evolved in an environment with arsenic and water in the ratios encountered on Earth. But if life had developed on a world with 6 million times the concentration of arsenic, critters on that world would be able to enjoy a lot more arsenic in their soup than we do. Indeed, if they didn't get enough arsenic, they might be in trouble.

The chicken-soup story gives us a key template for sifting out what really matters from what doesn't. When we want to judge the severity of any environmental intrudant, we should always first ask: By how much will the intrusion alter Earth's equilibria as a percentage of background levels of the intrudant?

One bowl of soup containing a teaspoon of arsenic will give one soup slurper a bad time. Yet, except for the immediate family and a few friends and colleagues, the rest of the world won't be much affected. But if suddenly every bowl of soup contained a teaspoon of arsenic, and an equivalent concentration were found in every bird bath, marsh and ocean, our world really would be disrupted.

So our principles of environmental intrusion need also to account for *spatial distribution*—the volume over which the intrusion occurs. If we want to consider the environmental impact of a specific intrusion on Earth's geophysiology, not just on our backyard, we need to ask: Over what *fraction of the planet* has the intrusion brought changes to normal flows and concentrations? Because what happens to the Earth's geophysiology will, in the end, surely arrive in our backyard.

Time is also a key consideration. How long will the intrusion, the mess we made, stay around before Nature's normal cleansing processes wash it away? If, for example, we stop injecting CO_2 and CFCs into the atmosphere, how long will it take for natural processes to lower their concentrations to their unperturbed levels? Engineers or scientists might put the question this way: What is the characteristic *residence time* of the intrudant? If a mess we make can be washed out in a few days, such a mess is less critical than one that won't be washed out for several hundred years.

We have now gone a long way toward developing the criteria for weighing the importance of different environmental intrudants—criteria that will help us sort out what matters, from what matters less, among the plethora of environmental issues that populate the daily news. The basic question comes down to this: *How much will the environmental disruption change Nature's equilibria and flows? And over what area? And for how long?*

12. Metastability: When You Can't Go Home Again*

In which we take our first look at climate disruption.

No one has the specialized knowledge to weigh *every* environmental threat. Reading the news, it seems impossible to determine which threats are nonsense, which trivial, which critical. That is why, in the last chapter, we began developing a template that anyone—scientist, engineer or layperson—can use to weigh the importance of the environmental challenges thrown at us daily. The template can be a prophylactic against panic. It can also warn us about very real dangers that we might otherwise dismiss.

We cast our template in terms of a single overarching question: What is the magnitude of the material intrusion upon Nature's equilibria? And to answer that question, we need to ask four more:

- How much does the intrudant push the environment away from normal conditions?
- How much of our planet is infected by the intrudant?
- How long will the intrudant stay around?

Yet, important as they are, these three clarifying questions are but a start. That's because perhaps the scariest question of all is this:

- Is the intrusion upon Nature's equilibria so large that, even if we stop causing the intrusion, natural processes will be *unable to return things to where they were?*

To answer this last question, we need the idea of *metastable* equilibrium. The meaning of metastability is illustrated by Figure 12.1, which shows steel balls sitting on different undulating surfaces. The figure shows three different equilibrium categories. All three systems are in some type of equilibrium. So

---

* *You Can't Go Home Again* is the title of a novel by Thomas Wolfe, published in 1940.

unless something pushes the ball to one side or other, the system will stay in its original position forever.

Stable   Unstable   Metastable

Figure 12.1 Examples of stable, unstable and metastable equilibrium.

But if something does nudge the ball, the three systems behave very differently. If the ball is in the stable equilibrium bowl, then no matter how big the force, the ball will always rolls back toward its original position when the force is removed. If the ball is in a state of unstable equilibrium, then no matter how small the push, the ball will roll away never to return.

However, if the ball is in metastable equilibrium, then it will respond in one of *two* ways, depending on the size of the push. A smallish force that doesn't shove the ball over the lip will take it away from its original position, but when the force is removed, the ball will roll back. But if we keep increasing the force, or give the ball a big push, at some point the ball will slide over the lip so it runs away—never to see its original position again.*

---

* I've often thought if something is worth knowing it's worth understanding in several ways. This jumped back into my mind when looking at the metastable case of Figure 12.1. As a young teenager, I read Arthur C. Clarke's *The Exploration of Space*[9] in which he explained the idea of an 'escape velocity' from Earth. His explanation used something that looked like a tall vase, but it had a lip. At lower speeds, whatever was thrown up the side of the vase would always roll back. But at some speed, the thrown up thing would fly over the lip and never return. For me, this was a profound revelation. For then I knew that, someday, space travel would happen because it *could* happen. One of my afflictions is an extremely poor memory for names, dates and numbers. But I've never forgotten that the escape velocity from Earth is 25,000 miles per hour (actually, just a bit

Sometimes systems have several metastable states. Figure 12.2 shows a system with three. When the ball is in any of these states, it can tolerate gentle forces that move it, left or right, without taking it out of its local equilibrium bowl. But if the force pushes the ball over the lip of its local bowl, it will run away, probably to a new metastable state, settling down to a new life tolerating little pushes. At least until the next big one comes along.

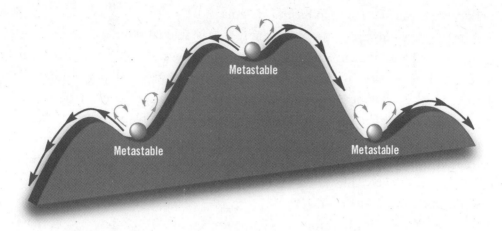

Figure 12.2 A simple system with three metastable states.

As this is a book restricted to two-dimensional pages, Figures 12.1 and 12.2 can only illustrate one-dimensional stability: The force on the ball depends only on the ball's *horizontal* position—because the force depends only on the *slope* of the surface, and the slope is determined by horizontal location (and, of course, gravity and the mass of the ball). When we choose a horizontal position, we precisely define the location of the ball on the undulating surface. Because we live in a world where motion is dominated by gravity, our mind's eye automatically took the height of the surface to mean elevation. So the force we were implicitly thinking about is the force needed to push the ball up an inclined slope against gravity—or to *hold* the ball away from its original neutral

---

less). This was long before I learned to think in Système Internationale (SI) units. And the idea that Earthlings would someday soar into space shaped my life from then on.

position, similar to the force the handbrake uses to prevent your car from rolling down a slope and into your neighbor's garden.

If the figures had been drawn three-dimensionally—showing, for example, the topology of a piece of thin sheet steel after it had been left outside during a violent hail storm—then a steel ball rolling over the banged-up sheet steel would represent a system of two-dimensional stability.

Let's carry these ideas over to Earth's systems.

Most geosystems are a mosaic of many metastable equilibria—tapestries of metastable states linked by interwoven forces and flows. Each force and flow can be considered a different dimension of stability. Nature's stability profiles are almost never neatly one-dimensional. Moreover, it is only from the perspective of our brief lives that we get tricked into thinking our planet's equilibria are static. Rather, they are moving targets that most often gradually, but sometimes suddenly, move to new states of equilibria. The realization that most of Nature's equilibria are metastable is essential to understanding civilization's impact upon the environment.

So how *far* we push things is vitally important. Too often people have caused environmental perturbations that *irreversibly* pushed the equilibrium of thousands of years into new circumstances. One example is when we change the environment to push species of flora and fauna into extinction.

Climate change is another. That's what I was thinking on the commuter flight between Victoria and Vancouver. It's not that I am obsessed with a few degrees of warming. Rather, what has me worried is that we don't know where the 'climate lip' *is*. And it is a lip we are rapidly approaching—if we're not already beyond it. Should that seem hyperbolic, remember that we have already increased atmospheric carbon dioxide to concentrations much higher than our world has seen for more than 600,000 years.* If, as looks likely, we push our climate over the lip, then we don't know where the next climate metastable state will be, or with what speed we'll be carried toward it.

---

* We will return to this data in the next chapter, 'Bubble, Bubble, Signs of Trouble.'

Climate is determined by so many linkages, so many dimensions of metastability, that predicting Earth's response to greenhouse-gas intrusion is extraordinarily complex. Perhaps it is the most difficult physical modeling task ever undertaken. So it is not surprising that there are uncertainties, which some newspapers (and all coal producers) are delighted to use as a basis for saying, 'there are even disagreements among scientists.'

Yet we must be clear on what the uncertainty is about. It's not about whether or not our climate is dangerously perturbed. That's for sure. Rather it's about the magnitude, timing and details of the changes that will result. Furthermore, today's attempts at climate prediction usually assume that the climate stays within its metastable equilibrium bowl. If we try to anticipate where the edge of the bowl is—if we ask, 'Where is the tipping point and what happens after we topple over it?'—then modeling difficulties are compounded many times.

We are not endangered only by those things we can predict accurately. If you take an afternoon stroll through a minefield in Kuwait or Cambodia, the fact that you can't predict *where* your foot might tread on a land mine will not change the way you will be spread about the landscape when your foot finds one. We should temper our feelings of smug well-being the next time we read that scientists can't agree about the details of what climate disruption will bring.

Sometimes, it's difficult to give numbers to our four measures of environmental intrusion—or to its flip side, environmental gentility. Yet even without numbers, the process of asking the right questions can bring a lot of understanding.

Often the things that count the most are not countable. As an engineer trained in quantification, and typically suspicious of claims without quantification, this is a truth I've learned slowly. Yet now, even without numbers, I realize that a good appreciation of what matters and what doesn't can be gained from even a cursory examination of our four criteria:

- the intrudant level compared to background level,
- the fraction of the global commons intruded,
- the residence time of the intrudant, and
- whether the disturbed equilibrium is metastable.

I'm frequently shaken by the intensity of public concern for issues that would be seen as trivial by a simple application of these criteria. On other occasions, I'm troubled when grave planetary environmental issues are dismissed because we haven't seen their effect in our backyard yet, but which should be flagged as critical.

Now comes a sting—our ability to judge good or bad by using our four criteria can depend on our perspective. Sometimes we might intrude upon the natural environment to make things *better*. For example, if the regional climate in the lee of a continental mountain range has caused a hostile desert, we might modify Nature's flows by building irrigation channels that change the desert into a lush, fertile Garden of Eden.* Many would claim irrigation improved the environment, and I accept that judgment. But we must temper it with the question: Improved for whom? The scorpions and cacti that see their homes destroyed would not be so pleased. And, if we're thinking *systems*, what happened to the land or rivers from which the irrigation water was taken?

Or consider our (almost complete) eradication of smallpox, one of Earth's 'natural' diseases. Was that eradication an environmental crime? I think not. But it certainly was an intrusion upon Nature's equilibrium among diseases.

Environmental gentility is in the beholder's eye, and we are the beholder. It's tough to acknowledge that neatly cut principles, even straightforward ones like those in this chapter, must be applied using human values. Some things can't be absolute.

On the glide path into Vancouver International I was thinking about climate instabilities. But my thoughts did not dig into the quantum mechanics of how tri-atomic CO_2 catches long-wave infrared radiation, nor to the complexity of atmospheric-oceanic-continental linkages that will determine how the details play out, nor to the physiology that characterizes how UV influences Earth's flora and fauna. Not even to the mechanisms of how all this will tie back to

---

* Recently, it's been suggested that the biblical Garden of Eden was located in the thousands of marshland hectares in southern Iraq that, on the orders of Saddam Hussein, were drained during the 1990s. As an example of 'unintended consequences,' if Saddam had not drained the marshes, the US drive up from Basra would have been *much* more difficult.

greenhouse-gas-induced climate instabilities. Rather I skipped the details and went in search of patterns.

For most people it's enough to know that the growth in atmospheric CO_2 caused by human activities is altering our planet's equilibria and disturbing Earth's flows of inbound and outbound radiation. But now we have a set of questions, the answers to which can give us a good indication of whether these threats are real, or hype.

Let's consider the metastability of climate systems.

The rapid changes between glaciation and non-glaciation periods, with comparatively long periods of relative stability in between, sharply demonstrate shifts from one metastable climate to another. Greenland ice core drilling has shown that 'rapid' can mean in as little as a few decades—very quick, indeed, when compared with the millions to billions of years typical of geophysical change. Moreover, there has been a long-term but steady *decline* in atmospheric CO_2 since the dawn of the Proterozoic Age, some 2.5 billion years ago. Gaia theory argues, persuasively for me, that living processes have been reducing atmospheric CO_2 as a means of keeping Earth at 'comfortable' temperatures as the Sun's temperature increased. The Sun now sends about 30% more energy to Earth than it sent during Earth's childhood more than 5 billion years ago.

From this perspective, by using Earth's atmosphere as our fossil-fuel waste repository and thereby pumping CO_2 up, we're pushing against Nature's much longer trend of pumping CO_2 down.

As a key industrial-environmental strategy, I've recommended changing processes to stop making pollutants, rather than adding collectors to catch pollutants. In the preceding two chapters we've carried these ideas a bit further and now conclude:

- Whenever we introduce technologies, the interaction of these technologies with the environment should, as much as possible, be coherent with Nature's flows and equilibria.

- And we must *never assume* we can balance the intrusiveness of one action with a *second countervailing intrusive action*.

Too often we look for technological fixes to clean up our messes, rather than looking for non-intrusive environmental pathways from the beginning. Too often we employ technologies that intrude on an already disrupted Nature in an attempt to somehow balance the original intrusion. To me, this is to play dice with a very complicated planet. People have grown distrustful of technological fixes—as well they should, because technical fixes characteristically fail with spectacular results. Unfortunately, this distrust sometimes spills over, indiscriminately, to all technologies.

The year 1988 may have been when concerns for the 'greenhouse effect' and the 'destruction of the ozone layer' first began seeping into the thinking of the person-in-the-street. In response, that year's August 14 'Science Times' section of the *New York Times* carried an article with the marvelous technological-fix headline, a classic of the genre: 'Scientists Dream Up Bold Remedies for Ailing Atmosphere.'

Fascinated by technological fixes, the article said, 'Work on futuristic cures for these pollution problems started in the 1970s and has accelerated in the 1980s as concern has grown about potential damage to the atmosphere.' The proposed futuristic solutions advocated in the article include:

- placing giant laser guns atop mountains. . .to blast apart one million tons of CFC per year;
- replacing depleted ozone with ozone manufactured on Earth and lofted into the stratosphere by rockets, aircraft or balloons. . .or firing aloft bullets of frozen ozone. . .
- increasing the reflectivity of Earth's atmosphere [by using sulphur dioxide] so that more sunlight is reflected back to space. . .about 35 million tons of sulphur dioxide would have to be transported to the stratosphere each year. . .but the method would have drawbacks. . .increase[ing] acid rain and give[ing] the blue sky a whitish cast;
- launching giant orbiting satellites made of thin films to cast shadows on Earth, counteracting global warming;
- fertilizing the oceans to spur the growth of phytoplankton [which eat CO_2] . . .and, when they die, will sink to the bottom and turn into limestone;

- covering much of the world's oceans with white Styrofoam chips which would reflect more sunlight back into space than regular ocean water, and painting the roofs of all houses white.

To give the article its due, Princeton University geoscientist Michael Oppenheimer is quoted as saying at the article's end, 'It's probably cheaper in the long run to rely on prevention rather than unusual cures.' This is a masterpiece of understatement—especially when there are obvious, straightforward ways to run our energy system, and the refrigerators that are a small part of that system, without emitting a drop of CO_2 or CFCs.

Some eighteen years after the *New York Times* article, technological fixes are still being proposed. The June 5, 2006 edition of the *Globe and Mail* carried a front page article 'Going to Extremes to Fight Global Warming.' Many of the earlier *New York Times* ideas were replicated and a few added, such as launching a fleet of 55,000 mirrors, each larger than Manhattan, into space to reflect sunlight—at a cost of US$120-billion with replacement costs of up to 40 times more. It was, however, pointed out that the mirrors would cause 'sunlight' to flicker, 'which could be distracting.' The piece also spoke of launching dust or other particles (let's say it straight, dirt) from naval guns into the upper atmosphere. Stand by for more nonsense to come.

When I look down from commuter flights linking Victoria, Vancouver and Seattle hoping to see a pod of killer whales coursing through the islands, I'd prefer not to see Styrofoam balls stuck in their blowholes.

13. BUBBLE, BUBBLE, SIGNS OF TROUBLE 🌶

In which we exhume ancient ice-entombed bubbles to analyze more than 600,000 years of Earth's geophysiology, and go on to examine other evidence that ratifies the prospect of human-caused climate disruption.

To maintain Earth's thermal balance, the energy arriving as sunlight must equal the energy leaving as infrared (sometimes called thermal) radiation. If more energy arrives than leaves, our planet heats up. If more goes out, the planet cools. This energy balance is called 'radiative forcing.' It's elementary accounting. When Earth's radiative forcing goes out of balance, we had better watch out. That's what climate disruption is all about.

Incoming solar radiation has been reasonably constant over millions of years, with minor fluctuations caused by solar flares or the wobble and tilt of Earth's orbit.* In contrast, the net outgoing radiation can change more rapidly and substantially—sometimes brought about by planetary processes that have been triggered by those wobbles and flares.

In 1824 the French mathematician Jean-Baptiste Fourier proposed that the atmosphere absorbs outgoing infrared radiation from Earth and reradiates some of it back to Earth. Today, well-established physics explains why gaseous molecules of more than two atoms like CO_2 and H_2O (molecules of three atoms) and CH_4 (a molecule of five atoms)—absorb outward-bound longer-wave radiation before it can escape into the furthest reaches of the universe. A very few diatomic molecules also have this property. We call these infrared-absorbing molecules 'greenhouse gases.' An increase in atmospheric greenhouse gases will trap an increased amount of outgoing radiation—reradiating some of it back to Earth, thereby increasing Earth's temperature. The reverse occurs if greenhouse gases decrease. So the mechanism that drives 'global warming'

---

* But not billions. Today, the Sun is giving off about 30% more energy than it did when Earth was born.

has been understood for more than a century. It is straightforward physics, 'Elementary, my dear Watson!'

The terminology of human physiology can be adapted to Earth's geophysiology. I attribute this approach to James Lovelock and his colleagues who developed the idea of Gaia—although others may have used it before. Using this language, today's climate disruption is Earth's most critical geopathology.* And carbon dioxide is the most important of the accused geopathogens. As often happens in the physical sciences, the cause of this geopathology was proposed long before there was experimental evidence. Then, more than a century after Fourier explained the mechanism, the geoepidemiological evidence began pouring in to confirm the mechanism. In this chapter we will focus on the confirming evidence, but only touch on some of the consequences. In Chapter 14 'Controversies, Conveyors and Consequences,' we'll expand on the physics and introduce more of the consequences.

Bubbles of our planet's ancient atmosphere, exhumed from polar ice laid down over eons, reveal more than 600,000 years of Earth's history. The bubbles expose the history of global atmospheric temperature and its corresponding carbon dioxide and methane content. Both Greenland and Antarctic ice caps have yielded mutually confirming ice core data. These bubbles forecast civilization's future even better than the witches presaged Macbeth's with their 'bubble, bubble, toil and trouble.'

The data of Figure 13.1 comes from the Vostok site[10] in Antarctica.[11] It's both an extraordinary record, and unequivocal writing-on-the-wall about the tight links between climate and greenhouse gases which, today, many still claim they can't see. In 2005, as these 1999 Vostok results settled into scientific minds, a different group of researchers reported ice core data from Dome Concordia (Dome C), another Antarctica site. The Dome C data confirmed and expanded the Vostok data, and took the time line a further 230,000 years into the past— to 640,000 years before today.[12] In early 2008, I learned that the temperature,

---

* I said Earth's geopathology, not civilization's, whose pathologies include religion, or tribally inspired conflicts.

CO_2 and CH_4 correlations have been extended still further. Now scientists have data going beyond 800,000 years into the past—all of which confirms the more recent data.

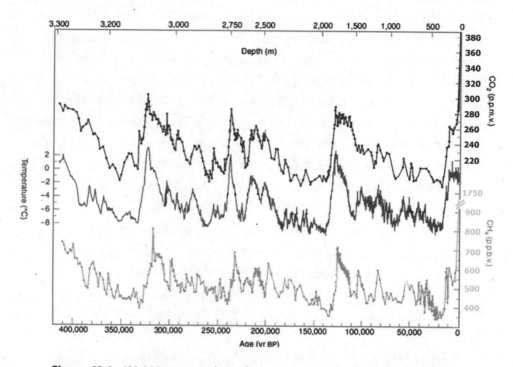

Figure 13.1 420,000 year correlation between Earth's temperature, CO_2 and CH_4.
Adapted and redrawn from Petit, J.R. et al (1999)[11]

I've used the Vostok data rather than the Dome C data in Figure 13.1 because, of the diagrams I was able to access, the Vostok presentation seemed easier to read. The correlation between temperature and the two dominant greenhouse gases (carbon dioxide and methane) over this sweep of time, is astounding. Let's start by considering the most obvious conclusions:

1. The cheek-by-jowl links between global temperature and the greenhouse gases brings to mind *three* dogs on short leashes, out for a walk with their master. This temperature greenhouse gas correlation can become a little unstuck for short excursions and brief times—just as one dog may wander briefly to sniff a shrub or hydrant. But for time frames of millennia, the

correlation is stunning. Look again at Figure 13.1. No rational person could argue the relationship is merely a stochastic fluke.

2. Over the last 420,000 years, climate has experienced four glacial cycles—times when the world was much colder than today, interspersed with times when it was much warmer. Both greenhouse gases have tracked the temperature upwards and downwards. Yet even in the warmest periods, atmospheric CO_2 levels never exceeded 300 ppmv (parts per million by volume) until today. Today CO_2 has already reached 380 ppmv.* This is more than 36% higher than at the start of the industrial age, and more than 22% higher than at *any time during the past 420,000 years* (or 640,000 years if we use data from Dome C). Moreover, geophysicists and climate scientists agree that there is zero prospect of stopping this increase before CO_2 levels reach 450 ppmv—and little prospect of stopping even then.

3. Atmospheric methane has followed much the same behavior as CO_2. Yet, as a percentage, the jump in methane is even greater than the jump in carbon dioxide. Today, methane has reached levels more than double those at any previous time during this same 420,000-year period. (I'm told that there is some evidence that the rate at which CH_4 is increasing has slowed in recent years.)

4. Looking at Figure 13.1 you will notice the data record shows increased 'noise' closer to the present (this is most obvious from the temperature data, which appears 'fuzzier' closer to the present and 'sharper' as we go further into the past). The reason is simple. As new layers of ice build up over the millennia, the deeper layers are compressed by the immense overburden. This deep-ice compression gradually smooths out short-term fluctuations in the air-bubble samples. Shallower, newer ice samples retain greater differentiation between years and decades. This is an artifact of when the

---

* The manuscript for the first edition of *Smelling Land* was on its way to the printer in January 2007, when the Intergovernmental Panel on Climate Change (IPCC) released its report *Climate Change 2007: The Physical Science Basis, Summary for Policymakers*. *Smelling Land* is entirely consistent with the IPCC's broad findings. The single caveat is in this chapter where I said that nighttime low temperatures were predicted to have greater increases than daytime highs. The IPCC report confirmed this prediction but noted that this phenomenon has not yet been observed worldwide.

bubbles were entrained—not a fundamental change in long-or short-term behavior.

5. One of the most important and unequivocal conclusions from this historical record is that climate systems are masterpieces of metastability.

There can be no doubt about these five observations. Nevertheless, some who would deny anthropogenic climate disruption choose to interpret the data of Figure 13.1 as proving, only, that climate has always jumped around, to argue, 'so what's the big deal about today?' What they miss is the magnitude and speed of today's human-triggered changes. To suggest that today's changes are just natural—not anthropogenic—is, in my view, to demonstrate either a unique immunization against scientific evidence or a stunning ability to select from and manipulate it.

Now let's consider a cause-effect postulate that, as far as I know, cannot be proven but which I find both interesting and plausible.

6. Anthropological studies indicate humans have changed little during the past 130,000 years. Yet civilizations (structured societies of reasonably large groups of people) first arrived no more than 10,000 years ago. Place this bit of *human* history against our *planet's* history. Within the context of the past 640,000 years, the most recent 10,000 to 12,000 years (the Holocene) have been an era of unprecedented climate stability. Many scientists believe it's probably unprecedented during at least the last million years. Putting these two histories together, some speculate that this exceptional climate consistency is what allowed civilizations to take root. If climate had continued to flip sharply back and forth, as it had throughout Earth's previous history, the agriculture that's required to sustain civilizations could never have gained a toehold. People would have remained trapped in small, primitive, hunter-gatherer groups with cruel, short lives of discomfort and uncertainty.

I'm unaware of a scientific explanation for why Earth is experiencing this climate-stable streak of 10,000 to 12,000 years. But whether or not we understand why, isn't it extraordinary that Earth's climate over the Holocene has enjoyed such unmatched stability? And aren't we damn fools to be jabbing this stability with such a sharp stick?

Let's explore in a little more detail how Nature laid down her record, and how today's scientists interpret it. When rain or snow falls onto the ice caps, air is entrained. This air later becomes entombed bubbles when precipitation compacts and freezes. Each year a new layer containing samples of that year's water and atmosphere freezes. Through millions of years, ice layers accumulate to preserve a record of Earth's atmospheric chemistry—not unlike the way a tree's growth rings record its history. Scientists in parkas have now drilled through the Antarctic ice cap to a depth of 3.6 kilometers (2.23 miles), bringing up ice-entombed atmospheric samples from the past 420,000 years.[14] Other scientists working at Dome C have brought up samples going back 640,000 years.

There is some smearing of the time-line correlation between the air bubbles and their surrounding water, because the air can migrate short distances up, or down, through the different annual ice layers for a decade or so, until it becomes firmly locked in bubbles. Still, the migration—even if over several decades—doesn't change the fundamental correlations demonstrated in Figure 13.1.

It's easy to understand how bubbles of ancient air can tell us the percentage of Earth's ancient atmosphere that was carbon dioxide or methane by using diagnostic equipment like a gas chromatograph. But how can ice core samples tell us about temperature? Clearly these bubbles don't host micro-thermometers that record temperatures over hundreds of millennia. Yet global temperature can be inferred, from the ratio of oxygen and hydrogen isotopes in the water that constitutes the ice itself. This process is analogous to using the ratio of carbon isotopes to perform what we call 'carbon dating' when examining old bones, manuscripts or the Shroud of Turin.

This is how it works for ancient atmospheric temperatures. There are two methods, both using the same principle. The first involves oxygen isotopes and the second, hydrogen isotopes. Isotopes are different forms of the same element, differentiated by the number of neutrons in the element's nucleus.*

The most common oxygen isotope is oxygen-16 (^{16}O). However, about one in a thousand oxygen atoms is the heavier isotope, oxygen-18 (^{18}O). Because oxygen-18 is heavier than oxygen-16, water molecules composed of the heavier oxygen isotope ($H_2^{18}O$) are heavier than normal water ($H_2^{16}O$). The lighter molecules evaporate more easily from the ocean surface, because being lighter, they jump into the air more easily. This causes the residual surface water (the liquid water left behind) to become slightly oxygen-18 enriched, oxygen-16 depleted. The differential between the ratios of oxygen18 to oxygen-16 in the residual water becomes more pronounced as temperature increases. Therefore, the change in isotopic ratios at different layers of the ice core mirrors the change in temperatures at the time the ice was formed.

The second method involves the same principle of isotopic-mass ratios, but uses hydrogen isotopes. The nucleus of a normal hydrogen atom contains a single proton and nothing else. But about one in every 6,500 hydrogen atoms also contains a neutron and we call this hydrogen isotope 'deuterium' (D). An atom of deuterium has twice the mass of ordinary hydrogen, so water containing the deuterium isotope is heavier than 'ordinary' water. Again, because it's lighter, ordinary water (H_2O) will evaporate more easily than water containing the deuterium isotope. And again the effect is more pronounced as the temperature rises. The changes in H/D ratios parallel the change in temperature.

The entombed air comes primarily from high-latitude atmospheres. Yet atmospheric mixing and the relatively short time it takes for the atmosphere to move around the world, ensure the air bubbles give a good approximation of worldwide atmospheric composition.† Evaporated oceanic water preferentially comes from high latitudes and—due to a multi-evaporative trail before it's finally laid down in the icecaps—the 'proxy temperature' from isotopic ratios

---

* For a discussion of isotopes, see Appendix B.3.

† The seasonal variation in atmospheric carbon dioxide samples taken at Mauna Loa on Hawaii, which reflects the northern and southern hemisphere growing seasons, might be an indication of the deviation of high-latitude atmospheres from a globally integrated atmosphere.

represents the temperature over the ice caps. Nevertheless, while the Vostok measurements of carbon dioxide, methane and (especially) temperature are weighted toward conditions in Antarctica, they are good indicators of global changes from one year, or decade, to the next.*

I've chosen to speak primarily about the Vostok data. But Vostok does not stand alone.

We've already identified the 640,000-year data from the Dome Concordia Antarctica site. If we move to the northern hemisphere, Greenland ice cores show the same behavior and follow similar time lines. Data from lower-latitude mountain glaciers provide further confirmation over corresponding time frames. So it's reasonable to assume the entire atmosphere followed similar time line changes. Indeed, an array of paleo-climatic indicators, drawn from tropical as well as high-latitude regions, corroborates the ice core data. These data include plant pollens buried deep in ancient bogs, lake and ocean bottom sediments, residues of aquatic life forms and corals.

Too often we refer to climate disruption as 'global warming.' Warming is only first in a series of cascading cause-and-effects. For example, planetary warming brings higher-temperature oceans, which increases atmospheric moisture content. And as water is also a greenhouse gas, this increased atmospheric moisture is a positive feedback that further reinforces temperature increases.† And so on. Some of the effects are not very important. Locally, some changes might even bring benefits. Others could bring human catastrophes.

More than a hundred years ago, scientists and mathematicians began predicting planetary warming if greenhouse gases increased. More recently, to anticipate features of our future on a warmer planet, a new generation of researchers has studied a variety of possible cause-and-effects. I've enjoyed following this science for more than a quarter-century and find it remarkable

---

* The temperatures represent comparatively local conditions, while the gases tend to represent global atmospheric conditions.

† Water is the largest contributor to natural greenhouse effects. But the question for us is: How large is the contribution of the *anthropogenic* increase in water vapor? Increased water vapor can also augment the amount and category of clouds which can reflect incoming sunlight off into the universe—a negative feedback to warming.

how, one after another, most of these projections have been validated. Here are a few examples, starting with the straightforward evidence of warming, and then proceeding to other predicted cause-effect phenomena:

Scientific Prediction: Atmospheric temperatures will increase.
- *Reason:* Greenhouse gas-induced warming.
- *Confirming Evidence:* Global average surface air temperatures have increased by about 0.6°C since the beginning of the Industrial Revolution. In Canada, south of 60°N, the temperature increase averages about 0.9°C. Further north in the MacKenzie basin, the temperature has increased by about 2°C and the summer Arctic sea ice has decreased by 30% during the last 100 years.

Scientific Prediction: Ocean temperatures will increase.
- *Reason:* Greenhouse gas-induced warming.
- *Confirming Evidence:* Ocean surface temperatures have increased around the world, mostly in the north and south Atlantic where warming has been measured down to 700m or more, but warming has been confined to the upper 100m in the northern Pacific and Indian oceans. Warming typically ranges up to a little more than 0.2°C. Moreover, observations show that about 84% of the total heating of the Earth has occurred in the oceans.[13]

Scientific Prediction: Increases in nighttime low temperatures will be greater than increases in daytime high temperatures.
- *Reason:* The increase in aerosols (particulate matter, which includes water vapor from a warmer planet) has both cooling and warming effects. During the day, aerosols reflect incoming sunlight and trap outgoing infrared radiation—so both cooling and warming results. At night, there is little to reflect but trapping continues, so nighttime temperatures are increased more than daytime temperatures.
- *Confirming Evidence:* World-wide evidence that confirms nighttime low temperatures have risen more than daytime highs is mixed—although it has been confirmed in my home town of Victoria, Canada.

Scientific Prediction: The high-salinity, warm, northbound North Atlantic Conveyor*will slow. (Many of us think of this as the Gulf Stream, but the Stream is only one component of the conveyor.)

- *Reason:* As the high-salinity northbound surface currents in the North Atlantic Conveyor flow north, they cool, which increases their density and decreases their buoyancy. Decreased buoyancy causes the water to sink, after which it joins deep water, southbound flows along and over the ocean basins. But today, Arctic warming is accelerating high-latitude melting of land-borne ice, which is adding freshwater to the ocean surface, increasing its buoyancy and further slowing its sinking.†

- *Confirming Evidence:* Oceanographic data shows that the Nordic seas are freshening (becoming less saline). There are indications that the slowing of the North Atlantic Conveyor has already begun.[15]

Scientific Prediction: High latitudes will warm faster than low and equatorial latitudes.

- *Reason:* As high latitudes warm, the retreating snow and ice lowers the albedo (ratio of reflected/scattered sunlight to absorbed sunlight).‡ Therefore, more of the incoming sunlight's energy, previously reflected by ice and snow, is absorbed by the ocean and land. This is a powerful feedback mechanism for warming.

- *Confirming Evidence:* Over the last few decades, Arctic warming has been greater than temperate and tropical warming.

---

* The North Atlantic Conveyor is part of the global thermohaline circulation. In the scientific literature, there are many names for the North Atlantic Conveyor including 'Atlantic meridional overturning' that may be more accurate scientifically but, for me, less evocative of what's happening.

† The northbound thermohaline circulation sinks at two (or three) major down-welling sites in the North Atlantic.

‡ See Appendix B-7, 'The meaning of albedo.'

Scientific Prediction: There will be an increase in extreme weather events.

- *Reasons:* There are at least two. First, warm ocean surface waters energize hurricanes. As surface waters (and trade winds) move east to west in tropical zones, hurricanes typically develop toward the end of the summer in the western sections of both the Atlantic and Pacific. Second, warmer temperatures evaporate more water, increasing the moisture content of the atmosphere—the energy-storage reservoir for storms.

- *Confirming Evidence:* Although, so far, there has been little confirmation suggesting that the frequency of hurricanes and typhoons has increased over recent decades, there is clear evidence that the severity of storms has increased. The infamous Katrina is but one example. Stand by for more.

Scientific Prediction: Atmospheric methane content will rise.

- *Reason:* There are several mechanisms, but one of the easiest to understand is that, as warming melts the Arctic permafrost, previously entrapped methane is released from the ice lattices of clathrates (sometimes called hydrates).

- *Confirming Evidence:* Over the last century, there has been an unprecedented rise in atmospheric methane. It seems likely that warming-induced permafrost destruction and wetland methane out-gassing are the primary causes, although some of the methane increase is the result of agricultural changes.*

These are but a few of the cause-and-effects that follow from greenhouse gas-induced climate change. My purpose is to demonstrate the many layers of evidence that validate the science consensus that anthropogenic climate disruption is real.

The heat capacity of the oceans (the ocean's ability to absorb and store thermal energy) exceeds the heat capacity of the land and, especially, of the atmosphere. The surface area of oceans is much larger than that of land, and its

---

* As noted earlier, there is some recent evidence that the increase in atmospheric methane has slowed. But I've been unable to learn why.

mass is many times greater than the atmosphere's. Moreover, unlike solid land, liquid oceans can roll about distributing energy within themselves, enhancing the rate at which they can absorb and store energy.

The ocean's ability to store energy is also enhanced because the 'specific heat' of water is higher than any other common substance.* So the oceans not only store immense thermal energy, they also absorb it with comparatively small temperature increases. Recognizing these truths, it's scary that we've already measured a significant rise in oceanic temperature.

The oceans continuously absorb and release carbon dioxide to and from the atmosphere. Over geological time frames, the net flow can be either in or out. If the partial pressure† of carbon dioxide in the atmosphere is greater than in the ocean's surface, the net flow will be into the ocean. The opposite occurs if the partial pressures are reversed. Today, because atmospheric CO_2 has increased to levels never recorded before, the partial pressure in the atmosphere is much higher than that in the oceans, so the net flow of CO_2 is from the atmosphere to the oceans. This makes the oceans CO_2 vacuum cleaners. Today, it's thought that about 50% of our anthropogenic CO_2 is taken up by the atmosphere, and the other 50% by the oceans.

But as the oceans warm they will be less able to draw down CO_2. Think back to a kitchen experiment you've inadvertently performed many times—opening a warm carbonated drink. Upon popping the lid, bubbling CO_2 frothed the contents over your hands and onto the kitchen floor. In contrast, when you open a cold carbonated drink, taken directly from the refrigerator, there is almost no frothing. This home experiment demonstrates that warm water doesn't hold carbon dioxide as well as cold water does. It also helps us remember that, as the oceans warm, they will be less able to vacuum CO_2 from our atmosphere.

---

* 'Specific heat' is the thermodynamic property that defines how much energy (for example, kJ) a given mass of material (for example, kilogram) can absorb for a given increase in temperature (for example, a change of 1°C). Compared to all other common materials such as rock, steel, wood, flesh or air, water requires the most energy to increase its temperature.

† The total pressure of a gas is the sum of the pressures contributed by each component of that gas. The contribution from each component is its "partial pressure." For example, the air pressure in your bicycle tire is the sum of the "partial pressures" contributed by the nitrogen, oxygen, water vapor, methane, argon and other trace constituents such as carbon dioxide that make up the air.

At some stage, rising ocean temperatures could even cause the oceans to pump CO_2 back into the atmosphere.*

Phytoplankton, which is at the base of the food chain for all pelagic life, eats CO_2, and uses it for carbonate formation. When phytoplankton die, some of this carbonate sinks to the bottom of the ocean to form material like limestone. So the question is: How will a warmer Earth influence the process of carbonate formation and sinking? This is important because limestone formation is one way Nature 'sequesters' carbon.

As our kitchen experiment showed us, rising ocean temperatures make it more difficult for the oceans to absorb carbon dioxide. But rising atmospheric CO_2 is keeping ahead of the game—by rapidly increasing the partial pressure of atmospheric CO_2. So the atmosphere continues pumping ever more CO_2 into the oceans. We're force-feeding the reluctant oceans with carbon dioxide.

Unfortunately, as we're often reminded, too much of a good thing is a bad thing. Force-feeding oceans with carbon dioxide is dangerous to oceanic life. As the oceanic CO_2 content increases, the waters become increasingly acidic. This poses a threat to all marine life, from coral reefs to the carbon-dioxide eating phytoplankton that constitute the broad base of all sea-borne life. Therefore, acidification will likely reduce the vacuuming ability of the oceans' most important residents. We'll talk more about ocean acidification in the next chapter, 'Controversy, Conveyors and Consequences.'

We now know that all Nature's CO_2 vacuum cleaners are retreating under a pincer attack. On land, expanding civilization and deforestation are rolling back photosynthetic life. And in the oceans, both warmer water and the acid-poisoning of aquatic photosynthetic life are further throttling Nature's CO_2 vacuuming processes.

Let's return to our template for environmental intrusion. We began this chapter by discussing the dominant physical processes carrying us toward climate disruption. Yet climate systems are probably the most nonlinear, complex and

---

* Whether CO_2 is pumped into, or sucked from, the atmosphere depends on the partial pressures of the gas in the ocean and atmosphere. With atmospheric CO_2 increasing at the rate it is today, the 'pumping' is from atmosphere to ocean—and it will be for a very long time.

multifaceted processes that the physical sciences must try to analyze. So I could only introduce some of the cross-linked phenomena driving climate change.

If your head is swirling from trying to assemble the many processes into an integrated whole, you will not be alone. But there is an escape: We can apply the four straightforward criteria set out in the last chapter to judge the power of any environmental intrudant. When applying these criteria to climate disruption caused by CO_2 emissions, the questions and answers play out this way:

- *What is the magnitude of the intruding material compared with its natural background level?* Pre-industrial carbon dioxide levels were about 270 ppmv. Early in the 21st century, CO_2 levels had reached 380 ppmv and, no matter what we do, they will climb above 450 ppmv.

- *What is the fraction of the global commons occupied by the intrudant?* One hundred percent of the atmosphere and the upper layers of the oceans.

- *What is the intrusion's characteristic residence time?* About 100 years for CO_2 in the lower levels of the atmosphere, up to 400 years when it reaches the stratosphere, and more than 1,000 years when it reaches the ocean's basement.

- *Is the equilibrium being disturbed metastable?* Climate systems have always been metastable, as can be plainly seen from Figure 13.1. In addition, with so many phenomena having positive feedback loops (positive feedback *reinforces* the change), it's easy to understand why climate has historically lurched from one metastable state to another or, in today's language, why climate history is a history of 'tipping points.' When we recognize that atmospheric CO_2 has grown far beyond its highest levels in the past half-million years, it's difficult to suppress the fear that we may already be beyond our tipping point.

The mechanism of greenhouse gas intrusion has been understood for more than 150 years. Although some still claim controversy, the evidence that it's happening is abundant, robust and unambiguous.

14. CONTROVERSY, CONVEYORS AND CONSEQUENCES 🌿

In which oceanic conveyors, acidification, collapsing ice shelves and the prospect of sudden change have us wondering if we still have a chance.

We sometimes hear, from the news media and corporate stakeholders, that it is still a matter of scientific debate whether today's climate change is the result of civilization's activities or just a result of natural processes. Yet within the scientific community, the debate is over. So why the persistent stories about debate?

One reason may be that people enjoy the argument culture.[16] The media obliges. I also suspect it's not just because conflict sells, but also because debate purportedly establishes impartiality. Unfortunately, such impartiality too often gives equal credence to the bizarre as to the rational. We should not glorify impartiality when judging between firefighters and fires. Worse, for issues like climate disruption, only a few journalists have science or engineering training, which makes it more difficult for them to distinguish firefighters from fires.

And sometimes we are bludgeoned by disingenuous stakeholder propaganda. The April 17, 2006 issue of the *New York Times* carried an op-ed piece titled 'Enemy of the Planet.' The article was about Lee Raymond, the former chief executive of ExxonMobil, who was paid $686 million during his 13-year tenure. But that's not the reason author Paul Krugman singled him out. It was for his role in having ExxonMobil lead an attack on climate science. I'll quote only portions of Krugman's article, leaving out most of his observations about individuals:

> To understand why ExxonMobil is a worse environmental villain than other big oil companies, you need to know a bit about how the science and politics of climate change have shifted over the years. Global warming emerged as a major public issue in the late 1980s. But at first there was considerable scientific uncertainty.

Over time, the accumulation of evidence removed much of that uncertainty. Climate experts still aren't sure how much hotter the world will get, and how fast. But there's now an overwhelming scientific consensus that the world is getting warmer, and that human activity is the cause. In 2004, an article in the journal *Science* that surveyed 928 papers on climate change published in peer-reviewed scientific journals found that "none of the papers disagreed with the consensus position."*

To dismiss this consensus, you have to believe in a vast conspiracy to misinform the public that somehow embraces thousands of scientists around the world. That sort of thing is the stuff of bad novels.

So how have corporate interests responded? In the early years, when the science was still somewhat in doubt, many companies from the oil industry, the auto industry and other sectors were members of a group called the Global Climate Coalition, whose *de facto* purpose was to oppose curbs on greenhouse gases. But as the scientific evidence became clearer, many members—including oil companies like BP and Shell—left the organization and conceded the need to do something about global warming.

Exxon. . .chose a different course of action: It decided to fight the science.

A leaked memo from a 1998 meeting at the American Petroleum Institute, in which Exxon (which hadn't yet merged with Mobil) was a participant, describes a strategy of providing "logistical and moral support" to climate change dissenters, "thereby raising questions about and undercutting the prevailing scientific wisdom." And that's just what ExxonMobil has done: Lavish grants

* Following this *Science* article, articles and blogs appeared questioning what constituted 'to be within' the 'consensus position.' If we use a narrow consensus bandwidth, it can be argued that some articles fell outside the consensus view. But this is using a fine-point, academic discussion to obfuscate the message the *New York Times* article laid out so well.

have supported a sort of alternative intellectual universe of global warming skeptics.

The people and institutions ExxonMobil supports aren't actually engaged in climate research. They're the real-world equivalents of the Academy of Tobacco Studies in the movie "Thank You for Smoking," whose purpose is to fail to find evidence of harmful effects.

. . .the fact is that whatever small chance there was of action to limit global warming became even smaller because ExxonMobil chose to protect its profits by trashing good science.

In spite of this sorry story, we must not vilify all corporate stakeholders. Most are responsible. From my own experience, I cannot confirm the kinds of morally challenged actions reported by the *Times*. Yet from the evidence I have, the article is entirely plausible. For me the *Times* piece became even more credible when it reported that Shell and BP had withdrawn from the Global Climate Coalition. I've found these two companies to be among a group of more realistic—let's deal with the issue, not deny it—corporate stakeholders.

Nevertheless, since Earth's climate change is such a multidimensional and nonlinear phenomenon with multiple positive and negative feedbacks, there is still much to understand and much to learn about the rate and magnitude of the consequences. The trouble is, these uncertainties are too often misinterpreted, wittingly or unwittingly, as a debate about the basic phenomenon and its cause—about which there is no uncertainty.

Nevertheless since I've introduced the issue of controversy it's appropriate to begin by outlining the arguments of those few skeptics.* As I understand

---

* I'm using 'skeptics' because that's how much of the news media describes them and they describe themselves.. But when I talk to climate scientists, they object—saying these folks aren't skeptics, they're 'deniers'—and go on to point out that skeptics are needed as part of the scientific process. Deniers spit in the face of the data. Those having suspicions about climate change and especially about its anthropogenic causes, have sent me several articles that are intended to make me a 'skeptic' too. What strikes me about these articles is how frequently they resort to ad hominem attacks—such as disparagingly referring to the overwhelming majority of climate scientists as 'true believers' and then using the acronym 'TB' throughout the rest of their article.

it, the skeptics' primary argument is based on the following: The deep-past correlation between carbon dioxide and temperature might represent some periods when the temperature rose before atmospheric carbon dioxide rose. In the past, Earth-orbit wobbling, or strong and extended solar flares, could have triggered warming in advance of a rise in CO_2. In turn, this modest warming could have caused methane bursts from clathrate (methane dissolved and trapped in ice) dissolution that would have reinforced the warming. Then, the warmer oceans would have out-gassed CO_2 (if the atmospheric and oceanic CO_2 partial pressures had previously been in equilibrium). So it is plausible that in some precivilization episodes, an initial warming may have caused the associated CO_2 rise, rather than the reverse.* Taking what might have happened in the past, the skeptics argue that the temperature increase we're witnessing today could be caused by the same mechanism. I consider this suggestion preposterous.

The International Panel on Climate Change (IPCC) and the overwhelming majority of climate modeling scientists tell us that, for the almost two centuries since the start of industrialization, the explosive growth in anthropogenic CO_2 has been the prime driver of today's climate change†. When I asked IPCC members why they didn't include the skeptics' warming-came-first scenario as one possibility, even if unlikely, they answered: we couldn't include theories that climate scientists have soundly refuted and for which there is neither supporting data nor empirical evidence.

Many reputable articles have disproved the few skeptics. Two examples are the work of Barnet et al[14] and Hansen et al[15] The many papers demonstrate that the warming of the world's atmosphere can only be the result of human activity and is not the result of natural variability. Moreover, substantially different climate models draw the same conclusion. Other articles point out that Earth

---

* Even if you give zero credence to the climate-skeptics' points of view, I think that reviewing the arguments can be useful—because the more ways we can look at an elephant, the better we'll understand what an elephant is. If you want to further consider some of the skeptics' arguments, you might try Jan Veizer, 'Celestial Climate Driver: A Perspective from Four Billion Years of Carbon Cycle,' *Geosciences Canada*, 32, 1 (March 2005).

† Land-use changes is the other—and they range from cutting down forests to build cities, to slash and burn tactics for creating farms in the Amazonian rainforests.

is now absorbing about 0.45 W/m^2 (watts per square meter) more energy from the Sun than it is re-emitting to space; that Earth has warmed by as much as 0.7°C since 1880; and that, even if we stopped all anthropogenic greenhouse gas emissions today, the inertia in global systems would cause the world to continue to warm for several more decades.

Now we'll return to the physics of what's happening. The atmosphere and hydrosphere are Earth's two great conveyors of energy (and entropy and exergy)* over our planet. Any realistic climate model must account for the coupling between these two great conveyors. While the atmosphere may be more important for day-today weather, the hydrosphere dominates climate. So let's zoom to the hydrosphere and its big players, the oceans and icecaps. We'll leave aside the bit players, such as lakes, rivers and wetlands, which primarily influence local weather.†

Surface winds are the primary driver of ocean circulation. When there are any north-south (meridional) components of the flow, the Earth's rotation-induced Coriolis forces bend currents in a clockwise direction in the northern hemisphere, and counter-clockwise in the southern.‡ Oceans are subject to thermohaline circulation, as shown in Figure 14.1 adapted from the work of Stefan Rahmstorf.[17] The adjective 'thermohaline' reminds us that a combination of temperature and saline differentials power these superimposed oceanic conveyors.

The Gulf Stream is the popular name for one component of the North Atlantic Conveyor, which is, itself, part of the global thermohaline circulation. The 'Stream' is the northern hemisphere portion of the northbound, largely wind-driven surface current. The southern hemisphere portion of the North Atlantic Conveyor surface current begins a little southwest of Africa's Cape of

---

* The meaning and significance of adding entropy and exergy to energy is covered in Chapters 22 and 24.

† The oceans contain more than a thousand times more energy than the atmosphere. However, atmospheric circulation is many times faster than oceanic, so most scientists say these two conveyors have about equivalent influence in moving energy about our planet.

‡ Because Coriolis yields a circulatory flow, the east-west flows curve too.

Good Hope. The complete conveyor includes a return southbound trip, called the North Atlantic Deep Water flow. Figure 14.2 gives a closer look.

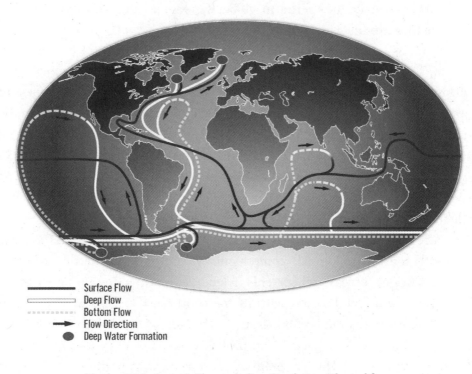

Surface Flow
Deep Flow
Bottom Flow
→ Flow Direction
● Deep Water Formation

Figure 14.1 Oceanic Thermohaline Circulation Adapted from
the Work of Stefan Rahmstorf[17]

There is growing evidence that climate change might weaken the North Atlantic Conveyor with consequences not just for northern Europe, but for the entire global climate because the Atlantic conveyor drives the thermohaline portion of all ocean currents. To understand why the conveyor might weaken, we must understand how it works.

Today the surface water density increases as it flows north from equatorial regions to the high latitude Atlantic. This causes a loss of buoyancy, and then sinking, of the water. The density increases for several reasons. First, the water cools as it enters higher latitudes.*Second, the process of ice formation as the water moves into the Arctic precipitates salt into the water immediately beneath the ice layer. This precipitation increases the salinity of the under-ice seawater, and thereby increases its density. High-latitude surface-water sinking is called 'ventilation,' because it pulls oxygen down into the deep waters.

Figure 14.2 North Atlantic Conveyor Adapted from the Work of Stefan Rahmstorf[17]

---

* Among common fluids, water has this unique property: As its temperature drops below 4°C, it expands. This is the reason ice forms on the top of lakes, rather than on the bottom. Good thing, too! If solid ice formed on the bottom, as it would if water continued to increase its density sinking toward the bottom until it froze at 0°C (as all other materials, such as steel, do), then ice would first form and then build up from the *bottom of lakes* (until lakes became a solid block of ice). Instead, ice forms an insulating blanket on the top of lakes. This is just one more reason water is magic— one more of Nature's tricks to build a universe in which some planets can come alive.

Figures 14.1 and 14.2 show the world's most prominent ventilation (deep water formation) sites. Ventilation provides a continuous supply of cold dense water to the bottom of high-latitude oceans. As this deep water leaves the North Atlantic sites, it has nowhere to go but south. It travels along the ocean floor after lifting over an undersea ridge that extends roughly from Greenland to Norway.

As this deep southbound ocean conveyor arrives at lower latitudes, it is gradually warmed by vertical mixing, thereby lowering its density. Working together, high-latitude ventilation and low-latitude mixing build up density differentials in the deep ocean; in turn, these generate forces that maintain the massive momentum of the thermohaline circulation in the Atlantic.

Today, global warming is weakening the sinking mechanisms at ventilation sites. Indeed, data indicate the flow has weakened by about one third since 1957.[18] I'll identify some of the mechanisms that are applying the brakes:

- Increased land-based summer-ice melting is escalating the freshwater run-off into the ocean. Freshwater forms the surface layers and is more buoyant than saltwater. The increased buoyancy hinders sinking.

- In addition, the slower formation of sea-ice as winter approaches changes the ocean's albedo (ice reflects more sunlight than does the liquid ocean), so more sunlight is absorbed to further warm the high-latitude ocean surface water, lowering its density and helping keep it afloat.

These are some of the slowing processes, but there are also countervailing ones. Floating ice acts as an insulating blanket. Thus, the absence of floating ice may allow the Arctic-bound warm-surface flow to cool more rapidly as it interacts with colder arctic air. However, this cooling effect seems insufficient to offset the effects of the increased summer-ice melting. There is evidence that the North Atlantic Conveyor is already slowing. Measurements indicate that the Faroe Bank Channel lower-level southbound currents decreased by about 20% in the 50 years between 1950 and 2000.[18] Data also show that Nordic Sea surface water is freshening.

Oceans profoundly influence the world's climate. So slowing the North Atlantic Conveyor will have an impact on everything—from the family life of polar bears to Polynesians as they depart their disappearing islands. Oceans cover about two-thirds of Earth's surface. And oceanic circulation, the key to redistributing the Sun's incoming energy over our planet, is being bent out of shape.

All greenhouse gases act to bring warming. But a single greenhouse gas, carbon dioxide, also affects the oceans' chemistry.

About three decades ago, when I first began tracking the theory and evidence of growing anthropogenic CO_2, I hoped the oceans might be one of our saviors—because they absorb carbon dioxide. It's thought that over the past 200 years, the oceans have absorbed about half the anthropogenic CO_2 emitted.[19] With this large fraction of our emissions sequestered in the oceans, we might have thought—a decade or so ago, I certainly thought—that we have a great place to stuff some of our excess anthropogenic CO_2. But now we've learned that stuffing carbon dioxide into the oceans brings its own problems. And, in some ways, these may be even more dangerous than the growth in CO_2 itself.

In Chapter 13, 'Bubble, Bubble, Signs of Trouble,' we spoke of a kitchen experiment—the different behaviors we get when opening warm or cold carbonated drinks—to help us remember that as water gets warmer it's less able to absorb CO_2. Now let's perform another kitchen experiment. This time, take a can of soda water. Before it's opened, all the CO_2 is dissolved in the water. After the can is opened, much of this carbon dioxide forms bubbles and is up and away. But some always remains. If we let the can sit around until it goes flat, we might expect the water to taste just like a can of room temperature water, but it doesn't. Rather, it tastes like stale water with a little bite. That bite results from the weak carbonic acid (H_2CO_3) that is formed whenever carbon dioxide is dissolved in water. A simple equation makes the process of carbonic acid formation easy to understand:

$$CO_2 + H_2O \rightarrow H_2CO_3$$

One molecule of carbon dioxide + one of molecule of water →
becomes one molecule of carbonic acid.

As atmospheric carbon dioxide levels rise, more CO_2 is forced into the ocean. This increases the carbonic acid and the ocean's acidity. In high school chemistry we learned that pH = 7 is neutral. Above 7 is basic (alkaline) and below 7 is acidic. Today, oceans are slightly alkaline, having pH between 7.9 to 8.5, depending on the oceanic region.[19]

Perhaps the most dangerous consequence of ocean acidification is the disruption of calcium carbonate ($CaCO_3$) formation, which is performed by both critters and photosynthetic organisms—from lobsters to coral. They use calcification processes to create themselves.

Scientists say it will take tens of thousands of years to return ocean chemistry to that of pre-industrial times. This long period results from oceanic-mixing time lines, which in turn determine the time that will be required to lift sediments from the ocean floor to buffer the increased CO_2 impact on surface chemistry. But now, as global warming slows the thermohaline conveyors, mixing will take even longer.

So we must think about what harm acidification is bringing today, and might bring in the future:

- Scientific evidence tells us that these changes in ocean chemistry will have an adverse effect on tropical corals, the reefs they build, and in turn, on the life they support. (Almost 30% of warm water corals have disappeared in the last quarter-century largely, it is thought, due to frequent and intense episodes of higher-than-normal sea temperatures, perhaps enhanced by acidification.)

- Acidification is disrupting the ability of marine life to build their carbonate skeletons and shells.

- Increased CO_2 and lower pH is thought to be especially stressful to squid and other high metabolism species.

- Photosynthetic marine organisms, predominantly free-floating phytoplankton, underpin 99% of the organic structure that constitutes

the base of the oceanic food chain,* which in turn constitutes almost half
of Earth's total food chain. Some uncertainty remains about the effect of
acidification upon the health of phytoplankton. But by acidifying the oceans
we surely gamble with the health of phytoplankton, the base of the pyramid
that forms all oceanic life.

Now I'll identify a few more, easy-to-explain, climate-related phenomena:

- As the oceans warm, their waters expand. This is one reason sea levels are
  rising. The melting, or collapsing, of land-supported ice is another. (Melting
  floating ice doesn't change ocean levels, just as a melting ice cube doesn't
  change the level of the drink you're holding in your hand. But dropping in
  another few ice cubes does change the level.)

- Temperate-region mountain glaciers are retreating. Seasonally delayed
  snow cover and earlier snow melt are decreasing Earth's albedo for longer
  fractions of the year, which is another positive feedback that reinforces
  warming. (Over the last few years, a scary series of photographs has
  recorded the retreating glaciers on Mount Kilimanjaro.)

- The global hydrological systems will be (and are) disturbed, leading to land
  decertification and increasing shortages of freshwater—already a growing
  world problem.

- Some other phenomena might bring either positive or negative feedbacks,
  to reinforce or mitigate warming. For example, rising ocean temperatures
  will increase the water content in the atmosphere which, in turn, will
  increase cloud cover. Clouds have a higher albedo than do oceans and land.
  So by reflecting away more incoming sunlight, increased cloud cover will
  mitigate warming. On the other hand, water vapor is a strong greenhouse
  gas. So while more clouds will reflect more sunlight from their tops, their
  water-vapor content will trap more outgoing infrared radiation. I'm unsure
  if science has determined whether the net effect will be a positive or
  negative feedback.

---

* The other 1% is contributed by the photosynthetic life in the intertidal zones—the seaweed and
  kelp that the cabin boy was probably smelling.

- The increased cloud cover originates not just from factories and surface-vehicle emissions. Aircraft contrails also reflect sunlight—but not necessarily enough to balance the warming caused by the CO_2 emissions from the same aircraft.

When I began to understand the physics of CO_2-induced climate disruption in the early 1980s, to me it was clear this would become a major threat to civilization. So my concern turned to our likely response. I feared we would be too slow waking up. Communities of people seldom respond effectively to gradual change, no matter how serious the ultimate change will be.* This truth is even more evident when the threat is environmental. The tragic story of the Easter Island peoples' self-destruction is one of the more heart-breaking.[20] So with this history of community behavior, I expected that a gradual, less-than-a-tenth-of-a-degree warming every ten years or so, would not be the stuff of action.

Then, in the early 1990s, evidence began to accumulate that climate has often changed abruptly—extraordinarily abruptly in the context of geological time. The reason for sudden change is rooted in the metastability of climate systems, a concept we introduced in Chapter 12. Let's consider one example of how the metastability of climate mechanisms might bring abrupt climate change.

When thinking about the straightforward physics of how melting polar ice would bring about rising oceans, I tempered my fears with the question, 'But how quickly can we *really* expect the Arctic and Antarctic ice caps to melt?' I knew the latent heat of fusion (the energy required to melt ice at 0°C to liquid water at 0°C) is very large.† So I knew it would take an awful lot of sunlight, and time, to melt the quantities of ice that form the icecaps, especially as incoming sunlight is weakest at the poles. But when I learned of the truly massive icebergs

---

* This is one of the reasons it's extraordinarily difficult for a thoughtful politician, with long-term concerns, to be elected. Our good fortune is that, despite this difficulty, we have had a few such legislators from time to time. Our bad fortune is that, during the critical first decade of the 21st century, in North America at least, they are not in command.

† The same energy input to melt ice at 0°C (32°F) to liquid water at 0°C would increase the temperature of liquid water from 0°C to 80°C (176°F). Please don't confuse the 'latent heat of fusion' with 'nuclear fusion' which, of course, is entirely different.

that were breaking away from Antarctica, I realized we didn't need to melt the icecaps *at the poles*. Large chunks could simply split off and then float about to be melted in the warmer waters of lower latitudes. So my question changed: How much ice could break off and how fast?

More answers began coming in. The Antarctic icecap contains more than 90% of the total ice on Earth. The east Antarctic ice sheet sits on land above (today's) sea level. But the base of the west Antarctic sheet sits on bedrock below sea level. Not only does much of this bedrock slope down toward the present coastline, most of the interface between the ice and bedrock is lubricated by water-saturated clay. This ice is ready to be launched, we don't know when; it could be any day.

Yet the denial continues. It continues to shape government policy and headlines. In January 2006, the head of the British Antarctic Survey, Professor Chris Rapley, warned that the huge west Antarctic ice sheet may be starting to disintegrate, an event (he said) that could raise sea levels by almost 5 meters (16 feet). Others have estimated it would be between six and fifteen 15 meters (20-50 feet).* Dr. James Hansen, the senior NASA scientist monitoring climate modeling, endorsed these concerns when he said, 'Once a sheet starts to disintegrate, it can reach a tipping point beyond which break up is explosively rapid.'

Then on February 9, 2006, one month after the UK report, the *New York Times* ran an editorial titled 'Censoring Truth.' The editorial began:

> The Bush administration long ago secured a special place in history for the audacity with which it manipulates science to suit its political ends. But it set a new standard of cynicism when it allowed NASA's leading authority on global warming to be mugged† by a 24-year-old

---

* See 'Antarctic ice sheet is an "awakened giant." ' NewScientist.com news service, February 2005.

† The *New York Times* article seems to have used the word 'mugged' to mean intellectually mugged. Deutsch, a young presidential appointee at NASA, instructed public affairs staff to limit reporter's access to Dr. Hansen.

presidential appointee who, quite apart from having no training on that issue [climate science], had inflated his résumé.

The NASA scientist was Dr. Hansen. The 24-year-old was George Deutsch, who had claimed to have earned a journalism degree from Texas A&M, a claim the university said was a fabrication.

So the danger is not just that our gradualist trip towards climate doom anesthetizes our response. It is also that some of our most powerful political commanders inject the anesthetic. Is Dr. Hansen one of our cabin boys?* Is George W. Bush our Admiral Shovell? History is replete with examples of societies unwilling to accept evidence of looming disasters before the disasters overwhelmed them.

Today, I fear a realistic analysis of our climate's status will conclude that, very likely, 'we're toast.' But it is also human to fight the odds. So we must give it our best shot. That's what the rest of *Smelling Land* is about: Our best shot.

---

* A more recent article by Hansen, 'Climate Catastrophe,' was published in the 28 July, 2007 issue of *New Science.*

Part Five

SUSTAINABILITY AND THE COMING HYDROGEN AGE

15. TEMPLATE FOR SUSTAINABILITY

In which we refine the meaning of sustainability and use our new understanding to fashion a template for sustainable energy systems.

I first heard the word 'sustainable' applied to energy systems during a lecture by Wolf Häfele. The year was 1980, when the world was still captivated by memories of the 1973 oil embargo, when people thought we were running out of energy—times, therefore, when everyone thought the solution was 'renewable' sources. It was thought we could never run out of renewables. Over several years of this pervasive mantra, I became uneasy about the acceptance of such a simplistic response—felt there was something more to this issue, felt there must be a better, more accurate word than 'renewable.' Wolf Häfele gave me that word.

Häfele was speaking at the University of Toronto as part of a lecture tour to introduce concepts that he and his team at the International Institute of Applied Systems Analysis (IIASA) had developed for their book *Energy in a Finite World.*[21] IIASA had been founded in 1972 as a consequence of the Cold War rapprochement led by Presidents Richard Nixon and Leonid Brezhnev. Located in a wonderful old Hapsburg castle outside Vienna, it is probably the world's foremost, non-governmental, multidiscipline, multinational, systems-analysis research organization. IIASA's goal was (and is) to assemble scientists from around the world to study issues of global importance, particularly issues that rise from scientific and technological development.*

Several years later, IIASA released *Sustainable Development of the Biosphere*[22] in 1986, because, in their words, 'Humanity is. . .entering an era of chronic, large-scale, and extremely complex *syndromes* of interdependence between the global economy and the world environment.' It is an impressive, readable

---

* At the time, IIASA was supported by seventeen national member organizations representing Austria, Bulgaria, Canada, Czechoslovakia, Federal Republic of Germany, Finland, France, German Democratic Republic, Hungary, Italy, Japan, the Netherlands, Poland, Sweden, the UK, the USA and the USSR. Four of these countries had ceased to exist as we entered the 21st century—symptomatic of nation fracturing and merging during the final decade of the millennium.

book that includes energy systems but places them within broader concepts of human development, world environment, social response and—something I found especially intriguing—the 'topology of surprise.' Both IIASA publications leapt ahead of the then-current thinking to introduce profound ideas. Yet they had little influence on the public.

Then in 1987 *Our Common Future*[23] was published. This report of the World Commission of Environment and Development was chaired by Norwegian Prime Minister Gro Brundtland. It set out criteria for global economic-environmental planning and finally planted the idea of sustainable development in the public mind. For more than a decade following its publication, the ideas of *Our Common Future* (often called 'the Brundtland Report') rolled from the tongues, word processors and TV scripts of every reporter concerned with our linked economic-environmental future. Of these ideas, the concept of sustainable development was among the most important, although, in my view, it had been articulated better, more comprehensively and with greater sophistication in the IIASA books.

'Sustainability' is now a fixture in everyday vocabulary. And to develop appropriate energy systems, it is essential that we have a strong understanding of what sustainability means.

According to *Our Common Future*, sustainability aims to '. . .meet the needs of the present without compromising the ability of future generations to meet their own needs.' That works, although I'm uneasy about the word 'needs' because it requires an answer to the ancillary question: By whose criteria? Moreover, it seems to me we should aspire to something more than mere existence, which 'needs' implies. Let's consider:

> *Sustainable development allows cultural and economic growth, embedded within environmental gentility, without jeopardizing the ability of future generations to live even better than us.*

My version is a bit longer, but I prefer the idea of cultural growth rather than needs. Still, I wondered if 'economic' was redundant. Without economic growth, cultural growth is normally foreclosed, although the converse (economic growth without cultural growth) is surely possible, both individually and nationally. In the end, I decided that removing 'economic' could be misleading, as it might encourage people to naively believe poverty-stricken nations can enjoy cultural growth without the underpinnings of economic growth. Recall Indira Gandhi's observation, 'Poverty is the worst polluter.'

We'll now adapt the general idea of sustainability to energy systems, where things become less abstract.

If we again evoke 'keep it as simple as possible but not simpler' to determine the sustainability of energy systems, I think first-rank metrics come down to evaluating:

1. The *quality* of the energy service.
   - Initially, the service cannot be inferior to an equivalent service provided by established systems.
   - The service must soon be significantly better.
2. The *environmental gentility* of the system chain providing the service.
   - Inputs and outputs to and from each link of the system chain must minimally intrude upon Nature's flows and equilibria.
   - Decommissioning (that returns occupied land to green space and recycles decommissioned material) must be both technically and economically feasible.
3. The *economic cost* of the service.
   - The initial cost of the service from new systems may, for a short time, be somewhat greater than the cost from established, competing systems.
   - The cost of services must soon be lower.
4. The *long-term availability* of the energy source.
   - Energy sources must be available into the foreseeable future (but there is no value in demanding foolishly long time frames, like thousands of years).
5. The *global distribution* of the energy source.
   - The distribution must be relatively uniform, so that geopolitical tensions that risk wars or economic blackmail are not aggravated.
6. The *resilience to surprise* of the system chain providing the service.

- As much as possible, the system must be resilient to geopolitical, technological, economic and environmental surprise—or any other surprising surprise.

Let's examine each of these concepts in turn. We'll treat the first four in the remainder of this chapter, skip the fifth because it's so obvious, and then devote the next chapter to the sixth.

The *quality of the service* may seem straightforward. Yet the apparent simplicity can be misleading, because we must ask: What is the *real* service?

Usually there are many facets to *what* you buy when you buy an energy service. If you invite a business colleague to dinner at an elegant restaurant where you're known by the maitre d', you're buying more than the food's energy content and trace elements. Beyond the pleasant surroundings, unobtrusive yet sophisticated music, artistically presented servings and excellent wine, I expect something else you're buying is the quiet, smug feeling of being recognized and called by name as the maitre d' sweeps you through the dining room to your usual table. What has an elegant meal got to do with energy services? Well, it is an energy service. The energy system produced the fertilizers and insecticides (if used); drove the tractors; shipped the farm products; processed, packaged and delivered the foodstuffs to the restaurant; heated the chef's stove—and washed the chef's white smock.

Frequently the service we buy is *saved time*. Consider travel. Elegantly appointed dirigibles, ocean liners and railway dining cars were certainly more comfortable than any of today's airplanes—even if booked in first class. Still, for most people, unless on vacation, the time saved by air travel outweighs the pleasant surroundings of earlier transportation modes. I like the answer the President of FedEx is reputed to have given when asked what he viewed as his company's business. I'm told he answered, 'procrastination.' So one of FedEx's services (products) is the option to procrastinate. It's a service delivered by modern transportation, high-speed communication and good management. Time saved is also delivered by computers, and hair dryers.

Hair dryers? In Chapter 24 'Exergy Takes Us beyond the Lamppost,' we'll use these dryers to help illustrate the difference between energy efficiencies (often

meaningless) and exergy efficiencies (*always* meaningful). My engineering
students had no difficulty calculating energy and exergy efficiencies for many
technologies especially for hair dryers, because an example was worked out in
their textbook.[24] But when I asked them to calculate how efficiently a hair dryer
delivers its *service*, it caused problems. Is the service dry hair? Not really. If we
stand around long enough our hair dries on its own. The service is time saved.
It took a while to get this answer. I pushed the issue to remind the engineering
students that it isn't good enough to be able to calculate correctly; it's even more
important to know what you should calculate, and *why*. Defining the right
question is often more important and usually more difficult than calculating
the right answer after specifying the right question.

After the students closed in on the time savings, the discussion became fun.
Some said, 'No! It's not time. It's fluffy, puffed-up hair.' When the men referred
to 'puffed hair,' it always got chuckles from the women. Then someone was likely
to say, 'I use hair dryers for shrinking shrink wrap.' At this point the discussion
would blow wide open and the students showed their creativity. They realized
that engineers must know more than how to design technologies. They learned
they must understand the service customers really want and what the service *is*
before judging its quality. In the end, they recognized that if you're selling hair
dryers, you're peddling more than hot air.

Environmental gentility was discussed in chapters 10 through 12. We know
energy services are delivered by a system. So when we analyze the potential
for environmental intrusion, we must go beyond individual technologies,
individual fuels (currencies), individual anything. We need to evaluate
environmental impact on the basis of the complete system.

While the importance of including the complete system should be obvious,
there are nuances. First, environmental intrusion may be exported from one
link to other links, either up-system or down-system.* Consider a hydrogen-

---

* This book avoids the descriptors 'upstream' or 'downstream,' which can be ambiguous: Upstream/
  downstream depends on whether we're watching energy flow or demand flow. Our perspective
  is that the system begins with a demand for service and responds by moving down-system to
  sources.

fueled city bus using a fuelcell powertrain. The bus will improve urban air quality because it releases only small amounts of water vapor, rather than the soot, nitrous oxides, carcinogens and tangy-things that otherwise would find their way into your nose—all of which were spewed from the diesel bus which the fuelcell bus displaced. But what if the H_2 were produced by electrolysis using electricity generated from a coal-fired generating station? If the station were far from the city, the urban environment will still benefit but the air quality near the generating station will be degraded. Moreover, carbon dioxide will be dumped into the global commons. By manufacturing the hydrogen from coal we push the environmental intrusion down-system push it from the service-delivery technology (buses) in the city, to the source-harvesting technology (coal-sourced H_2 production) on the outskirts.*

From Chapter 11 'Tidal Flats and Airports, Quantifying Environmental Intrusion,' we know environmental intrusion always involves material—either material *taken from* the environment, *diverted within* the environment, or *put into* the environment. Knowing that environmental intrusion is rooted in material, we set out four questions:

- What is the magnitude of the intrudant material compared to its natural level?
- What fraction of the global commons does the intrudant occupy?
- What is the characteristic residence time of the intrudant?
- Is the disturbed equilibrium fundamentally metastable?

The answers to these questions will give us a first estimate of the environmental impact from almost anything—from PCBs, to mine tailings, to spent nuclear fuel, to river dams, to electromagnetic radiation from overhead power lines, and so forth.

---

* Historically, a lot of H_2 has been manufactured from coal, but seldom by first making electricity and then using electrolysis to make hydrogen. Usually we use a process involving a reaction approximated by the equation $C + H_2O + (\text{thermal energy}) \rightarrow H_2 + CO$. Today, the $H_2 + CO$ mixture is called syngas. In earlier times it was 'town gas' or 'coal gas.'

The *economic cost* of a service must be measured against a clear understanding of what the service is, or what the services are. Often several services are delivered simultaneously, like the heating and lighting provided by skyscraper lights when they remain on throughout a cold winter's night—as we discussed in Chapter 3 'Turning out the Lights.' The danger comes when we mistakenly consider only one service—as when we think office lights uselessly illuminate vacated offices during the wee hours of winter.

We've said that, for short periods, the cost of services from an emerging technology can be greater than the cost of the same services from established systems, but soon the cost must be equal or lower, or it will fail. The question is: Why this initial reprieve? Earlier adopters bequeath the initial reprieve. These are individuals, institutions and (occasionally) political jurisdictions who want to be first. Sometimes they are motivated by the desire to lead the pack. More often, they have a vision for what the new technology will bring—like improved profits, a cleaner world or reelection.

Examples? In the late 1970s and early 1980s, early adopter office managers introduced word processors. In 1980, word processors cost more than mechanical typewriters, typists had to learn new skills, envelopes were trouble, and the printed copy wasn't a whole lot better. Whereas today, the cost of operating offices with mechanical typewriters would put businesses *out* of business. Turning to a different example, at the cusp of the 21st century, the early adopter cities of Chicago, Los Angeles and Vancouver introduced small fleets of hydrogen fuelcell buses. At the time, fuelcell buses were more expensive than diesel buses and the maintenance crews needed to learn new skills. But if we look to a distant tomorrow, the only businesses able to afford diesel buses will be theme parks showing what life was like back in the 20th century. But they'll need to operate within giant bubble-like structures so the quaint 20th century emissions don't escape to the surrounding environment.

Most early-adopter organizations are led by strong, free-thinking individuals—people in positions to make decisions without having to cajole teams of cautious administrators, bankers, accountants and lawyers. Metaphorically, early-adopters are booster rockets that lift new technologies through their critical embryonic phase so that, in a short time, they become less expensive than the established technologies. Once an emergent energy system outgrows its early-adopter phase, it *must* provide at least either substantially better services or more attractive prices. Preferably both. If it doesn't, it won't win widespread use. Established technologies have the enormous advantage of incumbency. Challengers with mere marginal advantages don't have the oomph to dislodge the established.

Sometimes there is a delay when a technology that doesn't make it within a ten-to-fifteen-year window lies in wait until circumstances change—but the wait can last many decades. For example, some hydrogen technologies that don't break out during their initial window may get a second shot several decades later when the imperative of reducing CO_2 emissions is driven further into the public consciousness and conscience.

As we consider the economic metric for creating a sustainable energy system, we must also evaluate the *internal* and *external** costs that we will discuss further in Chapter 35 'OK! Now Tell Me About Cost.' Some external costs can be quantified. The most important cannot.

The example without peer is the external cost of future climate disruption. If we choose to trivialize this danger as merely a degree or so of warming, then estimating the external costs of, say, running air conditioners a little more (and furnaces a little less) would be straightforward. But how can we evaluate the cost of a national insurance premium that would cover the economic damage of climate disruption, if the result were to shut down the North Atlantic conveyor or raise ocean levels by several meters? Impossible. Who would—could— underwrite the risk? Can you imagine an insurance company large enough to pay up? Indeed, insurance companies are already reevaluating premiums for damage caused by climate change and extreme weather.

Future legislation may internalize some of today's fossil fuel external costs. But this can only go so far and will surely come too late. Powerful stakeholders

---

* Internal costs are those paid by the producer or user of a product. But external costs must be paid by those who are not involved in the transaction but still pay—such as the cost homeowners pay for repainting their homes more frequently because they live downwind from a smelting operation.

are fighting long and hard—with an awful lot of money—to ensure that external costs are *kept* external. They'll fight even harder over the next decade.

Long-term availability is included in our template for sustainability, largely because so many people believe running out of such energy sources as oil is our greatest threat. Running out globally is not a danger, but the practical matter of uneven *global distribution* is important: Regional supply discontinuity has been the cause of many 20th-, and now 21st century military conflagrations. Therefore, reliable, long-term energy source availability, distributed with reasonable global uniformity, must be a requirement for sustainable energy systems.

Earth's coal, oil and natural gas will never be depleted. (Chapter 28 ' *Fossil Sources: A Lot Will Be Left in the Ground*' explains why.) But as we've said, the lumpy distribution of these resources around our planet leaves many regions without indigenous supplies.

Nuclear power will never be capped by the depletion of uranium ores. While some world regions are better endowed with uranium than others, there are three reasons this uneven distribution is unlikely to jeopardize economies the way fossil-fuel maldistribution might. First, uranium can be found just about everywhere, even in seawater—of course this becomes a matter of extraction cost, not availability. Second, the cost of uranium is a comparatively small fraction of the cost of producing nuclear power, so even if the cost of uranium jumped significantly, the impact on the cost of nuclear-derived electricity (or hydrogen) would be comparatively minor. Third, through the wonders of nuclear processes, we can design power plants—called breeder reactors—that *produce* more fissile fuel than originally supplied.

Renewables are constrained by the *rate* at which they can be harvested without causing environmental damage or depletion. And now another curious paradox. While we will *never* run out of nonrenewable (exhaustible) energy sources like coal and uranium, we *could* run out of renewable energy by over-harvesting—just as we're now running out of renewable fish, renewable freshwater and renewable forests.

So our criteria for long-term and global availability boil down to a few simple, perhaps surprising, realities:

- In a *global* context, depletion jeopardizes none of our nonrenewable (fossil and nuclear) sources.

- Geopolitical conflict does, however, threaten several of our nonrenewable energy sources. This risk is most critical for oil, less for coal and natural gas, and much less for uranium.

- Ironically, should renewables be called upon to deliver anything approaching the energy demand of modern civilizations, renewables will be much more vulnerable to depletion than the exhaustibles.

Now we'll move on to the next chapter, where we'll examine perhaps the most difficult of our metrics for judging sustainability.

16. THE TYRANNY OF SURPRISE

In which we search out ways to build resilience to surprise—an elusive, yet key attribute of sustainable systems.

How do we anticipate a system's response to surprise before we know what the surprise will be?

The difficulty of anticipating surprise is the reason why most predictions, and almost all popular futurist books, assume futures that are relatively smooth evolutions from today—futures free of the intellectual discomforts and the analytic difficulties wrought by unpredictable discontinuities. Yet *real* futures are determined more by discontinuities than by smooth evolution.

Unfortunately, many energy companies base their projections on a comparatively orderly evolution of business as usual, and mobilize their lobbyists to protect their future based on that concept. They often forget that 'business as usual' is frequently tripped up by sudden change, such as oil embargoes, the accident at Three Mile Island or the grounding of the *Exxon Valdez*.

Despite the difficulty of predicting surprise, indeed *required by* that difficulty, we must do our best to develop energy systems that will be resilient to the shocks of unanticipated, sudden shockers. During the 1970s, Ontario's energy system was brittle and so, when surprised by the oil disruptions, its economy cracked (at least briefly). So did the economies of the United States, the Netherlands and many western nations. Something else cracked too—clear thinking, which often cracks when surprised.

To a greater or lesser degree, Ontario represents a microcosm of all western economies. What if Ontario's transportation systems—and the transportation systems of all western economies—had been fueled by hydrogen? Then the electric-generation capacity having little to do throughout the long, cold winter's night—could have been redirected to making hydrogen for the cars and buses awaiting the morning rush. Had a mature hydrogen-electricity system been in place, the oil embargo could not have brought major disruptions. Our energy systems would have been comparatively surprise-resilient.

Pipeline history illustrates another side of surprise-resilience: Surprise opportunity.

During 1942, tankers carrying oil from the US Gulf States sailed up the eastern seaboard before turning east across the North Atlantic bound for England. Stalking this eastern seaboard route, German U-boats were sinking tankers at an appalling rate (appalling for the Allies, great for the Axis). For a time, east coast cities, notably Miami, insisted on keeping their cities brilliantly lit so as not to dampen tourism—ignoring the help they were giving the German submariners. Brightly lit cities provided a wonderful backdrop against which U-boat captains could frame their quarry. It's reported the fireballs from stricken tankers added tourist excitement.

Soon the Allies' strategists realized U-boats could not torpedo inland pipelines. So, in August 1942, one of the extraordinary engineering feats of World War II began: Construction of what came to be called the 'Big Inch' pipeline, running for more than 1,200 miles from Texas to the northeastern United States. Within eighteen months, the Big Inch was completed and carried five times more oil than the conventional pipelines of the day.[25]

Fortunately, wars end. So Britain and its allies needed much less oil for war machines. The Big Inch was mothballed to await either scrap metal dealers or someone with a bright idea for how it could be used. Several years later, a bright idea won out over scrap metal. Entrepreneurs began shipping natural gas from the Gulf States to the American northeast. And so began interstate commerce in natural gas.

The subsequent history of interstate gas trade is fascinating. We discussed one anecdote in Chapter 8 'The Gods of Energy Planning Foolishness.' But the point of the Big Inch story is this: A pipeline designed and built to ship oil began shipping natural gas during the pipeline's middle age. As a consequence, many smaller pipelines (and still smaller pipes running into homes and apartments, and ultimately into lamps and stoves) switched from coal-based town gas (a mixture of CO and H_2) to natural gas.

How does this relate to surprise? First, it shows that not all surprises are bad. Surprises jiggle the market, often shaking new opportunities into view. Ending the war introduced a new pipeline opportunity for the US. Second, many infrastructures—particularly pipelines—entered their mid-career, performing different tasks than those for which they were built.

I expect it's a pattern that will recur with pipelines being constructed today. Before the end of their useful lifetimes, many of these pipelines may have the opportunity to carry hydrogen. For this reason I've sometimes suggested to pipeline company executives that, when specifying new natural gas pipelines, they determine the marginal additional cost (if any) of assuring the pipe lining is hydrogen-compatible. To my knowledge, the suggestion has never been followed. So in a few decades a 'surprise' opportunity could be foreclosed for want of a few simple calculations when the pipeline material was specified.

Today's gas distribution companies should remember that the town gas distribution industry, which evolved into a natural gas distribution industry, is well positioned to become a hydrogen distribution industry.

The flip side of resilience-to-surprise is *intolerance*-to-surprise, a common vulnerability of technological monocultures. When people begin using new technologies—such as when airplanes, cars, radios and computers were emerging—many different designs and manufacturers compete for their place in the Sun. Later, after an early period of experimentation and innovation, one configuration usually prevails, then dominates. The winning configurations, safe in their market dominance, too often neglect all but the most trivial technological innovations and solidify into technological monocultures. This makes them surprise-vulnerable.

Before the 1974 oil disruption, the North American automobile industry was such a monoculture. Technological innovation was trivialized into little more than year-to-year body-style mutations. Technological innovation had moved to Japan and Europe. In North America, innovation (that might have led to improved efficiencies) was further impeded by the burden of selling cars into a market that enjoyed—*by far*—the world's least expensive gasoline. This set up a brittle monoculture waiting to be shattered. The oil-disruption surprise following the Yom Kippur War did the shattering. European and Japanese auto makers rushed in to fill the cracks. Some years later, North American automobile companies gradually returned to technological innovation—in part by forming alliances with offshore corporations. But for market share, they will never recover what they had before the fuel disruption. And now, what was once the world's largest corporation, General Motors, seems sometimes to teeter on the verge of bankruptcy. If GM hadn't been such a powerhouse, if oil prices hadn't been held so low in North America, if corporate strategies had prioritized innovative technological advances above chrome and marketing SUVs, things might have come out differently. But as it was, the North American automobile industry had painted itself into a surprise-intolerant corner.[26]

Or take computer software. As we entered the 21st century, computer users watched, in thrall, the unfolding struggle between the US Department of Commerce and Microsoft, especially over 'bundled' Windows operating systems. While I share the normal concern most people have about monopolies, this time my concern was not so much for the high prices that monopolies can charge and I must pay. My concern was, and remains, that the Microsoft operating system is a masterpiece of technological monoculture. I can't anticipate the surprise that might shatter it. But, at least for a short while, we'll all suffer when it inevitably shatters. After 2001, the US administration removed most of the pressure from Microsoft thereby unintentionally, but no less certainly, building the pressures for disruption that, I think, will almost inevitably occur.

So what can we learn about developing systemic surprise-resilience? It will always be tough, but it seems to me some general principles include:

- We must move toward hydricity, which will be resilient to the sudden supply disruption of any particular energy source, because both hydrogen and electricity can be made from any of the remaining sources.

- We must be alert to opportunities to use old infrastructures in new ways— which means clever, imaginative people must pay attention.

- We must be alert to the dangers of technical monocultures and to the circumstances that encourage them.

- To sharpen our ability to anticipate, we must continuously dream up a menu of 'what if' scenarios, and then test proposed new technologies and systems for their resilience to each 'what if.'

All these strategies point to opportunities for those who live by Shakespeare's maxim, 'the readiness is all.' Although oxymoronic, we could also ask: What is the surprise most likely to shock our energy system before 2025? The answer is a sudden, worldwide panic, resulting from some special event which shows us that the *extreme* consequences of climate volatility are already upon us.

17. Hydrogen: the Case For Inevitability

In which we see why our energy system must evolve to the Hydrogen Age.

People delight in skewering unfulfilled predictions. So as I became increasingly aware of the patterns that, it seemed to me, were taking us to a Hydrogen Age, I was cautious of predictions, mindful of skewers. Yet as the logic stacked up, it became difficult to escape the conclusion that civilization must evolve toward the Hydrogen Age, unless it first self destructs. So in this chapter I'll set out the layered logic that, for me, became impossible to set aside.

I first heard of hydrogen* in 1978 when, like just about everybody, I believed depletion would be the ultimate driver of energy system evolution. I had traveled to Ottawa to meet with Dr. Philip Cockshutt. He was then Director of Energy Studies at the National Research Council of Canada (NRCC). As the chair of the Department of Mechanical Engineering at the University of Toronto, I hoped to learn why our department's energy research proposal (which had nothing to do with hydrogen) had failed to be funded. My visit would reinforce, again, that we often learn more from failures than successes. I had not previously met Cockshutt. He had graduated from Toronto in mechanical engineering, and had then gone to the Massachusetts Institute of Technology. I hoped he would be willing to counsel his *alma mater* engineering department. I found Cockshutt to be a Canadian version of a soft-spoken, understating, English country gentleman. We sat in his office while he scratched out a series of energy options on his blackboard. I hadn't yet realized the importance of thinking in terms of energy *systems*; to me they were still just options, bits and pieces.

---

* This means hydrogen as a fuel for things like cars, trucks or airplanes—not just as a chemical. Anyone who took high school chemistry will remember their teacher using electrolysis to show that water was made of components, H_2 and O_2. By the mid-1970s anyone interested in the space race knew the main rockets were fueled by hydrogen and oxygen which, as it turns out, were precursors to the hydrogen future.

Cockshutt explained many wondrous things, from 'clever bugs that could be trained to eat shit and produce methane' (he has a gentle, poetic, yet pithy way to explain things) to something called 'hydrogen.'

I found three things about hydrogen intriguing. First, energy from *any* source can be used to *harvest* hydrogen by splitting water.

Second, the exhaust from *using* hydrogen is water. Third—and this is a deeper, less obvious idea—hydrogen is the ideal fuel for fuelcells.

I returned to Toronto thinking hydrogen was a sweet fuel clean, abundant and, in principle, easy to manufacture. Not yet having conceived of the idea of energy currencies, I could not slot hydrogen into its profound systemic significance.

In 1980, two years after Cockshutt wrote 'hydrogen' on the blackboard, I was invited to join a panel at the third World Hydrogen Energy Conference in Tokyo. Although I knew of the greenhouse gas threat, I had not yet broken out from thinking depletion would be the cap on fossil fuels. So my contribution to the closing plenary session was a five-step 'Case for Inevitability' based on depletion. The logic went something like this:

1. *Driven by depletion, civilization must move from fossil to sustainable energy sources.*

2. *Transportation requires chemical fuels which, today, are harvested exclusively from fossil sources.* Today, the Jet A for aircraft, the gasoline and diesel for automobiles and trains, and the bunker C for ships, can only be produced from fossil sources.

3. *In the future, we must be able to provide chemical fuels by harvesting sustainable sources.* Let's say it with an example: We must have a way for sunlight, wind, hydraulic or nuclear sources to make a fuel that can fly airplanes.

4. *Realistically, the only way sustainable sources can be harvested to make chemical fuels is to have them produce hydrogen.* By splitting water, any non-fossil source can be used to manufacture hydrogen, which can be used to fuel all transportation vehicles. There is no other non-fossil fuel that can serve this global role.

5. *Therefore, the move from fossil to sustainable sources can only begin with the increased use of hydrogen, and can only be completed with the preeminence of hydrogen among chemical fuels.* This is the only conclusion possible.

Because step four is probably the least obvious, the most difficult to accept without a lot of pondering, let's dig deeper. It took me many months to accept that hydrogen was unique. It seemed right, yet I needed to check 'seemed.' First I read a lot. Then I began asking informed people around the world: 'Can you think of *anything* else that might meet civilization's requirement for a non-fossil transportation fuel?' No one gave me a credible alternative. Perhaps ammonia (NH_3) came closest. But why use ammonia when you can use hydrogen? Ammonia requires that you first make hydrogen and then tie it to nitrogen, making it a much heavier fuel. And today, a key design strategy for efficient vehicles is weight reduction.* Finally, ammonia is prickly stuff. Have you ever got a whiff?

Over the years, at conferences and while serving on task forces, I continued to ask, 'Can you think of anything other than hydrogen that could fulfill this role?' Again, no realistic proposals came back. It was troubling. I disliked claiming something was inevitable, hoped for an alternative that would allow me to be less absolute, less strident, less vulnerable to skewers.

During these times, I also grew uneasy when people referred to hydrogen as an energy source, rather than something that *allowed* any energy source to power transportation, because it's transportation where the energy system has the fewest options. My need to get things straight in my own head led me to the idea of currencies. It seemed we were missing the systemic role of hydrogen because we had no linguistic way to differentiate the role of sources from the role of currencies. 'Running out' was the worry-of-the-day; finding new sources to replace the old, the obsession. No one asked, '*If* we develop a new non-fossil energy source, how can it be used to fly airplanes?'

---

* The single advantage of ammonia over hydrogen is that ammonia can be more easily carried as a liquid.

We must remember that fossil sources are used for more than transportation, so the need for hydrogen reaches beyond transportation. Fossil sources provide many of our material feedstocks—from building materials, to automobile, airplane and computer parts, to clothing, and to supplement some foodstuffs. Even when the end-product material itself contains little or nothing from the fossil source, fossil inputs are often used to produce the materials. Steel, for example. Chemically, iron ore is mostly iron oxide (Fe_2O_3). Today we use carbon, made in coking ovens from coal, to rip the oxygen off the Fe_2O_3 thereby leaving iron (Fe). A simple process equation describes what happens:

$$2Fe_2O_3 + 3C \rightarrow 4Fe + 3CO_2$$

This shows why carbon dioxide is the principal (albeit invisible) effluent from today's steel-making processes, and why steelmaking contributes to climate destabilization. But in the future we will be able to use hydrogen to do the ripping. Then the process is:

$$Fe_2O_3 + 3H_2 \rightarrow 2Fe + 3H_2O$$

Water is now the effluent and steel mills will become better neighbors.

In the early 1980s, we all thought the depletion of fossil sources was the danger. And this was the basis on which I set out that five-point case for hydrogen inevitability. We now know that a different peril has overtaken depletion as the decisive reason we must quit fossil fueling: The undeniable risk of climate disruption. And thus a parallel, five-step case for the inevitability of hydrogen now rests on its being the essential defense against climate change:

1. *Atmospheric CO_2 growth is bringing climate destabilization that, if unabated, will be catastrophic.*

2. *To eliminate anthropogenic CO_2 emissions, we need both non-carbon sources and non-carbon currencies.*

3. *There are many non-carbon sources.*

4. *But realistically, electricity and hydrogen are the only non-carbon currencies that, together, can supply the full menu of energy services.* Hydrogen cannot

do it alone, just as electricity can't do everything. The Hydrogen Age won't terminate electricity. Rather it will give electricity a wonderful new partner, a kindred spirit.

5. *Therefore, anthropogenic CO_2 emissions can only be slowed by the extensive use of hydrogen and can only be stopped with the preeminence of sustainable-derived H_2 among chemical fuels.*

Step two points out that to avoid CO_2 emissions we cannot use any carbon-based energy sources. Yet to be complete, I must temper the statement by noting that some argue there are circumstances, although limited, where it might be possible to use fossil sources. If, for instance, a carbon-based source is used in *stationary* applications, it's conceivable we might avoid releasing the waste CO_2 *if* we catch, process and sequester it. This will not be as easy as some suggest, but the principle is valid and it is discussed in Chapter 32 'Harvesting Hydrogen.'

If sequestering is to be at all feasible, carbon-based fuels cannot be carried on board transportation vehicles. How would we catch and sequester the CO_2 effluent from tailpipes? Aircraft cannot drag CO_2-catching sacks behind them, like the *fen dou* (manure-catching bags) I saw slung beneath horses' tails as they clip-clopped through Beijing during the 1980s.

The only chance for fossil sources to power transportation, without emitting CO_2, is if the source is used to *manufacture hydrogen* at a stationary setting where the waste CO_2 can be sequestered—and then the product hydrogen is the fuel carried on board. It seems a tortuous way to have fossil fuels hold onto their transportation markets. Moreover, then the fossil source would lose its only advantage: The fact that it can produce fuels that are liquids at standard temperatures and pressures.

So regardless of the path, even if we follow the twisted route of making the hydrogen from things like coal and sequestering the by-product CO_2, we'll still end up with Hydrogen Age vehicles. Seems to me we'd be better off making the hydrogen from clean sources from the beginning. At risk of restating what, by now, should be obvious: We'd be better off changing the process rather than adding collectors.

18. FOR BETTER OR WORSE

In which a nursery rhyme spotlights synergy, the imprimatur of the coming hydrogen-electricity age.

We now know a Hydrogen Age is our only chance to avoid calamitous climate instabilities. So if we're optimists, this means hydrogen is inevitable.

It can be scary when our future seems channeled towards the inescapable. Many apparently inescapable trends *are* scary—from severe overpopulation (especially in underdeveloped nations least able to cope), to the steady depletion of freshwater resources (including aquifers under North America's breadbasket), to the spreading destruction of primordial forests, wetlands and jungles, to the persistent over-harvesting of our oceans.

We already know the Hydrogen Age will bring a cleaner world. So the issues boil down to these: Will this future improve our service delivery or harm it, will the system have greater or lesser flexibility, and will the infrastructures be more or less resilient to surprise?

Therefore, let's probe our hydricity future, examine how it will differ from today, see if we're being pulled towards 'better' or 'worse.' Of one thing we can be sure: Special interests will argue tirelessly that to move 'prematurely' towards hydrogen will cripple our economy, that people will lose jobs, that we'll all be the poorer. What's more, these special interests will spend a lot of money to ensure their message is heard.

Earlier I said that hydrogen will give electricity a wonderful new partner and kindred spirit. We'll use this chapter to expand on this two-part claim. Let's start with the idea of a wonderful new partner. It's another idea that transports me to a nursery rhyme:

> *Jack Spratt could eat no fat,*
> *His wife could eat no lean,*
> *And so, betwixt the two of them,*
> *They licked the platter clean.*

Electricity-hydrogen synergies resemble the fat-lean synergies of the Spratt family:

- Hydrogen can serve as a transportation fuel or a material feedstock. Electricity cannot.
- Electricity can be used to transmit, process and store information. Hydrogen cannot.
- Hydrogen can be stored in enormous quantities. Electricity cannot.
- Electricity can transport energy without moving material. Hydrogen cannot.
- On Earth, hydrogen will be best for long-distance transport of energy. Electricity is not as good.
- In space, electromagnetic radiation (the proxy for electricity) will be best for transporting energy.

These six synergies deserve more explanation.

Hydrogen can be a transportation fuel or a material feedstock. Electricity cannot. Hydrogen's future role as a fuel is obvious. Yet we mustn't forget how much hydrogen is already used to manufacture commodities ranging from fertilizers to food, plastics to ultra-suede. Try to imagine how much more hydrogen will be used in the future when it becomes civilization's staple transportation fuel, while at the same time continuing to provide feedstocks for commodities.

Electricity can be used to transmit, process and store information. Hydrogen cannot. This second synergy is obvious. Electricity, not hydrogen, will be the currency that powers information systems.

Hydrogen can be stored in enormous quantities. Electricity cannot. This confers upon hydrogen its future as both the staple transportation fuel and the system's energy storage-sponge. Transportation is obvious. Storage is important to cover times of demand-supply mismatch—especially if we hope to harvest intermittent sources like wind and sunlight. And to store energy for seasonal fluctuations, and for national security in uncertain times of war or natural disasters that disrupt supply. Electricity cannot fill this role.

Electricity can transport energy without transporting material. Hydrogen cannot. As we've observed, material is always at the root of environmental intrusion. So it's interesting that when we deliver energy via electricity, we don't move material. Hydrogen is pretty good too, but it's not perfect. Still hydrogen carries more energy per-unit-mass than any other common currency.*

On Earth, hydrogen will be best for long-distance energy transport. Electricity will not be as good. For continental distances, transporting energy via gaseous fuels through pipelines is preferred to transporting it via electricity grids. There are three reasons. First, the fraction of the currency's energy that is consumed in pipeline transport (for pumping) is much lower than the fraction lost in electricity grids (to overcome impedance losses).† Second, as we've said, the pipeline itself is a load-leveling sponge. There is no equivalent sponge in an electricity grid. If you put electricity in at one end it instantaneously squirts out the other. Third, the security of energy transmission will be greatly improved, both because storms that knock down overhead electricity transmission lines can't dig up underground pipelines and because the electricity system network crashes that have periodically infected whole regions, such as northeastern North America, will be a thing of the past.

Across inter-planetary space, electromagnetic radiation (electricity's child) will be best for energy transport. Hydrogen (needing its sibling, oxygen) can transport energy only in comparatively small quantities. In the deeper future, should we want to move large quantities of energy from planet to planet—or farther—the energy will be carried by electromagnetic radiation, thereby avoiding the need to accelerate and decelerate the mass of the hydrogen-oxygen pair.

I've set out these dovetailed synergies because they can help us predict how each currency will capture different markets. Airplanes get hydrogen, computers get electricity. On the other hand, there are some tasks where both

---

* I have not included nuclear fuels among conventional fuels. Although nuclear fuels do power submarines and aircraft carriers, it seems unrealistic to think these fuels might become everyday currencies for civilian transportation. It's not just a matter of safety although that's an issue. It's also a matter of powertrain weight and volume. Nuclear power reactors are not easily built either small enough, or inexpensively enough, for a car.

† This energy is lost to the environment and is typically due to resistive heating in the case of electricity grids, and pumping energy to overcome fluid friction in the case of pipelines.

can compete—both hydrogen and electricity can warm living rooms. We can now pick any imagined task and, by applying the Spratt family template, see whether hydrogen or electricity is likely to win, or if both will have a shot.

Of course, faced with straightforward logic some will always turn their contrarian neurons to arguing exceptions. They might point to an airplane that can fly on electricity—perhaps for the length of a runway. Or, they might dream up computers that could run on hydrogen—albeit at a snail's pace and a snail's capacity. Yet to build a hydrogen-fueled computer as a high school science project, say in 2050, could be both educational and fun. And it would demonstrate that it's often possible to force-fit a currency into jobs for which it's genetically unsuited.* Sometimes human resource managers make this mistake with people too.

So now let's look at the kindred spirit claim. It seems to me kindred spirits share more than yin-yang synergies. They also share similarities. For hydrogen and electricity at least these three are important:

- Both hydrogen and electricity can be manufactured from any energy source.
- Both hydrogen and electricity are *interconvertible* (electricity can be made from hydrogen, hydrogen from electricity).
- Both hydrogen and electricity are *renewable*.

If you're at a social function and the conversation turns to our energy future, you might engage your friends with a logic that shows why *any* energy source can be used to manufacture hydrogen. While there are several ways to harvest hydrogen from any source, you could build upon the fact that most people know any energy source can be used to manufacture electricity. And because hydrogen can always be harvested from electricity using electrolysis, we will always be able to manufacture hydrogen from any source. While this

---

* During the early stages of a technology's development, inappropriate currencies can win for a while. As an undergraduate engineer, I was once offered a summer job working on control systems that used compressed air. Pneumatic schemes worked pretty well in the early days of control systems. Some still exist. But overwhelmingly, control systems use electricity.

simple argument demonstrates that any energy source can be used to produce hydrogen, it doesn't reveal the many ways the harvesting can be done.

For those few friends who haven't drifted off to find a beer or slice of cheese, you could add that natural gas—chiefly methane (CH_4)—is currently the most common fossil source of hydrogen. Hydrogen is produced from natural gas using a process called steam methane reforming (SMR). Although the process involves several steps, the (overall) process equation is:

$$CH_4 + 2H_2O + \text{energy (as heat)} \rightarrow 4H_2 + CO_2$$

Today, the heat in this energy-balance equation normally comes from burning natural gas. So carbon-dioxide emissions come from *both* the reformation by-product *and* the heat requirement. In the future, the heat could be produced by non-fossil sources, like nuclear. If we use nuclear, we'll still have CO_2 emissions as a by-product of the basic process, but no additional CO_2 from heat generation. It's clear, of course, that nuclear power can produce hydrogen without a natural gas partner, via electrolysis or thermochemical processes, and without producing a drop of waste CO_2.*

Another aspect of hydrogen and electricity kindred spiritness is that both are mutually interconvertible. Electrolyzers convert electricity to hydrogen. Fuelcells, hydrogen to electricity. Still, the important issue is not that hydrogen and electricity are mutually interchangeable—rather it is that gasoline and electricity aren't. Gasoline can be converted to electricity, the alternators in our cars do it every day. But electricity cannot be converted to gasoline. The path between electricity and gasoline allows only one-way traffic. It's a kind of diode that imposes a vicious rigidity on today's system. In turn, this rigidity has brought dangerous, systemic brittleness, and often the kind of planning foolishness described in 'Turning out the Lights.'

The third way hydrogen and electricity are similar speaks to our hankering for renewables. People have sometimes said to me, 'The good thing about hydrogen is we have so much water that we'll never run out of water to make

---

* We'll return to this topic in Chapter 32 'Harvesting Hydrogen.'

hydrogen.' That's true, but it misses the point. It's like saying 'The good thing about electricity is we'll never run out of electrons.'

Here *is* the point: Electricity is generated by separating electrical charges from their normal state of charge neutrality. Later, when the electricity is used, the positive and negative charges come back together giving charge neutrality—and we're back where we started. So electricity is 'renewable.' Analogously, hydrogen is generated by splitting water into its components, hydrogen and oxygen, which later, when the hydrogen is used, recombine to give water back. So hydrogen is also renewable.

In a previous paragraph I said, 'speaks to our hankering for renewables'—not to dismiss or praise. When we discuss the merits of renewable sources in Chapter 29 'Renewables and Conventional Wishdom,' we'll find it's an error to accept renewability as an unqualified imprimatur of goodness. But for currencies, renewability is an unequivocal blessing. Indeed, it's the core reason for the extraordinary environmental gentility of the coming Hydrogen Age. When we use either hydrogen or electricity, their renewability returns things to the way they were.

19. CAN SOMETHING BETTER COME ALONG?

In which, remembering how currencies differ from technologies, we come to appreciate that, once established, hydrogen and electricity will be used as long as civilization lasts.

The students were engaged, intent and enthusiastic, perhaps seeing the shape of personal opportunity. We'd been discussing the rationale for the inevitable Hydrogen Age and had moved on to the logic that, once established, hydrogen and electricity will be sustained as long as civilization lasts. Then from the back row came a skeptic's taunt, 'You mean until something better comes along!' A challenge had been fired across our bow.

It was a good shot. We've all grown to expect ever better *technologies*: Better cars, new improved soap powders, faster computers, surprising new IT widgets. That's been our lifelong experience. No reason for it to stop. But I had not meant until something better comes along. I *had* meant, until the end of civilization. But now back-row skepticism caused me to wonder if I'd been too hasty.

I needed to pause, to tighten arguments—or, perhaps, 'heave to' and surrender.*

Before we start, we must remember we're testing a claim for the immortality of two *currencies*, electricity and hydrogen, *not* for a technology. It's not a claim for the permanence of computers, telephones or fuelcells—or any technologies that might sprout up in time. It's reasonable to expect the future will bring wonderful as-yet-unimagined technologies.

---

* In previous centuries, when a sailing ship was overtaken by a more powerful enemy ship and realized she had no chance of escape, she would often 'heave to' and lower her ensign to indicate surrender. To 'heave to' is to set the sails so they counteract each other, causing the ship to lie with almost no forward motion. Sometimes, when friendly ships encountered each other on the high seas, they would heave to in order to exchange news, to 'gam.' On smaller ocean sailing yachts, heaving-to can be used to ride out a gale, or to make cooking easier and dinner more pleasant.

We've already learned that technologies and currencies play different roles. And to respond to this shot across our bow, we must go further: We must examine the properties of specific currencies that allow them to fulfill their roles. Repeating a diagram we've used several times, Figure 19.1 will remind us where currencies fit.

Figure 19.1 Components of the energy system chain with emphasis on currencies.

To cover the full menu of energy services, either today or tomorrow, we need at least one electronic and one protonic (chemical) currency. Let's consider the electronic currency and then go on to chemical currencies.

Electricity has always been our electronic currency. Its corpuscular bits and pieces are charge-carriers—normally electrons.* Electronic charge-carriers carry energy when they are not in equilibrium with their environment or, said another way, when they exist within an electromagnetic potential different from their environment. Atoms or molecules having a net electronic charge due to a surplus or deficit of electrons or ions, can also be charge-carriers. They are important in batteries and fuelcells, and in some industrial processes. Electricity is simply electric charges—of any type—moving from one place to another, usually driven by differences in the electrical potential of the two places. I can't envision a world that doesn't use electricity. You can go to Appendix A for more talk about electricity. But it should be clear that electricity is here to stay.

Chemical (protonic) currencies require more discussion. They are made of material, like today's gasoline, diesel, coal and natural gas—and yesterday's wood and dung. Just like electricity, a chemical currency carries energy when its corpuscular bits, its molecules, are not in equilibrium with the environment.

---

* In a transistor, an equivalent but inverse unit of charge is called a 'hole.' It represents the absence of an electron that, if there, would provide local charge neutrality. In storage batteries and fuelcells, charge carriers can be ions.

Throughout history, people have used many different chemical currencies. Looking to the future, we've set out the rationale why, over the course of the 21st century, hydrogen will push aside such earlier chemical currencies as gasoline. Hydrogen will win because it's lighter, cleaner, source-independent, inexhaustible, ideal for most fuelcells and better for internal combustion engines (when they are designed to exploit hydrogen's properties). We won't be going back to gasoline.

But the skeptical student's challenge was not that we could *return*. Rather it was that, after hydrogen, a still better chemical currency might come along.

If we want to seek a better chemical currency we must work within two constraints. First, we're restricted to the elements Nature gave us. As we know, an element is defined by the number of protons in its nucleus, because the number of protons determines the element's chemical properties. This brings us back to the periodic table we struggled with in high school. The teacher would normally start at the simplest element, hydrogen, with its single proton. Stepping up the periodic table, we next arrived at helium with two protons, then lithium with three. . .on to carbon with six. . .oxygen with eight. . .and so on. We're not going to find more elements.*

The second constraint is even more restrictive. An overriding criterion for environmental gentility is that the amount of a material's waste product be insignificant compared with its environmental background level. This is especially true for any currency considered for universal deployment— remember the arsenic-laced soup in Chapter 11 'Tidal Flats and Airports.' This means that to avoid environmental intrusion, chemical currencies can be made only of elements that exist in environmental abundance. If a universal currency were to include trace elements, the waste products would quickly impose an unsustainable environmental burden.

In the atmosphere, three atomic species dominate: Nitrogen (N), oxygen (O) and hydrogen (H). The hydrogen is tied to oxygen in water (H_2O). A chemical

---

* We can't rule out synthetically manufactured unstable elements having even more protons than, say, Lawrencium with its 103. But because of its very rarity, no such element could be a practical currency, or part of a chemical compound that could fulfill the role of a currency.

currency needn't be a single element; almost all of today's chemical fuels are molecular combinations of at least two, commonly hydrogen and carbon. Although carbon is plentiful in the earth's crust (and us), it's a trace element in the atmosphere—so carbon must be eliminated as a candidate ingredient of any 'forever fuel' whose waste products will exhaust to the atmosphere.

Now we're down to H, N and O. When we set out to look for combinations of these three candidate elements, we're pretty much back to currencies we already know—either ammonia which uses hydrogen and nitrogen, or pure hydrogen. Of these, for the reasons we've discussed, including weight and toxicity, hydrogen wins. We know the menu. Nothing else will come along.

Before leaving this chapter, we should remember that there is a subset of currencies which I'll call 'embedded' currencies. These are currencies employed over short distances *within* technologies. Take coal-fired electricity generation to illustrate the principle. Coal is the currency input to the combustor that produces the embedded currency, heat. Heat is the embedded currency input to the turbine that manufactures the next embedded currency, rotating-shaft power. Rotating-shaft power is the embedded currency input to the alternator producing electricity. Now we've reached the prime systemic currency, electricity. Embedded currencies will run through the veins of machinery, but they can't be viewed in the same context as the two prime systemic currencies, hydrogen and electricity.

Once wedded, the bond tying civilization and hydricity will be until death do us part. The vital components of our forever currencies will be single electrons and single protons. The challenge is to get civilization to hydrogen, not to stay there once we get there.*

---

* Some of my former students had fun trying to poke holes in the 'forever' claim to hydrogen and electricity. Usually they dreamt up something like 'dark matter'. But, even if we were able to harvest dark matter, it would play the role of a source, not a currency.

Part Six

EARTH'S ENERGY SYSTEM

20. AFTERNOON ON A HILLSIDE

In which, reminded that hydroelectric stations don't consume water, we realize Earth doesn't consume energy. So we embark on a quest for what Earth does consume.

It's late summer afternoon. I'm sitting on a nurse log in an old-growth forest.* The log lies across a granite outcrop that lifts it above the surrounding trees allowing me to see down into the valley. Relaxing, gazing, mindlessly breaking off pieces of bark, I watch a river course through the valley on its way to the Pacific. The river widens as it flows into a lake created by a dam. Below the dam, the water runs through the machinery of a hydroelectric generating station and then out into the lower valley, whence it travels around a bend, vanishing behind hills, seeking the ocean. Electricity lines march away from the generating station toward distant towns and farms, delivering power to light bulbs, laptops and milking machines.

It's an afternoon for reflection. The hydroelectric station *extracts* energy from the water flowing through its machinery. But it doesn't consume water. Easy enough to understand. When upstream of the dam, the water has more gravitational potential energy than when it leaves the station's tailrace. The turbines extract most of this energy difference, scrape it out from the water passing through their innards and send it away as electricity.

I smile at something fundamental. The hydroelectric station does not consume water. Rather, it *extracts* energy from the water that flows through its machinery.

To the east, the Moon a few days before full, floats above the mountains, ephemeral against a dimming sky. From the reality of my nurse log, the sight carries me into a reverie. I am lying back comfortably, weightless, in a space capsule drifting somewhere between Earth and its Moon, gazing back at our

---

* For those who haven't enjoyed climbing through the old-growth rainforests along the northwest coast of North America, a nurse log is a fallen tree that lies proud on the ground for hundreds of years, while fungi, ferns and new trees take root in its fallen majesty. The fallen log 'nurses' the next generation of trees into adulthood. An inspirational witness to rejuvenation.

planet. Through the viewing port I watch sunlight stream toward Earth. It carries inbound energy at the rate of some 180,000 terawatt (TW)—of which about 58,000 TW is immediately reflected by the upper atmosphere, bounced away to the depths of space. The remaining inbound energy, roughly 122,000 TW, is absorbed within Earth's material.*

After entering Earth's material, this 122,000 TW zaps through the intricacies of Earth's machinery, where it powers things like tomatoes growing, oceans circulating, termites wiggling. As it moves through Earth's machinery, the energy often changes form, from electromagnetic to kinetic, to potential, to chemical, and back and forth again—an energy polymath. The stunner is that, in spite of this twisting, transforming route through Earth's bits and pieces, the energy always ends at the same place and in much the same form as when it reached Earth, because ultimately, all this energy is exported to the universe as infrared radiation. So the 122,000 TW that the Sun continuously delivers to Earth is matched by an equivalent 122,000 TW that Earth continuously sends back to the universe. Earth has an import-export trade balance in energy.

If you haven't yet read Chapter 13 'Bubble, Bubble, Signs of Trouble,' it may seem eerie that the rate of energy entering the Earth is the same as the energy leaving. But it is. It must be, or Earth would heat up if more energy came in than went out, or cool down if more went out than came in.†

Struck by the parallels between water flowing through the generating station and energy flowing through Earth, I recognize something else that is fundamental: *Earth does not consume energy.* Rather, it *extracts something of value* from the energy that flows through its machinery.

In the next several chapters, we'll embark on a voyage-within-a-voyage, to find that something of value. The ideas we'll explore are essential to a true

---

* The prefix, tera-, means 10^{12}. So a terawatt is 10^{12} watts, the energy needed to feed 10,000,000,000 (ten billion) hundred-watt light bulbs. If you dream about being rich, you can use the prefix 'tera' to dream large—of having terabucks, or terayen, or teraeuros in your bank account.

† Some people might ask, 'What about energy contributed from geothermal, tidal sources or other things?' It turns out that all non-solar-sourced energy flows are trivial compared to solar-sourced flows. Much the largest of these non-solar sources is geothermal, and the magnitude of geothermal-sourced energy flow is about 2/10,000th the magnitude of solar-sourced energy flow. Tidal energy is trivial, even compared to geothermal. Still, if you want to be precise, to account for geothermal

understanding of how energy systems work. Although some of the chapters will be a bit demanding, the adventure will allow us to better navigate through energy systems toward a comparatively uncharted future.

In these chapters, we'll dig into the reality that we don't consume energy. We'll reflect on how mantras such as 'we must conserve energy' have led public policy astray. On the other hand, we'll see how Nature's law of energy conservation is the most profoundly useful principle for understanding how our universe works. Then we'll introduce *entropy*, the prime girder in the intellectual scaffolding that will take us to *exergy*—which is the something of value we *do* consume. Exergy is what powers our planet, our energy system and us.

The ideas we'll uncover are needed to understand how our energy system ticks, and are essential for designing the advanced technologies that must be developed over the next century. Understanding exergy and its role can also be useful for strengthening business planning, national policy or industrial-process optimization.

Astonishingly, these same ideas can tell us something profound about what it means to be alive, hint at why we might like music, or food exquisitely set out on a dinner plate.

---

and tidal contributions, the energy *leaving* Earth as infrared radiation must be a teensy bit larger than the energy *entering* as sunlight—*and* starlight if you're picky about the 'little bits.' (After all that, you might ask, 'OK, but why is he saying 'solar-sourced,' rather than simply 'solar'? Answer: Because there are all kinds of energy flows like wind and biomass—and, indeed, oil and coal—that *are* solar-sourced. But we don't describe coal or even wind as solar energy, so I've called them solar-sourced.)

21. CONSERVATION, CONFUSION AND LANGUAGE

In which, watching how the careless use of words confounds understanding, we invoke the Chinese saying, 'the first step to wisdom is getting things by their right names.'

Sitting on that hillside, we realized Earth doesn't consume energy. So it should be no surprise that neither do we. Nor does our energy system. Nor does anything. So why do we confuse our children? Schools preach the righteousness of 'energy conservation', and simultaneously teach that 'conservation of energy' is an inviolate law of Nature.

Over the past 200 years, we have learned that Nature designed its universe with surprisingly few fundamental rules. I think of these as the four laws of classical mechanics, plus Maxwell's four laws of electromagnetic theory, joined together by the three constitutive relations that link them all.* We call these eleven rules Nature's laws. They govern how the universe behaves. The most sweeping of them have exquisite simplicity. And of these, one of the first we discovered is the law that energy is conserved. Repeatedly, we have proven that Nature designed our universe to work that way. We didn't set the rule. We merely learned what Nature decreed.

Let's take a moment to say, as straightforwardly as possible, what the law of energy conservation means. It means that energy can never be used up, never destroyed, never consumed. But, while never consumed, energy can certainly change from one kind to another, because it exists in many guises. Sometimes it stays in one form for billions of years. But under the right circumstances, it can change from one form to another in a wink.

---

* These eleven laws of classical physics are explained more completely in the next chapter, 'Entropy and Living Planets.' Beyond classical physics, there is, of course, all the fun modern physicists are now having looking for unifying theories that, it's hoped, will apply to everything from the very small (quarks, leptons and so on) out to the vast reaches of space. So by saying 'very few' I mean those laws of classical physics which govern almost all of engineering.

The century between the 1830s and 1930s was a time when people rapidly uncovered Nature's fundamental rules. During this period, our faith in the law of energy conservation was sometimes shaken. But whenever that happened— whenever some observation made it seem that energy conservation was violated—we always found we'd made a mistake. These mistakes were usually rooted in not knowing that Nature had another trick for storing energy, a method our ignorance had hidden from us. Then, having learned about this new mode of energy storage, the law of energy conservation came out unscathed, sitting there smugly, in supreme control.

This happened, for example, when we discovered that matter, itself, is a form of stored energy. That truth led to our understanding of nuclear energy, which is simply the energy released when material is annihilated. Material disappears and another kind of energy appears, such as thermal energy and light. It is the energy conversion process of the stars—and, therefore, the process that powers our universe.*

There are many forms of energy, of course, which is the same as saying there are many modes by which energy can be stored:

- *Kinetic energy* is the energy something carries by virtue of its speed—like the kinetic energy of a speeding bullet, car or train.

- *Potential energy* (due to gravity) is the energy something has by virtue of its elevation—like the potential energy a famous apple had while clinging to a tree and which, after the apple lost its grip, caused it to bonk Newton's head.

- *Electromagnetic energy* is stored in an electromagnetic field by virtue of the field's intensity—like the electrostatic energy stored in the capacitor within your car's ignition system, which is spent to make the spark.

- *Chemical energy* is stored in material when the material is not in chemical equilibrium with its environment—like the chemical energy in gasoline, which is changed into thermal energy when it is burned.

---

* Understanding that mass could be converted to more usual forms of energy, such as thermal and light, led to our hopes for the peaceful use of nuclear power and to a myriad of often invisible benefits to human well-being—benefits now becoming indispensable.

- *Thermal energy* is the vibrational energy of atoms within material.* People often call thermal energy 'heat'—which will help you recognize what I'm talking about, but it is, in fact, incorrect terminology. As we shall see, heat is one of Nature's three ways to *move* energy, not store it. Thermal energy is sometimes called 'internal' energy by engineers. But I decided to avoid the term 'internal' because it could be confused with chemical energy† which is also stored within material but is identified separately.

Nature has still more strategies for storing energy, but the five listed above are probably the most important for our journey. I won't be surprised if people someday learn that we haven't yet found all the tricks for storing energy that Nature has up her sleeve. For instance, during the first years of the 21st century, astrophysicists are considering the possibility that, throughout the universe, immense energy may be stored in 'dark matter.'‡

But even if we haven't yet discovered all the ways Nature has to store energy, we surely know that her overarching energy law will always be this: *Energy is conserved.* In contrast to the many ways Nature has for storing energy, she has only three ways to move energy from place to place:

- *Heat* is the first method. Heat is the process of energy flowing from one place to a second because the first location is hotter than the second. For example, heat carries energy from your stove top to your soup, because the stove is hotter than the soup. When heat moves through materials, through the stove's heating element and the pot, it's called *conduction* heat transfer. But heat can also move energy from hotter to colder locations without the use of intervening material. Then it's called *radiation* heat transfer. Sunlight is the most obvious example—moving energy from a hotter Sun to a cooler Earth. The important thing to remember is that heat is energy moving from

---

* Thermal energy includes the energy of intermolecular forces and some other forms of microscopically stored energy. Yet it's probably easiest to think of it as molecular vibration energy.

† Besides chemical energy, several other energy forms can be stored within material, such as *magnetic* energy if the material is magnetized, or *strain energy* that is stored within a stretched elastic band or a steel spring.

‡ See, for example, "The once and future cosmos," *Scientific American* (Special Edition). December, 2002.

one place to another caused by temperature differences. It is not energy stored within material.

- *Work* is the second way energy can move from one location to another. It *always* results from force pushing through a distance.* Your car's transmission transfers work from the engine to the wheels. A ski lift does work carrying a skier to the top of a ski run, thereby pushing energy from the bottom of the mountain to the top because skiers at the top have more gravitational potential energy than do skiers at the bottom. Work transactions are by far the most significant deliverable of today's energy systems, and include the work of mining ore, flying airplanes, pulling trains or pushing ships, and running computers, TV sets and MRI machines.† Here I'm using 'work' in its technical sense of lifting or pushing, a force times a distance—not how we use the word in everyday speech—to describe the effort of thinking, for example.

- *Material movement* is the third method of transporting energy. Material always contains thermal energy, but it may also contain chemical, kinetic, electromagnetic or other forms of energy. Whatever energy is contained within the material will be carried along as the material moves. (A special case of energy carried within moving material is called *convection*, or convective heat transfer. Convection is the transport of thermal energy by a moving fluid—either a gas or liquid—between locations of different temperatures. An example is the mechanism your car uses to transfer excess thermal energy from the engine's hot cylinders to the radiator. In this case, the moving fluid is the engine's coolant. Then, once the thermal energy is in the radiator, convection by the passing air carries it away to

---

* Technical readers may find it fun to note that work is always the product of an *intensive* property (a generalized force) multiplied by the change in its cognate *extensive* property (a generalized displacement). This is most familiar when what we consider an 'everyday' force is multiplied by its cognate 'everyday' displacement—usually written in vector format as: $W = F \cdot \Delta r$. But it can be a product of more abstract generalized forces, like pressure, multiplied by change in its cognate displacement; volume, when we have $W = p \Delta v$; or when surface tension (generalized force) is multiplied by the change in its cognate displacement, surface area and so on.

† This may be difficult to understand now, but I hope it becomes clear after reading Chapter 23 'It's Exergy!'.

the surroundings. Another example is the mechanism a cool breeze uses to carry heat away from your body. It's especially effective if you're bald and hatless on a cold winter's day.)

Now that we understand the mechanisms of both energy transport and energy storage, it can be fun to watch Nature ensure that energy is never consumed—to watch her move energy to here, to there, and then plunk it down somewhere else.

A car comes around a bend speeding toward a stoplight. It contains a lot of kinetic energy by virtue of its velocity. Some of that energy could smash you if you were foolish or unlucky enough to get in its way. In this story, there will be no unpleasantries. The light goes red. Brakes bring the car to an orderly stop. To slow down, the car's kinetic energy must be removed—changed into some other form and sent somewhere else. The first step is done by friction between the brake discs and pads, which change the ordered kinetic energy from the car into disordered thermal energy (random molecular vibrations) at the rubbing surfaces. Heat transfer then moves this thermal energy from the friction surfaces to the interior of the brakes. Soon after, the energy is pushed out to the surrounding air (by convective heat transfer) with some slithering off into the axles and wheels (by conductive heat transfer). Later, as the car waits responsibly at the stoplight, its kinetic energy is gone. But none of that energy was consumed. Rather, it has simply changed its type (from kinetic to thermal) and its location (from totally within the car to mostly within the surrounding air, with some remaining in the brakes and wheels).

The increased thermal energy in the brakes means the brakes' molecules are wiggling and waggling more vigorously: 'Hot molecules.' You can sense this by putting your hand on the brake disks or pads, because then some of the brakes' molecular exuberance will be transferred to the molecules in your hand, making the molecules of your hand exuberant too. Next, information from your hand's agitated molecules is telegraphed to your brain so your brain can tell you that you've just touched something hot, something made up of feverish molecules.

Let's return to our car at the stoplight, to continue following the energy now in the wheels and air. We know what it *can't* do. We know it can't be changed back into the car's kinetic energy. We can't take thermal energy from the warmed air, wheels and axles, and magically use it to accelerate away from the stoplight. That's obvious from everyday experience, and we'll speak to why in the next two chapters.

Engineers don't want this thermal energy to stay in the brake material. If it did, the brakes would get hotter and hotter with each stoplight, jaywalker or skateboarder. To help cars tolerate erratic drivers who roar away from one stoplight to screech to a halt at the next, engineers design brakes with fins and other means to speed the transfer of thermal energy to the outside world. Moreover, the brake drums, disks and pads are made of materials that can survive high temperatures. By tolerating high temperatures, the brakes have more time to push their thermal energy into the air, extra time to squeeze out heat before the brakes fail from thermal overload. This thermal energy must go other places and do other things.

The energy we've been following, now stored as thermal energy in warm air, joins other energy already in the atmosphere. Carried by this marvellous atmospheric conveyor belt—'convection' belt if you like—this energy embarks on voyages around the world, sometimes helping to build winds or waves that, in turn, can power sailboats or windmills or knock down trees. After all these adventures, the energy we began following when it was the car's kinetic energy is exported to the universe as infrared radiation—sent off on a voyage to engage different worlds, in different corners of the universe, where it sets about doing different tricks.

We've been following where the energy went, or might have gone. We haven't thought about where it came from. You can do that yourself. But you must be prepared for some heavy-duty speculation because, before it became the car's kinetic energy, it had been traveling from one place to another, changing from one form or another, for a very long time. Indeed, it had been traveling hither, thither and yon since the Big Bang brought the universe into existence. It's thought that happened about 15 billion years ago. That little bit of energy has an impressive résumé.

It's all very simple. Similar energy transitions are happening all the time. It is what Nature decreed when she established a universe where energy is conserved. Walking through our world watching energy jump could have inspired one of my favorite authors, Dr. Seuss,[27] to say:

> *From there to here,*
> *From here to there,*
> *Funny things are everywhere.*

Which brings us to people who forget that energy is conserved.

These include inventors of new machines that, the inventors claim, produce energy. They assert their machines can produce more energy than the machine receives. With the glee of someone about to save the world, the inventors are heard to exclaim, 'It will put the oil companies out of business.'

After having difficulty raising money for the development of their machines, these inventors are often drawn to mechanical engineering professors as a court of last resort. On those occasions when the professor is me, I feel a discomfiting mixture of sorrow and despair. I don't like to discourage. But encouragement could be a license to drain yet more of some family's modest savings into bits of useless wire and wheels scattered about the garage. Almost always they don't have a patent. Almost always the inventors are unwilling to explain to me how the thing works—for fear I'll steal their idea.

I wish these inventors had understood Nature's principle of energy conservation before they purchased their first box of screws. They are most often bright, have the best entrepreneurial spirit, and have families standing behind their dream. If they had understood what Nature allows and what she doesn't, their inventiveness might have really paid off.

Let's search out the origins of the phrase 'energy conservation' to see if we can explain, or at least better understand, how it came to be hyped into a moral objective—when all the while it's one of the most fundamental laws of Nature.

As so often happens, the first origin is the careless choice of words. When people use the phrase 'energy conservation,' I expect most mean 'energy efficiency.' I expect they want our energy system to operate as efficiently as possible—in the belief that efficiency means we will use our natural resources sparingly, to provide the best possible services with the least possible environmental intrusion, at the lowest possible cost. Stated this way, the objective is appropriate and important. Indeed, it is the only path that can lead to a brighter 22nd century. Although it's a distinction difficult to keep clear in our heads, conserving a resource is different than conserving energy.

The second origin of 'energy conservation' comes from persistently thinking of energy as material, such as coal or oil. If you believe we could use up all the oil in the ground, then it's reasonable to use it sparingly—to 'conserve' oil in the ground by leaving it there. There are good reasons to use it sparingly—the most important being to reduce CO_2 emissions. But that idea can be better expressed by the words 'consume less oil.' We don't need 'energy conservation.'

When we have a simple, direct phrase like 'energy efficiency,' why not use it?

Early in his book *Consilience*, Edward O. Wilson speaks of the Chinese saying, 'The first step to wisdom is getting things by their right names.'[28] Upon reading this, I was impressed that the objective was wisdom, not mere knowledge. We could easily trivialize 'naming correctly' as little more than rote memorization of arbitrarily assigned monikers. Yet as all wisdom is context, we must not dismiss 'getting things by their right names.'

A dandy example of getting things by their wrong names is the admonition that 'conservation is an energy source.' This is a double-barreled shotgun of nonsense. Pellets from the first barrel splatter muddy thinking all around. Because energy conservation is a fundamental principle, it has no more relation to an energy source than to a bumblebee. It's a principle that must not be hijacked to serve as a political slogan. It must be kept firmly on its throne as a central law of Nature. It's not good enough to say, 'Oh! Everyone knows what we *really* mean by "conservation is an energy source." They know it means to use energy efficiently.' It is simply not correct to say; 'Everyone knows what we really mean.' Everyone doesn't.

As for pellets from the second barrel, they shatter understanding of what energy sources are and are not. But the vital point is that conservation has no business being identified as *any* component of the energy system—not a source, currency, technology or service. Naming conservation as part of the energy system chain confuses what energy systems *are* with how they *work*.

We must reserve 'conservation' for where it's meaningful, where it is the right name. Let's try to conserve justice, freedom, and above all, rationality. But let's rid ourselves of 'conservation is an energy source.'

I was talking about this with my now-on-his-cloud friend Ken Hare, one of the world's foremost climatologists and the first non-US citizen to be president of the American scientific society, Sigma Xi, when he said, 'Muddled language is a specialty of English. You wouldn't find it happening in French.'

I thought Ken's an interesting idea. I love the English language, with its freewheeling willingness to adapt, change and unabashedly import words from other languages. Most English-speakers have few hang-ups when our language evolves—often so quickly that the meaning of English words can morph within decades and lose recognition over one's lifetime. ('Gay' had one meaning to my parents, another today. Same for 'cool.') Sustaining the rivalries between English and French that have lasted from the Middle Ages, English speakers sometimes chortle at what they consider French language paranoia—sooner or later huffing: English doesn't need an Académie Anglaise. I enjoy the flexibility of English, enjoy its willingness to borrow foreign words to help express nuances, enjoy how the great magazines such as the *New Yorker* and the *Atlantic Monthly* expose us to new and creative phraseology, enjoy rereading William Zinsser's chapter on evolving usage in *On Writing Well*.[29] But sometimes, precision in language, getting things by their right names is critical. Sometimes French fussiness helps.

I scorn words that dishevel thinking. But I don't disparage the underlying objective of those who intend 'conservation' to mean using less energy. Whenever appropriate, using fewer energy services can both clean up the place and enhance lifestyles.

I like biking to work, to downtown, to wherever I can get away with it. Besides the exercise, little things make biking more fun. One route takes me past an elementary school; often about the time the kiddywinkles are hopping, skipping and jumping off to a new day. On my bike, I exchange smiles and nods of recognition with the old salt who controls the crosswalk. If I drive, I nod to the crosswalk guard from my steel, plastic and glass cocoon. But he never sees me. I'm saddened to miss this bit of human contact. Biking reduces my call on civilization's energy supplies, but not my quality of life.

But when reduced energy services do diminish the quality of life, I feel different. There will always be energy. I believe our mission is to deliver services ever more efficiently, with minimal environmental intrusion—not to shiver in the dark.

You might ask yourself, 'Why has this guy been going on for so long about this need for precision in our language?'

I'll tell you why. Nature's law of energy conservation is the most important and robust guide we have for designing energy systems that will work and that make sense. We must never mush up our understanding of this fundamental law. Because words *do* shape actions.

22. ENTROPY AND LIVING PLANETS 🍃🍃

In which, contrary to what some people think, we see that the laws governing entropy are what keep both Earth and us alive.

Have you ever wondered if something deep in our primordial brain attracts us to a manicured garden, a gracefully set out meal—most of all to music? Have you ever sensed that these experiences encapsulate the essence of being *alive*? Have you ever wondered why?

I suspect the answer lies in *structure*.

Metabolism is the process of living. But the downside of metabolism is that it destroys structure. The ability to maintain the structure within our body is what keeps us alive. So to stay alive we must continuously replenish the structure that metabolism destroys. We do this by mining and importing new structure from our surroundings. And to do *that*, we must live where there is structure to mine—the more abundant the better. We get our structure from the foods we eat. Carrots and trees get theirs from sunlight.

Until now, I haven't mentioned entropy. That's because entropy is not a measure of structure—rather it is a measure of disorder, or lack of structure.

I'll start with an admission. I'm unable to give a nice linear, step-by-step explanation of entropy, and of its storage and transport. To get a good feel for the subject, it seems most scientists and engineers must use an approach that's a bit like round and round the mulberry bush. Read and discuss once, then go back to read and think again. Having thought about and researched entropy for many years, I've come to believe that anyone who says they understand entropy on first explanation hasn't quite got it. Yet the good news is that with one circle of the mulberry bush we can get a sense of what it's all about. And that sense is all we'll need for smelling land. So let's get started.

As we've said, entropy is a measure of disorder, or lack of structure. The thermodynamicists who uncovered this thermodynamic property came to—*had* to—define entropy in terms of disorder, rather than its opposite, structure. That's unfortunate because it's structure that living things need, and that our energy system uses. Yet what we've named is lack of structure. It's as if we wanted to describe something beautiful, but the laws of physics didn't allow us to define 'beauty.' Saying 'It's a low-ugliness sunset' doesn't quite work, does it?

Although entropy may *seem* to be an upside-down property, it is not. Entropy is defined as zero for perfect structure (order)—but there is no upper limit to *dis*order. So the entropy scale begins at zero—which represents perfect structure—and increases as disorder increases. In this way, the entropy scale mimics an absolute temperature scale. Zero temperature on the Kelvin scale means as cold as we can get, nothing can get colder. But there is no upper limit to hotness.

All processes produce entropy. Any time anything happens, entropy is created. Sometimes only a little. Sometimes a lot. For instance, when you push back your chair in puzzlement (or wonderment) from reading these last few paragraphs, you will have produced entropy—some caused by the friction within your blood vessels carrying the extra blood to your agitated brain, and some by friction as the chair rolls across the floor. Pushing your chair back produced just a little entropy. In contrast, if your car hits a tree, much more entropy will be produced.

The entropy law is unique because, unlike *all* other laws of Nature, it is not a conservation law. The eleven laws of classical physics are:

- The four laws of classical mechanics (conservation of matter, of momentum, of energy and the non-conservation of entropy),
- The four laws of electromagnetic theory (elegantly wrapped in Maxwell's four equations), and
- The three force laws (gravitational, acceleration and electromagnetic, which link the eight laws of mechanics and electromagnetics).

Ten of these eleven laws are built around the idea that something is conserved.* The striking exception is the *non*-conservation law of entropy—because entropy is continuously produced.

This is why Nature's entropy law aims the arrow of time—the barb of higher entropy pointed to the future, the feathers of lower entropy leaving the past. Any process that would increase the entropy of the universe is a possible route to the future. Any process requiring the entropy of the universe to decrease is impossible. All of Nature's other laws—such as her 'conservation of momentum' law—would allow time to run either forward or backward.

When I introduce engineering students to the ideas of entropy, entropy production, and the arrow of time, we often speak of movies. When we want to anticipate how much entropy a process will produce, because all processes will produce some, it's useful to imagine a movie running *backwards*. If backwards can seem realistic—like a backwards movie of a pendulum swinging in a vacuum—the process produces little entropy. Conversely, if backwards appears preposterous—as in a backwards movie of an egg thrown at a brick wall—the process is strongly entropic.

We often describe a process that generates very little entropy, such as a pendulum in a vacuum, as a reversible process. No process is ever *purely* reversible. But the idea of reversibility and irreversibility is a good way to differentiate between pendulums swinging (almost reversible) and eggs hitting walls (strongly irreversible). It's an important engineering optic when designing energy technologies.

Now I must deal with a widespread misunderstanding. Many people understand that the total entropy of the universe is always increasing. But some believe that, because all real processes produce entropy, entropy must increase *where* it's produced. That's the mistaken part of things—because entropy can be moved

---

* It's easy to recognize that most laws are conservation laws, like energy. But the three force laws are also conservation laws, although more difficult to visualize. Should you be interested in thinking more about these eleven laws, you might look up my article, 'Engineering and Classical Physics,' *International Journal of Hydrogen Energy*, 25, No. 9, (2000): 802-806.
  Or go to www.smellingland.com or www.h2.ca.

from one place to another. We can dig entropy out from where it's made, and ship it away as fast, or faster, than it is generated. The often-forgotten truth is this: Entropy can be imported and exported.

So although the total entropy of the universe is increasing, locally it can stay constant, increase, or decrease. An example of local entropy decrease is when liquid water changes to ice. (The internal distribution of ice is more structured, because the molecules are frozen in place, not free to go wandering about as they can in liquid water.)

The storage and transport of entropy is related to, but not the same as, the storage and transport of energy. We know energy can be stored as *kinetic, potential, electromagnetic, chemical* or *thermal* energy. And we know it can be transported as *heat,* or *work* or *within materials* that are themselves moving. Purely structured energy, like kinetic energy, does not contain or move entropy. In contrast, thermal energy is disordered and is therefore a repository for entropy. (Thermal energy is the disordered energy of random molecular motion.)

The key to *which* energy storage and transport modes participate in the storage and transport of entropy depends on whether the energy has a disordered distribution, or whether it's perfectly structured. The mechanisms of entropy storage and movement are related to a *subset* of energy storage and movement mechanisms. Entropy is stored in *chemical* and *thermal* energy within materials, and is transported by *heat* and *material movement.* Every time thermal energy moves, entropy goes along for the ride. The *amount* of entropy stored or transported is the amount of energy divided by its absolute temperature in, say, degrees Kelvin.

There are many tough ideas in these last few paragraphs some of which will become easier to accept when we've covered the ideas of exergy. But for the purposes of this book, let's summarize:

- Entropy is produced by any process, so the entropy of the universe is continuously increasing;

- Entropy is stored in material—its quantity is the amount of thermal energy divided by its absolute temperature; and,

- Entropy is transported by heat or moving material—its quantity is the amount of thermal energy flowing (as heat or material) divided by its absolute temperature.

If we forget the last of these three truths, we'll be unable to develop energy systems properly—and we'll never understand how Earth is kept alive.

Now let's consider entropy and people. A good place to begin is by thinking about thinking.

The metabolic process of thinking produces entropy in our brain. If our brain couldn't shed entropy, then entropy build-up would increasingly disorder our brain, turn it to mush. Fortunately, we can export entropy from our brain. Heat departing the top of our head carries entropy with it. (Baldness has advantages.) Blood carries away still more—exporting it to other parts of our body where we pitch it out as heat or high-entropy material.

You might now ask: If entropy is so important, why isn't it more familiar? The answer may be that, unlike temperature or pressure, we can't *feel* or *see* entropy. What's more, while we have thermometers for measuring temperature, gauges for measuring pressure and weigh scales for mass, there will never be an entropometer that can directly measure entropy.*

Now let's treat ourselves to an out-of-body experience. Just as we had an out-of-world experience when looking down from our imaginary space capsule to watch Earth work, let's look down upon ourselves to watch how we work.

Gazing down we're struck by the fact that whatever we're doing—eating or fasting, sleeping or waking—one thing is constant, one thing never stops: We continuously produce entropy throughout our innards. Blood circulating generates entropy. So do muscles twitching, stomachs digesting, lungs breathing, brains thinking, hormones percolating. All these activities dump disorder into us, break down defined molecular structures, grind neat arrangements of atoms into slightly more random distributions. All are working to turn us into mush.

---

* I attribute 'entropometer' to James Lovelock. We can't measure entropy directly. But thermodynamicists have developed explicit relationships between entropy and other common properties, like temperature and pressure. So by measuring these other properties the value of entropy is easily calculated.

So to *stay* alive we must rid ourselves of the entropy we produce by *being* alive. We must flush out disorder or we'll croak. To lose our ability to shed entropy is to die. (While a corpse has lost its ability to shed entropy, it still contains structure—which is valuable to micro-organisms. They mine this structure to keep *themselves* alive and, in so doing, speed us through the process of 'ashes to ashes, dust to dust.')

People have two ways to flush entropy. One way dumps it in batches. The other pushes it out continuously—using a process that, within limits, can be speeded up or slowed down to match the rate we're producing entropy. So the two entropy-shedding mechanisms are synergistic.

We batch-shed entropy when we leave material waste behind in rooms marked Gents or Ladies—or, as the Brits say, 'in the loo.' The stuff we leave behind contains more entropy, is much more disordered, than the food we ate. Our excrement is entropy-rich. The food we eat is entropy-poor, structure-rich.

To augment batch shedding, we continuously flush entropy by expelling heat. The amount of entropy carried away by heat is equal to the amount of heat, divided by the temperature of your skin. (To make this calculation, we must use an absolute temperature scale like Kelvin.)

When we are especially active, we ratchet up our entropy production. So we better ratchet up our entropy shedding. During these activities we often don't want to call 'time out' to visit the loo. But we *can* increase the rate at which heat leaves our bodies. Sweating helps and brings evaporative cooling. Evaporative cooling substantially increases heat rejection, and therefore entropy flushing.

Food brings structure, energy and material to our lives. But of these, I expect structure is the most important—then energy. Material is the least important. Indeed, most of us would be better off with less material; I would certainly benefit from less material around my middle when lunging after a cross-court tennis shot.

So far we've focused on us, on people. Let's get a little less parochial. Let's think of how we fit in with our fellow travelers in what we call the biosphere.

People are structure parasites. And so are those lions, monkeys, fish and caterpillars. We mine structure from things that were once alive. This is very different than the trees and phytoplankton who mine their structure from sunlight. All the structure that ends up in any of Earth's living species was originally harvested from sunlight by photosynthesis. So we should look at life's frontline troops, photosynthetic life.

Let's start with a tree. I view a living tree as a factory whose product is more tree—and sometimes fruit and seeds, that perchance, will start more trees. The factory metaphor leads us to ask, what are the material inputs? What are its energy inputs? What are its manufacturing technologies? What are its waste products and, of course, its products?

Water and carbon dioxide are the primary material inputs from which the tree scoops out hydrogen and carbon. The tree also mines trace elements, but hydrogen and carbon are the big guys. The energy input is sunlight, which the tree uses to do the mining. Photosynthesis is the mining technology. Oxygen is the waste product—the tailings from mining hydrogen and carbon from water and carbon dioxide. The tree also keeps some of its oxygen and imports a little nitrogen. The product, as we said, is more tree and nuts to start the next-generation trees—or to feed squirrels and us.

That's a preliminary look at our tree. Yet something seems to be missing. Mining hydrogen and carbon is one thing. But a tree is much more complicated, much more *structured*, than just a bunch of hydrogen and carbon atoms—which could also be a bag of methane or a lump of coal. So we must ask: Where does the tree get the structure it uses to arrange the carbon and hydrogen into the exquisite molecular assemblies of walnuts or maple leaves? Bet you know. The tree harvested structure from sunlight.* So while sunlight brought energy to the tree, even more importantly it brought the *structure* the tree used to assemble the wonderful molecular arrangements of its roots, trunk, branches, leaves, flowers and fruit. (The *technology* the tree uses to choreograph the structure of the leaves, et al., fills volumes of biology books. But a good engineer sweeps the

---

* The route is via hydrogenation of carbon, oxidation of the hydrogen, and then (with a little phosphate and nitrogen) the formation of the energy currency of life, ATP (Adenosine Triphosphate).

details away, to focus only on the issue at hand—in this case, understanding the role of entropy in our living planet.)

To keep itself alive, our tree uses its photosynthesis technology to mine both energy and structure from sunlight. Both are required. Yet to me, structure seems the more important.

Energy is a means to an end. Structure is the end. Our ability to replenish structure is what keeps us alive.

I invite you to return with me to the space capsule we first entered during our afternoon on a hillside. This time we'll train our binoculars on the hierarchy within Earth's biosphere. We look down to watch petunias and apple trees, forests and blue-green algae all dipping into the incoming sunlight, all scooping out structure with a scoop called photosynthesis. The petunias and forests pour this structure into themselves—where it appears in the exquisite architecture of leaves, stamens and all the floral wonders you see about you.

Then what happens? Let's turn our binoculars to lions and eagles, termites, cows and people. Herbivores eat flora to get their energy, material and structure. Carnivores eat herbivores to get theirs. Omnivores and maggots eat them all. Then there are the fungi, bacteria and protoctists. . .and the whole canopy of life that envelops our Earth. In his book *The Strategy of Life*,[30] Clifford Grobstein wrote:

> Life—macromolecular, hierarchially organized, and characterized by replication, metabolic turnover, and exquisite regulation of energy flow—constitutes a spreading centre of order in a less ordered universe.

It seems to me this gets close to the core of it—at least that bit about 'constitutes a spreading centre of order in a less ordered universe.' Merging Grobstein's statement with the fact that the entropy of the universe is growing, we can add one word so the last clause becomes, 'constitutes a spreading center of order in an *increasingly* less ordered universe.' In Chapter 26 'From Steam Engines to Symphonies,' we'll listen as a student changes one more word.

In a universe where universal entropy growth gives time's direction, the essence of being alive is that all living things must prevent their own metabolism from causing entropy growth within themselves. It seems to me that: *Life swims upstream against the arrow of time, propelled by absorbing structure and disgorging entropy.*

Now we'll expand our view again, to consider Earth's import-export trade in entropy.

Because the temperatures of Earth's incoming sunlight and outgoing infrared radiation are very different, they carry very different amounts of entropy.* We can bundle these ideas like this:

---

* Engineers often use the symbol J to represent the flow of a commodity, using a subscript under the J to indicate *which* commodity. If we assign Js to entropy flow, Jm to material flow and s to specific entropy (entropy stored per unit mass), the equation for entropy carried by material is simply, $Js = sJm$. Entropy transport by heat conduction is proportional to the amount of heat divided by the local *absolute* temperature. First consider conduction heat transfer. If *conduction* heat flow is $Jq(c)$, the associated entropy flow $Js(c)$ is: $Js(c) = Jq(c) /T$. That's OK for conduction, but when we talk about *radiative* heat transfer (like sunlight), it gets trickier. For radiation, the macro-*shape* of the equation is similar, but we need to put a constant in front of the $Jq(r)/T$ to account for frequency distribution of photons. Typically, this constant is about 4/3, or slightly greater, depending upon the distribution of radiation frequencies. So entropy transport by thermal radiation becomes: $Js(r) \approx 4/3 \ Jq(r)/T$.

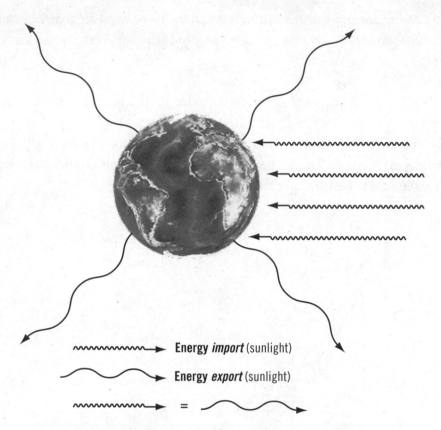

Figure 22.1 Earth's Import and Export of Energy

Import and export of energy are equal, so energy is neither produced nor consumed on Earth. Shorter wavelengths symbolize ultraviolet radiation. Longer wavelengths symbolize infrared radiation.

- The Sun's temperature, which sets the temperature of Earth's solar energy import, is much greater than the temperature of the Earth, which sets the temperature of Earth's infrared energy export.

- Therefore, solar radiation delivered to Earth carries much less entropy than does the infrared radiation leaving Earth. Earth accepts low-entropy streams from the Sun and throws out high-entropy streams to the universe.

- Therefore, because Earth's entropy is constant (at least as constant as any other of Earth's thermodynamic properties), Earth must produce entropy at a rate equal to the difference between the rates of outgoing and incoming entropy.*

Figure 22.1 shows Earth's incoming and outgoing energy flows. Energy entering via sunlight is equal to the energy leaving via infrared radiation. Figure 22.2 shows Earth's incoming and outgoing entropy infrared radiation, so a lot more entropy leaves via infrared than enters via sunlight. Incoming entropy is carried by low-entropy sunlight. Outgoing entropy is carried by high-entropy infrared radiation.

By comparing Figure 22.1 with Figure 22.2, it's obvious that although Earth hosts no *energy* consuming (or producing) factories, she does host a myriad of *entropy* producing factories.

I like the image of Earth consuming structure and sending its entropy excrement off on a speed-of-light trip through the universe. The infrared energy that Earth pitches out to the universe is at least as important as the incoming sunlight. Perhaps more so. For it pours into the universe all the entropy produced by our planet's exuberant lifestyle. We can think of Earth as eating low-entropy sunlight, discarding high-entropy infrared radiation and hosting a myriad of versatile, multifaceted, and sometimes even intelligent entropy-producing factories.

Except for the 'sometimes intelligent part,' every planet does much the same. I've been asked, 'How do you know every planet behaves much the same?' Well any planet, anywhere in the universe, must be cooler than the star it's orbiting, but not as cold as deep space. So the planet will receive low-entropy (high-temperature) radiation from its star and (in order to maintain constant temperature) must export higher entropy (cooler) radiation to the cold universe.

---

* If you want to dig deeper into entropy transport by radiation, Adrian Bejan wrote an excellent book *Advanced Engineering Thermodynamics* that clarifies much.[31] Nevertheless, unlike the expressions for conduction entropy transport (which has been well-defined for some time) the formulation(s) for radiative entropy transport still have ragged edges.

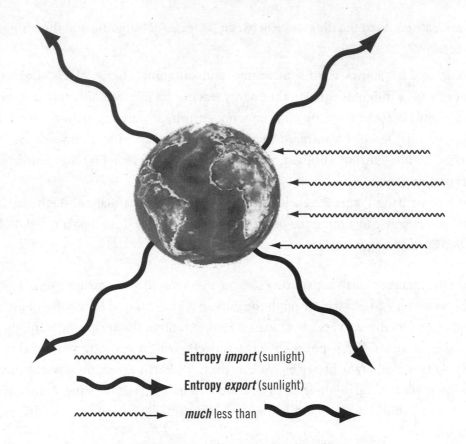

Figure 22.2 Earth's Import and Export of Entropy

Entropy export is much greater than entropy import, so a lot of entropy is produced on Earth. Shorter wavelengths symbolize incoming ultraviolet radiation; longer wavelengths symbolize outgoing infrared radiation.

Back to Earth. Overwhelmingly, most of Earth's entropy is generated by non-living things—tides, winds, rivers, earthquakes, volcanoes and so on—most of which are likely to occur on any planet. So what distinguishes Earth from dead planets is not entropy production. Rather, it's those living things that produce entropy by swimming upstream against the arrow of time.

On Earth, life is chemicals having fun—as they vigorously paddle upstream against the currents of time.

Because life must shed its entropic waste, as it cannot survive within an increasingly entropy-fouled nest, it seems to me that every living system must also develop strategies to place itself where there are both sources of structure from which to graze, and conveyor belts for removing entropic waste. Therefore, for life to thrive on *any* planet, I expect it to have fluid (gaseous or liquid) conveyor belts. Fluid conveyors spread life's entropy waste over the living planet as a key step toward pitching it out to the universe as infrared radiation. A solid planet couldn't do it.

The hydrosphere (oceans, lakes and rivers) and the atmosphere are Earth's fluid conveyors. They give more uniform planetary temperatures than would exist without fluid conveyors. And these more uniform temperatures aid the process of throwing high entropy infrared radiation out into the universe.

These same conveyors also deliver the ingredients for life support, such as material nutrients. But I'm not persuaded these are as fundamental to life as the need for entropy shedding. What is, or is not, a material nutrient is *specific* to a particular form of life. Yet what *must* be common to all life, on any planet, is the need to eat structure, shed entropy and thereby to keep swimming upstream.

With these ideas burbling, I'm drawn back to the Gaia hypothesis developed by James Lovelock, Lynn Margulis and others. The Gaia perspective argues that there is great value in viewing Earth itself as a living organism. Two of Gaia's insights are:

- A living planet is unlikely to be merely weakly alive—having only tiny bits and pieces of life scattered over an otherwise dead ball in the sky. If life develops on a planet, it will almost certainly bloom quickly—in geological terms—to be robustly alive, spreading to every nook and cranny as it has on Earth.

- A robustly alive planet almost certainly will have an atmosphere that is *not* in chemical equilibrium with the planet's bulk material. Rather, its atmosphere will be held out of equilibrium by life's gaseous waste products.

Now I wonder if a third requirement for a planet to come alive might be to have an extensive network of fluid conveyor belts for removing entropic wastes. Moreover, it's probably best if one conveyor is liquid and the other gaseous, because then synergies blossom.

Any planet, alive or dead, will have one of the ingredients necessary for life—incoming low-entropy (high-structure) starlight from their neighborhood star, although for outlying planets it might be a pretty weak stream. And all planets will also have a second ingredient—a location within a cold universe into which the planet can pour its entropic waste. And now I think a third, complementary, requirement for life seems to be an extensive network of fluid conveyors to vacuum out entropy-fouled nests. So if we ever find another living planet, anywhere in our universe, I expect a robust network of fluid conveyors will course across its epidermis.

For those who want to learn more about the wonders of entropy, many excellent books get things right. For entropy's importance to living systems, two books by the same title, *What is Life?*[32,33]—the original by Erwin Schrödinger, the second by Lynn Margulis and Dorion Sagan—are a must. I believe Schrödinger's ideas should be one of the introductory courses in medical schools. Lovelock's *The Ages of Gaia*[34] is hard to top for a great overview of our living planet. In 1990, Peter Coveney and Roger Highfield published *The Arrow of Time*.[35] More recently, Brian Greene wrote *The Fabric of the Cosmos*.[36] These and several other books can be a joy.

If you're the sort of person for whom understanding can be clarified by watching how some people get things wrong, Jeremy Rifkin wrote an exquisitely incorrect book, *Entropy: A New World View*.[37] On page 37 you'll find this: 'The point that needs to be emphasized over and *over* again is that here on Earth material entropy is continually increasing and must ultimately reach a maximum.' The italics are Rifkin's. Fortunately, he's dead wrong.

Let's take one more quick trip round that mulberry bush. We need to do that because it's easy to think that, in the context of understanding entropy, the opposite of disorder is order. For example, in Grobstein's definition of life, he used the word 'order'; whenever possible, I've used the word 'structure.' I've a reason for preferring 'structure' to 'order'—which I'll now explain.

Using everyday ideas, disorder is one way to *visualize* entropy—while structure is one way to visualize low-entropy. That's OK. But it's probably safer to think of structure and disorder as *manifestations* of low-entropy and entropy. Because equivalent they aren't.

That's an important warning. There are many nuances lying about. To help illuminate these nuances I cannot do better than to quote from Schrödinger's *What is Life*:[32]

> The difference in structure [as manifested by low-entropy] and lack
> of structure [as manifested by higher-entropy] is of the same kind
> as that between an ordinary wallpaper in which the same pattern is
> repeated again and again in regular periodicity [which is ordered but
> not structured] and a masterpiece of embroidery, say a Raphael tapestry,
> which shows no dull repetition, but an elaborate, coherent, meaningful
> design traced by the great master.

These ideas *do* need a lot of marinating.

23. IT'S EXERGY! 🌶🌶

In which we find exergy—the stuff that powers our planet, our civilization and us.

When I was a boy, my dad enjoyed telling me of coming home from his evening shift at the paper mill when he encountered a fellow 'three sheets to the wind' stumbling about a lamppost, looking for something he'd lost.* The drunk seemed gentle, so my dad offered help. Turns out the fellow had lost his glasses. After looking about together for some time, the glasses still unfound, dad asked, 'Are you sure you lost them here?' 'Oh no!' slurred the drunk, pointing unsteadily down the dark road, 'I loshed them down there, but therish no light there, sho I'm looking here.' Later in life I realized other fathers must have told their sons similar stories. But that doesn't matter. Every father should tell that story. And it is the perfect introduction to exergy.

People often speak of energy as if it's what our energy system consumes to deliver services. But as we realized during our afternoon on a hillside, we don't consume energy. That realization set us off on this mini-voyage within-a-voyage to find the something-of-value we do consume, which is 'exergy.' All energy systems—from our planet's, to our civilization's, to those that allow you and me to walk, talk and think—harvest exergy from flows of energy. In fact, just as a hydroelectric station scrapes out energy from flowing water, we and our energy system scrape out exergy from energy.

It might be comfortable staying under the lamppost where, of course, we'll find a few insights. But we will never have the joy (and profit) of really seeing what's happening. For that we must find those missing spectacles.

---

* 'Three sheets to the wind' is another seafaring expression my father seemed to favor. The sheets are the lines (ropes) that hold a sail in position. To have three sheets a-flopping, surely means you're out of control.

I'll start by giving a definition of exergy: Exergy is the maximum theoretical work that can be delivered by bringing an energy source or currency into equilibrium with its environment.

'Availability' was an earlier name for what we now call exergy. 'Availability' emerged from steam-engine language. In those early days, engineers knew that even a perfect steam engine could only convert a fraction of the steam's energy into work—and they called that fraction the steam's 'availability'. Made sense then. But later, some people realized there were two problems with 'availability'.

First, the term 'availability' could be ambiguous. If I say, 'the availability of steam is 50%,' how will you know whether I mean 'it's available for use 50% of the time' or whether I mean '50% of its energy could be converted to work by a perfect steam engine'. The second problem arrived when people learned that, for some energy sources or currencies, the work delivered could be greater than the energy supplied to do the work. Seemed weird! How could the availability of something be greater than 100% of that something? For these reasons, I much prefer the word exergy to availability. Happily, exergy is catching on almost everywhere.

Slovenian professor, Zoran Rant, introduced the word 'exergy' back in 1953. He mined Greek and Latin for the bits he needed for his better word and published the reasons for his choice in German. When placed at the beginning of a word, the Latin *ex* means 'out' or 'out of'. The 'erg' part of both energy and exergy came from the Greek word *ergon* meaning 'work.'* So exergy means that fraction of the total energy that we can extract (excavate, excise, expel)† from the energy-containing currency or source to deliver as work.

We do the extraction by bringing an energy-containing medium, say a source or currency, into equilibrium with its environment. Yet, in some

---

* By the time Professor Rant came along, energy had been assigned a precise, scientific definition—having a broader, more inclusive meaning than work. So, in that sense, the *erg* Zoran Rant used in 'exergy' is more faithful to its roots than is *erg's* use in 'energy'.

† *Ex* words come from both Latin and Greek, including several I've used here. Of course, many of what we consider Latin-based words originated with the Greeks. These words were imported into Latin after Roman soldiers conquered Greece and returned with Greek words, wives, mistresses and slaves. In time, the wives, mistresses and slaves died. But the words lived.

circumstances, the idea of extracting exergy from the energy can be misleading because sometimes the energy to do the work is not taken from the medium, but from its environment—as we'll illustrate later. (From this point on, I'll use the single word 'currency' to encompass all energy-containing stuff.)

Now we need another idea. Energy grade is the name many engineers have given the fraction of energy that can be converted into work, if we could use a perfect energy conversion technology. 'Perfect' means without producing entropy or, what is the same thing, using perfectly reversible processes. (Later in this book, I'll use 'exergy grade' to describe the same concept, because it seems to better remind us what we're talking about.)

Historically, the energy grade of things like steam or wood was always less than unity. (The exergy was always less than the energy, as measured in, say, watt-hours.*) But when electricity came along, we found the energy grade was unity. Even more recently, as we shall soon explain, some currencies that contain almost zero energy can still deliver a lot of work. That made the energy grade much greater than one, sometimes approaching infinity. At first this seems bizarre. But as we dig deeper, we will come to see it's a natural result of the concept. For that deeper digging, we'll build on four ideas:

- The exergy associated with a currency† is the maximum work that can be harvested (by a perfect technology) by bringing the currency into equilibrium with its environment.

- Therefore, exergy is a function of both the currency and its environment, unlike other thermodynamic properties such as pressure, temperature and so on, which are properties of only the currency.‡

- The direction of energy flow to bring the currency into environmental equilibrium can be either from the currency to the environment, or from the environment to the currency.

---

* Energy grade, being a ratio of two things having the same units, is a 'non-dimensional' quantity.

† Exergy can be associated with either material or with radiation.

‡ Some chemical engineers may think exergy is nothing more than a new name for free energy. But free energies are a property of only the material, not of both the material and the environment, as is exergy.

- The mechanism of harvesting exergy, using energy flowing from the environment to the currency, is what sometimes allows the energy grade to be greater than unity.

These four bullets contain tough ideas. So we'll now seek explanations.

An energy currency possesses exergy when—and only when—it is *not* in equilibrium with its environment. Because it's such an important idea, we should review what we mean by environmental equilibrium. It helps to remember that Nature always moves things toward equilibrium. So let's consider the different forms of equilibrium:

- *Mechanical equilibrium.* Nature moves rainwater from your roof to be in equilibrium with water in your yard, and then moves the water in your yard to be ultimately in equilibrium with water in the oceans. The process of bringing the rainwater from your roof into equilibrium with the ocean is a process of achieving mechanical equilibrium. But elevation is only one part of mechanical equilibrium (or non-equilibrium). Other components include pressure or tension.

- *Thermal equilibrium.* This is particularly simple. If the temperature of a substance we're considering is the same as the environmental temperature, then the substance is in thermal equilibrium with the environment. If the substance is not at the same temperature as the environment, it's not in thermal equilibrium with the environment.

- *Species equilibrium.* Consider methane (CH_4). Methane is the primary component of natural gas. But methane is not a constituent of air, so it's not in chemical equilibrium with air. However, when Nature oxidizes methane—or we burn it—the products are water and carbon dioxide. Because these are part of the environment, they are in species equilibrium with the environment. So the process of oxidizing methane is a process of bringing its chemical constituents (C and H) into environmental species equilibrium. A simple equation makes it obvious. $CH_4 + 2O_2 \rightarrow CO_2 +$

$2H_2O$. Environmental reference materials include H_2O, O_2, CO_2, and nitrogen (N_2).*

- *Concentration equilibrium.* To be in concentration equilibrium, a reference species must exist in the same concentration as it does in the environment.

We often refer to the sum of thermal and mechanical equilibrium as thermo-mechanical equilibrium, and the sum of species and concentration equilibrium as chemical equilibrium. Exergy is always associated with a currency out of equilibrium with its environment. It results any time there is either thermo-mechanical or chemical non-equilibrium or, most often, both.

There remains the question: What environment? For our purposes, I'll be taking the environment to be our atmosphere, or to keep it simple, air. But if we were thinking of exergy on the Moon, or Jupiter or in the depths of the ocean, the environmental reference materials and conditions would be much different—and so would the calculation of exergy.

To extract work we put cleverly designed technologies *between* the exergy-containing medium and its environment.

Think back to our afternoon on a hillside. When we extract work from a river's mechanical exergy to make electricity, we do so by placing technologies in the path Nature uses to pull the river's water toward equilibrium, toward its oceanic fate. Using *Smelling Land* terminology, we employ source-harvesting technologies such as dams, turbines and generators. The higher the water behind the dam, the greater the exergy available for harvesting. After the transformer technologies have extracted the exergy—or at least most of it—the water is released to continue its voyage seeking the ocean. A couple of things should jump out from our hydraulic example:

- The amount of exergy depends upon how much the energy deviates from environmental equilibrium (for example, the elevation of water behind the dam relative to the water below the dam). This reminds us that we need to know more than just the properties of the energy; we must *also* know the

---

* The technical literature lists an array of chemical species and their concentrations which define standard environmental (atmospheric) equilibrium.

properties of its environment (such as the elevation of the water below the dam), to know how much exergy is available.

- Nature's entropy law guarantees that energy will eventually degrade to environmental equilibrium, no matter what we do. But left to themselves, equilibration processes don't yield useful* work. So to deliver civilization's energy services we intervene in equilibration processes with technologies that can harvest exergy—and its cousin, structure.

Dams with their hydraulic turbines, heat engines (energy conversion devices that use a heat-to-work step, such as jet engines and nuclear power plants) and fuelcells are all intervention technologies. In the language of our five-link systemic chain, they are either source-harvesting or service-delivery technologies. So, too, are mitochondria, which dip into our body's energy flows to extract exergy to power our running, jumping and thinking. The same applies to photosynthesis technologies, which catch incoming sunlight to direct its exergy to bringing Earth alive. Not all cleverly designed technologies were designed by people.

We know Nature stores energy as kinetic, potential, electromagnetic, chemical and thermal. Electromagnetic and kinetic storage are the least important for civilization's energy system. (Electromagnetic storage is restricted to capacitors and, via electrochemistry, batteries. Kinetic storage might be flywheels. But compared with other storage forms, batteries, capacitors and flywheels hold trivial quantities of exergy.)

We are now left with three modes of mechanical exergy storage and two modes of chemical storage:

- Thermomechanical non-equilibrium (*gravitational, thermal* and *pressure*†
differences between the currency and its environment), and

---

* I've used the adjective 'useful' to mean useful to *us*. Unconstrained rivers can certainly do work—the work of moving riverbed boulders or scouring out the landscape. But that's not useful to us. Indeed, scouring may cause us problems, especially if we've built homes by the riverbank.

† Positive- and negative-pressure exergy exists in solids as well as gases and liquids—for example, as tension or compression in a spring.

- Chemical non-equilibrium (*species* and *concentration* differences between the currency and its environment).

For what comes next, I need to use a few symbols or the ideas will suffocate beneath the verbiage needed to explain them. I'll give two examples, one each of mechanical and thermal storage, and then make a short comment on chemical exergy storage. I'll start with mechanical exergy—a subset of thermomechanical exergy. Let's look at a few cartoons of piston-cylinder combinations, simplified versions of what you'll find in your car's engine. We'll designate atmospheric pressure as p_e (where the subscript 'e' reminds us that it's the environmental pressure), and the pressure inside the cylinder as p_c. To ensure we consider only mechanical non-equilibrium effects, we'll assume the piston-cylinder is thermally insulated from, and doesn't exchange material with, the environment. We're left with pressure non-equilibrium (one form of mechanical non-equilibrium) as the only possible source of exergy.

We'll consider our piston-cylinder combo in three different situations. Figure 23.1 shows that the gas within the piston-cylinder is at atmospheric pressure. In Figure 23.2, it's at a higher pressure. It's lower in Figure 23.3.

In Figure 23.1 the pressure inside the piston-cylinder is identical to the outside environmental pressure; that is, $p_c = p_e$. So although the gas within the cylinder contains energy, this energy (which includes molecular motion) has zero exergy because the inside and outside pressures are in equilibrium—they're perfectly balanced. The piston will just sit there, impotent, unable to deliver any work and unable to extract exergy because there's none to extract.

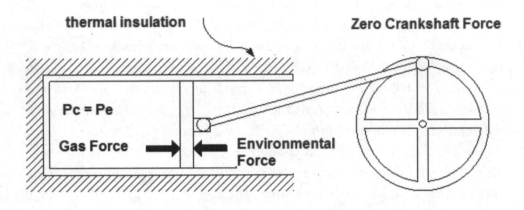

Figure 23.1 Cylinder and atmospheric (environmental) pressure are equal.

Figure 23.2 illustrates what happens when the inside pressure is greater than the outside pressure, that is $P_C > P_e$. This time there is an inside-to-outside pressure differential that can push the piston out, to turn a crankshaft or lift a weight—doing work. At least it can push the piston out until the inside pressure drops (as it must) to equal the outside pressure, when $P_C = P_e$. Then the work stops. These examples reinforce two features of all forms of exergy:

- The exergy existed because, initially, the gas in the cylinder was not in equilibrium with the environment.
- After we harvest all the exergy, the gas within the piston-cylinder still contains energy—the energy of the gas molecule's thermal motion.

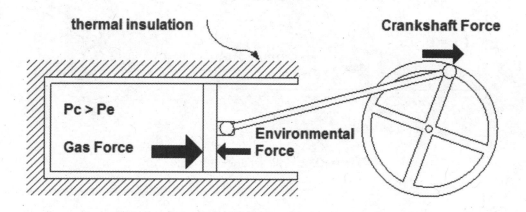

Figure 23.2 Cylinder pressure is above atmospheric pressure.

The big surprise comes with Figure 23.3. This time the cylinder contains a perfect vacuum. All the gas is out. Nothing is left, which means there is zero energy within the cylinder. But lo! There is a differential pressure across the piston. Our equation for how pressures relate is now $P_c < P_e$. (in this example, $P_c = 0$). So this time the atmosphere can push the piston in, until it reaches the end of its stroke or hits the cylinder head. Three things are of particular interest:

- Although there is no energy in the cylinder, either at the beginning or at the end of the piston's stroke, we were still able to harvest exergy.
- Exergy existed because, on Earth's surface, a vacuum is not in equilibrium with the environment.
- The energy within the cylinder was zero, but the exergy was substantial. This is a case where the energy grade (exergy/energy ratio) is infinity.

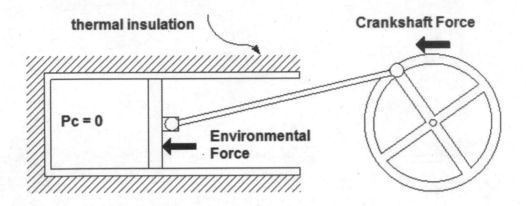

Figure 23.3 Cylinder pressure is zero.

When introducing the idea of exergy, I often ask my engineering students, 'How much exergy is contained within a piston-cylinder combination when it's filled with air at atmospheric pressure?' When you throw a class any question, most students are silent—most fear stepping up to be the designated hitter. Some suspect a trick question. One or two might risk an answer, usually 'zero.'

Of course, the proper answer is not an answer. It's another question: *Where* is this cylinder of gas located? Because exergy is always a function of both the currency and the environment, if you don't know the environment you can't know the exergy.

If the piston-cylinder is sitting in our backyard (where the environment is at atmospheric pressure), the exergy is zero—which is why most students would guess 'zero.' But if it's on the Moon, the air in the piston is not in equilibrium with the vacuum on the Moon's surface. On the Moon, there is no atmosphere to prevent the piston from moving outwards, doing work. So the piston-cylinder contains quite a bit of exergy.

Now imagine the piston-cylinder immersed in the dark depths of the ocean. The immense surrounding pressure will create an equally immense differential pressure across the piston—and we can harvest a corresponding amount of exergy as the sea pushes the piston in. In the ocean's basement, the cylinder not only contains exergy, it contains a lot more exergy than energy.

This should reinforce what I said earlier. Exergy exists because a currency is out of environmental equilibrium, but that doesn't require the currency to do the work. In some cases, the environment does the work. When our piston-cylinder sits in an ocean trench, it's the ocean overburden (not the air in the cylinder) that does the work. Yet the ocean couldn't have done the work if the cylinder had not been sitting on the bottom. The ocean did the work. But the piston's exergy allowed the work to be done. Earlier it may have seemed strange that a medium can contain more exergy than energy. But at least now it shouldn't be mysterious.

Next consider thermal exergy. Just as mechanical exergy results from mechanical non-equilibrium, thermal exergy results from thermal non-equilibrium. If an energy currency is at a different temperature than its environment, then the currency contains thermal exergy. Normally we expect the exergy-containing substance to be hotter. Steam is certainly hotter. But thermal exergy will also exist if it's colder. For example, liquid hydrogen contains thermal exergy because LH_2 is colder than the environment on Earth (but perhaps not on Pluto).

As you've probably now come to expect, the magnitude of thermal exergy is set by how much hotter or colder it is in relation to the environment.* So to know the thermal exergy content, we need to know the temperature of both the environment and the energy currency. If we use an absolute temperature scale (like the Kelvin scale), quantifying thermal exergy is eerie in its simplicity.† What could be simpler than to have the thermal exergy be proportional to the difference between the material temperature (T) and the environmental temperature (T_e)?

---

* In this chapter, we repeatedly use the same principle. That's engineering. If we work with fundamental principles, these *do* recur repeatedly. It's just a matter of applying them to different circumstances. In this case, we're applying the principles of exergy to the different forms of exergy storage. Are you becoming persuaded that engineering is a great career for the logic-gifted but memory-challenged?

† For the Kelvin scale, K = °C + 273.

Now consider energy grade. Because $(T—T_e)$ is proportional to how much exergy the currency contains, and because T is proportional to how much energy it contains, the thermal energy grade (the ratio of thermal exergy to thermal energy) is $|(T—T_e)/T|$. The symbol '| |' means the absolute (always positive) value of the quantity within. Here the absolute value is required to take care of the situation when T is below T_e, because without the | | the contained quantity would be negative. But that is physically impossible, because energy grade can never be negative.

We harvest thermal exergy from sources that provide heat such as coal, gasoline or uranium—by placing heat engines between the thermal source and the environment. A heat engine takes heat from the source, converts up to a maximum of $|(T—T_e)/T|$ of it into work, and dumps the rest into the environment as waste heat. The word 'maximum' is important because this is the heat that would be converted to work by a perfect engine. To appreciate the work we can get from a real heat engine, we must shave off all the exergy losses caused by technological imperfections. These are the inefficiencies. Unfortunately, those who don't understand exergy often use energy to calculate efficiencies and then the results can be terribly misleading—which will be one of the topics covered in the next chapter, 'Exergy Takes Us beyond the Lamppost.'

I'll wrap up this section about exergy storage by discussing chemical exergy—in part because it will be needed when we consider how to use exergy for quantifying environmental impact. We've talked about how fuels like methane contain exergy, because the fuel is not in chemical equilibrium with its environment. To determine the chemical exergy of a fuel—or, for that matter, of any material—we need to know the chemical composition (both species and concentration) of both the material and the environment.

So to evaluate the chemical exergy of a fuel like methane, we first determine how much thermal energy would be released if we were to oxidize the methane (usually by burning) to H_2O and CO_2—thereby bringing its elements into species equilibrium with the environment. We must then calculate the exergy of the thermal energy as we bring it into equilibrium with the environment. Finally, we must calculate the concentration exergy, because the products of oxidation (water and carbon dioxide) are unlikely to be at the same concentration as the water and carbon dioxide in the environment. After all that, we've got the exergy of methane.

We know exergy is the *maximum* work that we can get *from* an energy currency by bringing it into environmental equilibrium. But exergy analysis can also tell us the *minimum* work we must *put in* to extract a pure material from the environment. That last sentence contains a difficult idea.

Take the example of oxygen. The common procedure for producing pure oxygen is to mine it from air, using oxygen-separation technologies. The exergy of pure O_2 (which has concentration exergy, but not species exergy) is the minimum work required to separate oxygen from air. So by comparing the actual work of oxygen separation with the minimum possible work defined by exergy analysis, we know the real efficiency of the machine. This helps block manufacturers from pulling the wool over our eyes.

The same idea will give us the minimum energy needed to harvest iron (Fe) from iron ore (Fe_2O_3), or clean water from dirty. Useful design targets.

Some think it's a close-run thing whether climate volatility or the scarcity of clean freshwater will turn out to be the environmental issue of the 21st century. I expect climate disruption will overwhelm that curious race toward infamy. Still a dearth of freshwater will be critical in large regions of the developing world—and perhaps in some regions of the developed world, like California. Increasingly, a big part of our energy services will be used to provide freshwater—or, as people in countries such as Germany and the Netherlands say, 'sweet' water. The exergy of clean freshwater (relative to a seawater environment) is the minimum work needed to harvest potable water from the oceans.

Exergy can also tell us the minimum work required to clean up polluted water—or to harvest aluminum from bauxite.

Earlier we chuckled at the sanctimonious exhortations to conserve energy. It struck us like exhorting Gibraltar to stand up proudly like a rock. Yet something seemed amiss. Surely we *do* consume something of value when we drive cars, fly airplanes, turn on the TV.

Indeed we do: We consume exergy. In fact, when most people speak of energy they usually have in their mind something that behaves more like exergy than energy. Yet knowing that exergy behaves the way many people think energy behaves is not good enough. We must dig deeper. Not just for academic purity. Rather, if we're blind to exergy (staying only with the idea of energy) we will:

• Miss opportunities to greatly improve efficiencies—especially when using such emerging currencies as the cryofuels: Liquefied hydrogen and liquefied natural gas.

• Waste money and time chasing after non-existent opportunities—like trying to harvest energy from low-temperature reservoirs, which the media and proponents often tout.

• Fail to recognize innovation opportunities that can only become clearly visible using the exergy optic.

Most importantly, by understanding the exergy optic and then using it, we'll be better able to design and implement efficient, clean energy systems—just a little beyond the lamppost.

24. Exergy Takes us Beyond the Lamppost

In which the exergy optic shows us what's to blame for consuming exergy, how to calculate meaningful efficiencies, how heat pumps work and how to select the best targets for improving systemic efficiency.

How would you define the efficiency of the postal service, of studying for exams, of generating electricity? We use the idea of efficiency all the time. For many activities, like studying, the answer will always be squishy, opinionated, hard to measure.

Efficiency means the degree to which real performance approaches ideal performance. The idea is clear, but *quantifying* efficiency is often difficult. We all know we can't find a meaningful, numerate measure of efficiency when studying for exams. But for energy technologies, such as power plants or light bulbs, most people think efficiency is well defined—as the ratio of how much energy comes out divided by how much energy is put in. Yet efficiencies calculated this way are often meaningless—and the trouble is, many people don't realize these calculations can give false results. It's an error that can bite both ways. Sometimes the actual performance is overrated; at other times, it's underrated. And on both sides the errors can be very large. For true efficiency, we must use the ratio of *exergy* out to *exergy* in.

True efficiencies can be used to determine both the exergy and financial costs of delivering energy services. They can also tell us much about environmental intrusion, because if we improve the technical efficiency of delivering a service, we proportionally reduce environmental damage.

Leaving aside environmental intrusion for now, if we properly quantify the efficiency of an energy system, this efficiency tells us at least two things:

- How well the system is operating today, and
- How much improvement might be possible tomorrow.

With this background let's begin to apply the ideas.

Refrigerators and heat pumps behave strangely. These are technologies where energy ratios would *over*estimate performance, giving efficiencies between 200 and 300%—implying these technologies could be up to three times better than a perfect technology. Yet the refrigerator's true efficiency, which we get by using exergy ratios, is seldom greater than 30%, and often less.

Heat pumps are just refrigerators—except they refrigerate the outdoors in order to heat your home, while refrigerators heat your kitchen in order to cool your beer. So heat pump efficiencies, if defined by energy ratios, would give the same kind of silly numbers as refrigerators—often sillier.

The good news is that manufacturers of heat pumps and refrigerators don't claim these meaningless efficiencies for their products. Usually they use different metrics. Heat pump manufacturers sometimes use the term 'coefficient of performance' (COP), which is the heat energy delivered, divided by the energy put in (usually as electricity). Refrigerator manufacturers usually avoid the issue entirely.

The conversion of steam to electricity is a case where energy-ratio efficiencies *under*estimate performance. These underestimated efficiencies are sometimes used to falsely accuse electric utilities of wasting more than half their energy by throwing it out as heat into lakes or cooling towers. (Cooling towers are those large structures giving off a slowly rising cloud of steam beside any thermal generating station that is not located beside a river, lake or ocean into which the waste heat can be dumped.)

Correct but meaningless numbers are vicious weapons in the hands of those whose purpose is to mislead—including those environmentalists driven by political dogma rather than careful thought. Later in this chapter, we'll find that the *value* of the heat rejected (the heat's exergy) is seldom more than 3% of the total exergy input—nowhere near the 'more than 50%' sometimes implied by those whose avocation is to attack large utilities.

In spite of the misleading nature of these accusations, we can often find appropriate uses for this otherwise 'waste' heat—regardless of its low energy grade. The adjective 'appropriate' is key. We can use it for warming our homes, growing tomatoes, sometimes even to provide process heat for manufacturing. And it points to a supplementary prize: Exergy analysis greatly simplifies finding appropriate services for energy currencies that, today, we frequently throw out as waste. Often currencies unsuitable for one task are suitable for another. If we train ourselves to think in terms of exergy, we will often find opportunities staring us in the face.

Exergy efficiencies can identify where we should direct our efforts for improving tomorrow's designs.

Consider two energy conversion technologies, one operating at 30% *true* (exergy) efficiency and another at 90%. Imagine technical improvements that take the performance of each technology 'halfway to perfect.' It's reasonable to imagine technical advances that improve the efficiency of the first technology from 30% to 65%, halfway from where it was toward perfect. That would increase output by more than 100%, and reduce the energy use and environmental intrusion by more than 50%. These large margins point to opportunities for engineers to engineer and, if they are successful, for investors to invest.

For comparison, an equivalent halfway-to-perfect improvement in the second technology would bring the efficiency from 90 to 95%, increasing output (or reducing cost) by about 5%. Not as attractive a goal. Plus, it's typically very difficult to improve performances within the 90-95% range.*

So clearly, efficiencies give valuable information for both engineers and investors—*if* these measures correctly tell you what efficiency *is*. 'Aye, there's the rub!' And that is why exergy analysis should always underpin energy system designs.

We'll now look at four technologies to compare energy and exergy efficiencies: Niagara Falls electricity generation, home heating, coal-fired electricity generation and electric hair dryers.

Niagara Falls is one of Nature's more spectacular equilibration processes— water tipping over the lip to crash on the rocks below. The temptation to intervene with penstocks, turbines and generators was irresistible. Together,

---

* In practical terms, the true efficiencies of modern energy conversion technologies are usually above 50%, but seldom higher than 95%.

these technologies harvest exergy from rushing, dropping water to send it across the countryside delivering energy services.

As we observed back in Chapter 5, during the 1850s the energy from Niagara was harvested locally to saw logs and grind wheat. But after the invention of electricity, the energy could be shipped farther afield, to pull streetcars and light street lamps in Toronto and Buffalo. Table 24.1 is a snapshot of the two different ages at Niagara—the first before significant human intervention, and the second after hydroelectric generation. Both entropy production and exergy destruction are shown in the table.

Throughout this time frame, and for thousands of years before, the total rate of entropy production caused by water tumbling from Lake Erie to Lake Ontario remained constant. However, before civilization built power plants, all the entropy production happened between the two lakes. Afterwards, some entropy production moved away from the falls to be spread over the countryside, particularly within cities. Indeed, because using the electricity also dumped entropy into small towns, these towns bloomed to become cities.

Now that we understand energy, exergy and entropy, there should be nothing surprising in this observation. Yet this business of moving the location but not the amount of entropy production still intrigues me. We're learning that producing entropy is not all bad, depending on what it's produced for! Frankly, I hope to continue producing my personal quota of about 0.34W/K for several decades yet.*

---

* The average rate at which people produce entropy, if their metabolism rate is 100W.

	Dumped in River and Falls	Used for Energy Services	Total
Before Civilization Intervened			
Exergy Consumption [MW]	5199	0	5199
Entropy Production [MW/K]	17.7	0	17.7
Percentage	100%	0%	100%
After Hydroelectric Generation			
Exergy Consumption [MW]	2095	3104	5199
Entropy Production [MW/K]	7.1	10.6	17.7
Percentage	40%	60%	100%

Table 24.1 Entropy Production and Exergy Consumption (Niagara)

Comparison of exergy consumption and entropy production rates, before and after hydroelectric generation (based on mean annual rates, not hydroelectric installed capacity).

Two more things about Niagara. First, compared with most fossil energy systems, using hydraulic energy to produce electricity is very efficient. Most of the exergy taken from Niagara gets to the energy services—although a bit is consumed by friction in penstocks and turbines, and a little more as the electricity passes through transmission lines, voltage transformers and so on.

Second, the prime reason for high efficiencies is that there are no heat-to-work steps. All the input, output and intermediate currencies have energy grades of unity. (Recall, energy grade is the ratio of exergy-to-energy in a currency or source.) When all input, output and intermediate carriers have unity energy grades, the exergy and energy efficiencies will be identical—when designing electric motors, for example. But if you are designing technologies for warming your living room or cooling your beer, exergy really helps.

Let's look at warming your living room. Natural gas furnaces, electric-baseboard heating and heat pumps are three technologies for keeping us snug through cold winter nights. Table 24.2 sets out typical performance characteristics for these technologies.

Home Heating Options	Efficiencies		Exergy Wasted (% of input)
	Energy (meaningless)	Exergy (meaningful)	
Typical, Modern Natural Gas Furnace	85%	7%	93
Electric Baseboard	100%	8%	92
Electric Heat Pump (ideal)	1,330%	100%	0
Electric Heat Pump (real)	350%	26%	74

Table 24.2 Home Heating Options
Comparison of exergy and energy efficiencies.
Outdoor temperature: 0°C (32°F). Indoor temperature: 22°C (72°F).

Starting with the column for energy efficiencies, we see that at 85%, the performance of modern natural gas furnaces looks pretty darn good; at 100%, the performance of electric-baseboard heating appears unsurpassable; when we get to heat pump efficiencies of 350% things have become spooky—talk about getting something for nothing. If we accept the efficiencies given by energy ratios, we have every right to be smug, little need to think further. No need to seek improved designs.

Yet the message of this chapter is that energy efficiencies are often misleading, often don't represent what we *mean* by efficiency at all. So let's be careful. Better look at exergy efficiencies, see if they give us a different message.

Now move to the column for exergy efficiencies, where we find the performance of gas furnaces is about 7%, and electric baseboards about 8%. Atrocious! The thing that keeps gas furnaces in business is that natural gas is inexpensive compared to electricity.

For baseboard heating there is neither an efficiency justification, nor one based on operating cost—just low initial capital cost. That's why some 'spec' home developers in my area of southern Vancouver Island install baseboard resistive heaters. Although resistive heating destroys about 95% of the electricity's exergy, the capital cost is low. Builders 'speculate' that new home buyers will be lured by the lower capital cost and not think about operating cost because many people live by the rule: Never put off until tomorrow what you can put off until the next day. Resistive heating takes high-energy-grade electricity and squashes it into low-grade energy to warm a living room. And in the end, it will cost the home buyer more.

We'll look more closely at heat pumps, because it will be easier to understand how we can better use natural gas and electricity. 'Better' means to capture more of their exergy. We spoke of heat pumps before. But we never got around to explaining how they tick. We can better hear the ticking if we start with an analogy between pumping water and pumping heat.

Imagine you want to pump water from a lake up to your cottage located ten meters (33 feet) above the lake. By pumping we're increasing the water's gravitational potential energy from what it was in the lake to what it must be in order to pour from the tap in your cottage. Of course, the lake water already had a lot of gravitational potential energy, because it was already far above the center of the Earth (where its gravitational potential would be zero). So to get the water from the lake to your cottage you need only lift it that last little bit—that last ten meters above the more than 6 million meters of gravitational potential it already had (above Earth's center). Yet it's an essential few meters because, added to the lake elevation, it allows you to fill your kettle.

The temperature of thermal energy is analogous to the elevation of water. The higher the elevation of water, the greater its pouring potential. Analogously, the higher the temperature of thermal energy, the greater its warming potential.

Even when the outdoor temperature is 0°C (32°F), the environment still contains an awful lot of thermal energy. The amount of this energy is proportional to the environment's *absolute* temperature; so, in this case, it's proportional to 273K.

Compared with the thermal energy you need in your living room, the environment contains unlimited amounts. The problem is, the temperature of the environment's thermal energy is below the temperature of your living room. That hurdle prevents you from pouring any of the environment's thermal energy into your home. To jump the hurdle, you must increase the temperature of (some of) this outdoor thermal energy to (a bit above) the temperature of your living room. Then the pouring is easy.

This is no different than lifting the water from the lake those last few meters to your cottage. In the water's case, we need to add ten meters to the more than 6 million meters of gravitational potential it already has. For thermal energy, we need to add 22°C (72°F) to the 273K of thermal energy it already has. Then we'll be able to pour *all* the thermal energy into your home. When warmth is poured into your home from a heat pump, you get both the energy you lifted from the environment *and* the energy you used to do the lifting.

Home heating choices illustrate an important principle: We can always reduce energy use by substituting knowledge and capital. For home heating, we can displace energy with the knowledge of how to build heat pumps and the capital to build them. We've often spoken of the continuing evolution toward technologies that provide more service for less energy—something that's has been happening since the early days of the Industrial Revolution. The delivery of more service for less energy comes from using knowledge (and sometimes capital) in place of energy.

Table 24.3 sets out both the overall performance of a typical coal-fired utility, and the performance of the plant's most important components. I've normalized the data to a coal *exergy* input rate of 100 MW, so the entries in the exergy column also represent a percentage of the coal's input exergy.

I have not included this data because I think coal-fired generation will be important during the Hydrogen Age. Rather, the figures demonstrate how exergy analysis provides much more realistic insights into where the internal components of energy systems have opportunities for improved efficiency, while energy analysis can be quite misleading.

Component	Efficiencies of Component		Exergy Destroyed within Component (MW)	Exergy Delivered by Component (MW)	Emissions to Environment from Component	
	Energy	Exergy			Energy (MW)	Exergy (MW)
COAL (input)			100			
Combustor & boiler	**96%**	**54%**	**46.2**	58.7	**5.4**	**4.3**
Turbine (incl. steam extraction)	55%	81%	7.5	47.4	2	0
Steam condenser					**54.5**	**0.8**
Generators and transformers	98%	98%	0.8	36.7	0.7	0
ELECTRICITY (net output)				35.8		
OVERALL (output)	37%	36%		35.8	62.6	5.1

Table 24.3 A Typical Modern Coal-fired Electricity Generation Plant
Although this table may be difficult to follow, the text below sets out the key observations.

In the two far left columns, this table compares the energy and exergy efficiencies of the generation plant, and of its key components. The exergy destroyed within the plant is in the third column. The exergy delivered to the different parts of the plant is set out in the upper rows of the fourth column. The exergy consumers receive as electricity is set out in the bottom rows of the fourth column. The energy and exergy of emissions from the plant are given in the last two columns.

Because coal (the utility's input) and electricity (the output) both have energy grades close to unity, the *overall* plant efficiency, calculated from either energy or exergy ratios, is much the same. The value in exergy analysis comes when we study the generating station's internal components, its bits and pieces—the turbines, boilers, condensers, pumps and so forth. This is where exergy efficiencies become much more informative, more meaningful and more useful than energy efficiencies.

Look at the row showing the combined combustor and boiler performance. The first of two highlighted cells compare energy and exergy efficiencies. By energy-efficiency criteria, the performance looks good—nothing shabby about an efficiency of 96%. Yet its exergy efficiency is only 54%. So the *true* performance of the combustor-boiler is very poor—worse than all other components combined.

Now let's consider emissions, which are set out in the right-most two columns. The highlighted cells show the exergy and energy that escapes to the environment—mostly out the stack. Of the stack emissions, about half is chemical exergy and the rest is thermomechanical exergy.[*] The thermal exergy lifts the plume and gets it downwind. As we'll see in a moment, the chemical exergy usually causes most environmental damage.

If we look at environmental emissions from the steam condenser, we get a very different message. Astonishingly, through the condenser, the plant tosses out more than half the plant's total input *energy* to the environment. But hold on. These emissions from the condenser, in fact, constitute less than 1% of the exergy input to the plant.

So when we set out to prioritize research, development and engineering, it's important to know which step (in a series of process steps) has the largest margin for improvement. And for that, we must know which step has the lowest exergy efficiency.

Finally, let's take a look at those electric hair dryers. Table 24.4 shows the performance of these popular gadgets. One way or another, all the electrical energy goes to warming air. Some energy will first go to the fan, which gives the blown air a tiny bit of kinetic energy. But the air soon slows, transforming the kinetic energy to thermal energy. Therefore, all the input electrical energy ends up warming air and the energy efficiency is 100%.

---

[*] The data for these figures were developed by my colleague Marc Rosen, with very little help from me, and may be found in Rosen, M.A. and Scott, D.S., 'Entropy Production and Exergy Destruction- Part II: Illustrative Technologies.' *Int. J. of Hydrogen Energy* Vol. 28, No. 12 (2003): 1315-1323.

Hair Drying	Energy Efficiency	Exergy Efficiency	Service Efficiency
Electric Hair Dryer	100%	9%	Return to chapter 15

Table 24.4 Hair Drying
Comparison of energy and exergy efficiencies applied to an electric hair dryer.
Input air is 22°C (72°F) and output is 83°C (180°F)

But the hair dryer's exergy efficiency is a mere 9%. The reason for such terrible exergy efficiency is that the hair dryer degrades high-energy-grade electricity into low-energy-grade warm air. The energy grade of the warm air is a mere 0.09. So the hair dryer destroyed 91% of the exergy carried by the input electricity.

The column for service efficiency tells us to look back to Chapter 15 'Template for Sustainability,' because it's always important to be clear about the product our technology delivers.

So far, we've aimed exergy at strengthening performance. But, as implied when we were looking at emissions from the coal-fired generating station, we can also use exergy to flip things over, to help us better understand environmental intrusion. We've already spoken of energy and exergy escaping to the environment, so let's see what exergy can tell us about the magnitude of an emission's environmental impact.

We start by reminding ourselves that exergy is 'the ability to do work.' Up to now, we've been thinking of this as work to push streetcars, fly airplanes or drive computers. But it can also be the work of eating the paint off your shiny new car, or your house. So although exergy is a measure of an energy currency's value, it is also a measure of an effluent's danger. Exergy is a proxy for an effluent's environmental bite.

The exergy bite of emissions from the combustor-boiler of Table 24.3 comes from both thermal and chemical exergy. The thermal component is low-energy-grade—and does little but carry the plume into the sky. The chemical exergy results from unburned hydrocarbons, acids that will form acid rain and (among other toxins) the level of radioactivity in the fly ash.* The low-grade thermal exergy is quickly dissipated, while the bite from chemical exergy is saved until it has something to bite.

When we first spoke of exergy back in Chapter 2 'Charting the Course: Toward a Cleaner, Richer Hydrogen Age', exergy may have seemed an exotic and difficult concept because it had an exotic name. An exotic name it has, but an exotic concept it isn't. And, except when we dig deeply, it isn't very difficult. Before we (incorrectly) spoke of consuming energy—we can now (correctly) speak of consuming exergy. Still, we'd probably be wise to revert to 'energy' at cocktail parties.

---

* Often, more radioactivity is dispersed to the environment from coal-fired generating stations than from nuclear stations. Don't worry, to the surprise of many, this bit of radiation probably keeps you healthier, as we'll describe in Chapter 31 'You've Got to Be Carefully Taught—Know Nukes.'

25. A Whiff of Earth from Afar

In which we realize oxygen may be the true fossil fuel of our moist blue planet.*

My friend Hans-Holger Rogner and I were strolling through Cambridge, Massachusetts, talking about the recent drilling into the Siljan crater in central Sweden. The Swedes had been drilling through non-sedimentary rock that had been shattered by an asteroid during ancient times. This was 1986 and they were looking for abiogenic methane.

As we know, methane is the primary constituent of natural gas. But abiogenic methane is special. It has not been derived from biogenic origins—like dead dinosaurs and, of course, life that lived long before and after dinosaurs. In contrast, abiogenic methane would have existed as a primordial constituent of planet Earth. The Swedes had been tweaked to invest in the drilling, in part, by the late Thomas Gold, who spent much of his life smelling land, sometimes to be wrong, sometimes to be right, but always to make a contribution to looking at things the other way round. The Swedes were interested in the prospect of abiogenic methane giving them a boundless source of energy—a source under *their* ground. There was a lot of scientific interest and even more controversy.

I was interested in what to call the stuff.

It seemed to me that abiogenic methane could not be called a 'fossil fuel' because it was not fossil-sourced. Moreover, the theories describing why abiogenic gas should exist also led to the conclusion that much of what we now call the heavier fossil fuels, such as oil and coal, might also be derived, at least in part, from polymerizing abiogenic methane as it outgassed over the eons from the Earth's interior. So are *any* of what we now call fossil fuels, pure fossil fuels? Probably not. Does a *pure* fossil fuel exist?

Then it struck me that free oxygen is the fossil fuel of Earth. I'll explain.

Free oxygen means *molecular* oxygen, oxygen atoms tied together in molecules of O_2. When oxygen atoms are tied to other elements like hydrogen

---

* I attribute the wonderful description, 'our moist blue planet,' to Lewis Thomas.

to make water, or carbon to make carbon dioxide, or sulphur to make sulphur dioxide, or to anything except another oxygen atom, they are not described as free. Of course, not all the free oxygen is in the atmosphere. Some is dissolved in our oceans and lakes. Some infuses the soil.

Now why might we consider oxygen to be the fossil fuel of planet Earth? *Webster's Encyclopedic Unabridged Dictionary** tells us that 'fossil' pertains to 'any remains, impression, or trace of an animal or plant of a former geological age, [such] as a skeleton, footprint, etc.' Other dictionaries say similar things.

Recall our tree. The tree's primary raw materials are water taken from the ground, and carbon dioxide taken from the air. The energy the tree uses is sunlight, the technology is photosynthesis, and the product is more tree. *But the waste product is free oxygen.* So it turns out that free oxygen is a footprint of life. It is the footprint left by the processes life uses to mine the raw materials it needs to build itself—whether trees, algae or petunias, or the billy goats that eat the petunias, or the tigers that eat the billy goats, or us who often seem to eat them all. Photosynthesis is the technology used to mine all the hydrogen and carbon in biological material. The mine tailings are the free oxygen. All of Earth's free oxygen is the excrement of life.† So Earth's free oxygen is a fossil gas. But why is it a fossil fuel?

Rewind to *Webster's*, which tells us that a 'fuel' is a 'combustible matter used to maintain a fire, [such] as coal, wood and oil.' But fire is a tango. Coal, wood and oil will not burn by themselves. They need oxygen to burn. That means oxygen is also a combustible material. Most of us know that. Certainly any firefighter knows that. The engineers who redesigned the Apollo systems after the astronauts Roger Chaffee, Ed White and Gus Grissom died because the space capsule was filled with pure oxygen, rather than air, surely knew that. So why do we commonly speak of wood and coal as fuels, but not air and oxygen?

Webster doesn't mention that air is a partner in the tango. Webster's attention seems to be only on the partner we *pay* for. We don't pay for air because there

---

* *Webster's Encyclopedic Unabridged Dictionary of the English Language.* Portland House, NY, 1989.

† If you're wondering how early life got started without oxygen you might turn to the first few paragraphs of Chapter 44 'Our Sliver of Time.'

is so much more of it—at least so much more of it on the Earth's skin where we live. The difference between gasoline and air is their relative abundance and their ease of acquisition. Air is everywhere. Gasoline isn't.

The gasoline and oil that we call fuels exist in trace amounts compared with their combustion partner, air. There is a lot less gasoline in your tank than air in our world. And while we know it is impossible for gasoline to push our car without air, we say it burns gasoline, not air, because we know the gasoline was consumed. We never think about the oxygen which was also consumed.

We now know that any fuel is a fuel *only* because it's *out of equilibrium* with its environment. We covered that when discussing exergy. We harvest the exergy in a fuel by arranging when and how the fuel comes into environmental equilibrium, so that the equilibration process can serve our needs. The common way for two materials to be chemically out of equilibrium is for one to be a *reducing* material, and the other an *oxidizing* material.

I'll give an example. When we leave an axe head outdoors for several months, it rusts. The steel in the axe head is a reducing material and the air is an oxidizing material, which means that the steel and air are not in equilibrium with each other. Because it happens slowly, you don't notice the energy being released, but you do see the waste product, which is rust (Fe_2O_3). Rust is the ash left by slowly burning your axe head in air.

Oxidizing materials are often called 'oxic.' They include oxygen or hydrogen peroxide. Reducing materials, which include hydrogen, methane and the iron in your axe head, are sometimes called 'anoxic.' Oxic and anoxic materials like to form partnerships that bring them closer to equilibrium. They like to tango. In the right circumstances they like to 'burn.' The progeny of their union, the ash, is neither anoxic nor oxic. I'll call it stable stuff. It includes water and rust.

When we stand back from Earth, to see it from afar 'floating free beneath the moist, gleaming membrane of a bright blue sky,'* our planet is overwhelmingly a ball of *anoxic* material. The great mass of molten iron and other stuff at our Earth's core are all anoxic, and in this way Earth resembles every other planet

---

* 'Floating free. . .' is one of the magnificent phrases from Lewis Thomas, *Lives of a Cell*.[38]

in our solar system—but with a difference. Enveloping Earth's reducing body is a thin oxic atmosphere.

Let's glue these ideas together. A veil of free oxygen envelops Earth and some of it diffuses down through the soil, oceans and lakes. This oxygen has several characteristics. First, it is a fossil gas. Second, it is out of equilibrium with Earth's anoxic bulk and can burn in any of that bulk—in the oil, coal, or molten iron. Third, the quantity of this oxic gas is trace compared with Earth's immense anoxic body. Taken together, this all means that Earth's oxygen epidermis can be considered, perhaps should be considered, the fossil fuel of planet Earth.

An incoming alien space traveler, taking a whiff of Earth from afar, perhaps using a whiffer like a spectral analyzer, would see Earth's thin oxidizing atmosphere as a fuel. She may even go to her commander to say, 'Captain, my data indicate that the planet is enveloped in oxygen. We don't want to fly through a blanket of fuel. I think we should turn about.'

What is the difference between a cabin boy spending most of his life at sea who, smelling the intertidal zone, reports he smells land, and a space traveler who, whiffing the oxic atmosphere of Earth, reports Earth is surrounded by a fuel? The folks are different. What they whiff and how they interpret their whiffery are different. Both the cabin boy and space traveler are right. The cabin boy's thinking is strange to landlubbers. The space traveler's thinking is strange to Earthlings.

The eerie part is that the similarities go deeper. The most profound insight our space traveler could have upon smelling oxygen has nothing to do with fuels. Her most profound insight would be that she was smelling *life*. Not just life in the intertidal zone—that jumble of crabs and mollusks, seaweed and kelp—but the extraordinary plethora of life with all its grandeur and modesty, failings and successes, simplicity and complexity, that makes its home within the oxic skin of our moist blue planet.

To me it's the same story. I hope she won't be hanged.

Part Seven

INTERMEZZO

26. FROM STEAM ENGINES TO SYMPHONIES

In which we discover new ways to think about aesthetics. Then a student suggests the purpose of life, and it's all glued together during an ocean passage on a small sailboat.

In this chapter, I'll succumb to my fascination with possible links between entropy and aesthetics. I imagine these links as a kind of metaphysical corpus callosum, connecting the left-brain ideas of the 18th century steam-engine designers to the right brain's eternal question, 'What is beautiful?' Let's explore this corpus callosum, searching for what might link the entropy ideas developed by steam engine designers to Beethoven's Sixth or to a dew-draped garden in the morning.

Having come this far, having plowed through the concepts of energy, entropy, exergy—and how these relate to life, living planets and to designing better technologies—let's lighten up. Let's cast off the constraint of speaking only of things about which we can be pretty damn sure. Let's treat the caverns of our minds to some speculation—drift through intertwined ideas from aesthetics to baldness.

Our drifting can begin anywhere, so let's begin by wondering about our likes and dislikes. We know that to stay alive, all living things must have sources of order and structure from which to graze. Therefore, I suspect living things are programmed to seek out highly structured foods and environments. And so we might ask:

- Does this explain why our enjoyment of dining is enhanced when food is exquisitely arranged on a plate?
- Is this why we enjoy environments of orderliness and structure, like a garden, a forest or a tidy living room?
- Is this why we are repulsed by messes like garbage dumps, sewage outfalls or inner city decay?
- Is this the reason we're troubled by children who, when bored, cantankerous or over-tired, take to stirring the food on their plates? We admonish, 'Don't

mess up your food like that!' Whining, they rebut 'But it all goes to the same place.' We're frustrated, believe something is wrong, but don't quite know what. The feeling that something is wrong may be based on our reptilian brain telling us the structured plate is somehow *better*.

- Do the child's messy food and the teenager's messy room mean that, as much as our reptilian brain seeks out order and structure, some aspects of creating or maintaining order must be learned?

Staying with our likes and dislikes brings us face to face with aesthetics. The aesthetics of *settings* and the aesthetics of *processes*.

Have you ever played with one of those toys which has five identical stainless-steel balls, hanging side by side, each ball suspended by two strings, one string each from two cross bars like the parallel bars in a gymnasium? Few of us can resist pulling one of the end balls up and away from its colleagues. Then we release it to swing down, striking its former neighbor. We watch with glee as our dropping, accelerating ball strikes—causing an *orderly*, sudden transfer of momentum from the original ball, through the next three balls, until the fifth ball swings up *almost* mirroring the trajectory of the ball we initially released. Once started, the process keeps *almost* repeating itself until, gradually, the balls slow to a few spasmodic wiggles. Then we're saddened; we'd have liked it to be perfect, to have bounced back and forth forever. Are we sad because, had the balls bounced back and forth forever, it would have been a zero-entropy-production process?

Sometimes, on the first bounce it doesn't bounce at all. Rather all the strings get tangled and the motion twists itself to a quick stop. Then we're really upset. Is our dismay because the tangling has rapidly destroyed the orderly motion? No realistic backwards running movie here—which makes it a great childhood demonstration of irreversibility and high entropy production.

Let's dream up other scenarios that might demonstrate our response to processes that are close to reversible (low entropy production) and processes that are strongly irreversible (high entropy production):

- Most of us find joy watching an India-rubber ball hit the hard wall of, say, a handball court, to ricochet off to the floor and to other walls, almost forever—but we feel only disgust when an egg is thrown at that same wall where it goes splat. The first is low entropy production. The second, high.

- Could the ideas of entropy and structure be one way to separate good art from poor—one of the most subjective tasks of being human?

Shortly after I moved to the west coast of Canada, a Vancouver artist's unique form of 'performance-art' was to tie a live rat to a concrete block. He then dropped another concrete block upon the tied rat so the rat, like the egg, went splat. I expect many folks, like me, felt repulsed. But I also expect we felt trapped, barred from passing judgment because who are we—who am I, a mere engineer—to judge the artistic value of anything, even a splattered rat? Might we now have a way to judge this nonsense? Rat-splattering is a *strongly* entropy-producing process.

Goodness! Let's get away from splattered rats. Let's escape into music and poetry.

Is our desire for structure the reason we prefer music to noise? Is this why we enjoy Mahler's Second Symphony, the *Resurrection,* or Beethoven's Sixth Symphony, the *Pastorale,* or the Beatles' *Lucy in the Sky with Diamonds,* or Paul Simon's *Diamonds on the Soles of Her Shoes?*

To me, the link between music and low-entropy is the wonderment of wonderments. We are born loving it, evolve our taste for its different forms through life until, lying abed, dementia wormholing our mind, a smile crinkles our face when an old friend hums, for us, a childhood tune. Some part of music's magic must be rooted in music's nuanced structure.

These ideas lead to more thoughts, this time about language and poetry:

- Does low-entropy (exquisite structure) lie at the root of our joy in poetry?
- Is this why Shakespeare gets through to us better than a legal description of the same idea?
- Is this why cadence is so important in writing and speech making, why it carries us along whether we understand the literal message or not?
- Is that part of the power and danger of cadence-canny demagogues?

Occasionally high-entropy processes are highlighted for their beauty—literally—as when spotlights illuminate Niagara Falls at night. Yet perhaps we're seeing through the high-entropy processes to see the structure within. I often think the most mesmerizing part of Niagara is not the violence of the water's crashing turbulence below the falls. Rather, it's the ominous majesty of ordered, high-exergy water sliding over the lip, turning down toward its exergy-destroying doom. Lean over the railing, a few meters from that lip. You'll understand.

Now let's wander back to the business of being alive. In Chapter 22 'Entropy and Living Planets,' we saw that by shedding heat, animals have a wonderful mechanism for getting rid of entropy. Could this be another advantage of being warm-blooded?*

We've also learned that the *amount* of heat shed is proportional to the temperature difference between the object giving off the heat, and the place receiving the heat. This means that in very hot climates, because the human body is much closer to the environmental temperature, it's more difficult for people to employ heat rejection for entropy-shedding. On the other hand, bitterly cold climates require an unusually large fraction of metabolic exergy consumption just to keep warm. Is this why the most vigorous societies seem to have arisen in what we (with body temperatures about 37°C) consider temperate climates—climates where most symphonies have been written, most inventions invented, most new political processes tried?†

We also know that every creature has two ways to shed entropy. The first, voiding entropic *material*, suffers a key limitation: intermittency. Conversely, voiding entropic *heat* has the advantage that the rate of entropy shedding can be exquisitely matched to the rate of entropy production. We've noted that when fighting or running, it can be dangerous to pause in order to shed entropic material. But you can increase sweating, which raises the rate at which heat carries entropy away from you and into the environment.

---

* By providing the means to sustain regulated, high exergy-consumption rates.

† For those holding commissions in the politically correct police, please note that this reflection is about the quality of thermal environment in which people live, not about the people.

The absence of hair (fur) can be a disadvantage to mammals not smart enough to make clothes or set fires to keep warm. But if you *are* smart enough, being hairless has a great advantage when you want to shed entropy quickly during times of crises. We sweat. Dogs, having fur, pant. Sweating sheds entropy faster. More dogs are in more trouble during Manhattan's dog days of summer than are people. And why is it that among land-based mammals people have so little hair?

Now we'll leave what keeps us healthy, wealthy and full of spunk, and return to what keeps our planet healthy, wealthy and full of delight. As we've said, life will want to migrate toward locations where there are conveyor belts to deliver highly structured food *and* to remove entropic waste. Is this the reason life is so profuse within tropical and temperate rainforests—and in ocean upwellings and tidal estuaries, those exuberant linkages between our atmospheric and oceanic conveyors?

If the temperature of Earth's epidermis increases, the entropy departing Earth will be reduced—because entropy transport is proportional to energy transport divided by temperature. (Recall that $J_s \sim J_q/T$, where J_s is entropy flux and J_q is heat flux.) That might give us another anxiety twinge about climate disruption:

- Could global warming reduce Earth's ability to shed entropy?
- Might greenhouse gases constipate Earth's entropic waste disposal?

Most residents of our biosphere, cats and rats and elephants among them, take structure from their environment and put it only into themselves. The structure a tomato plant harvests from sunlight stays in the tomato plant, until it's eaten or oxidized. Some animals and birds put a portion of the structure they harvest back into the world around them, most often applying it to their nests.

Beavers go further. A family of beavers puts structure into their home and their dam, thereby reconfiguring the landscape (for better or worse).*

Now to people. When hunter-gatherer societies build structure into tools, tents and weapons, they don't go much beyond beavers. But modern civilizations have poured immense structure into cathedrals, highways, railroads and canals, into the skyscrapers of Manhattan and Shanghai, into the temples of Ramses II, into the Library of Congress, and the art treasures of the Louvre, the Tate and the Metropolitan Museum of Art. Into manuscripts, compositions and encyclopedias, into compact discs and hard drives, into equations.

It's amazing the amount of structure people have built *outside* their own bodies. In this, we are unique among living species:

- Is this one reason to think better of ourselves?
- Is this one way to differentiate greater civilizations from lesser?
- When lying abed during the last few days of our life, would it bring some peace to reflect on the amount of structure we've created and left behind— one way to measure whether it was all worthwhile?

The mind wanders to Lenny Bernstein, Wolfie Mozart, Willie Shakespeare, to the great architects, to the builders of canals and railroads. . . and then. . . to the great destroyers.

Toward the end of my thermodynamics course for engineering students, I sometimes slide in a few ideas about how thermodynamics can be used, not just to design steam turbines, but also to get more joy watching our living planet. And perhaps as another way to see the dangers of highly entropic, anarchic societies. At this point, I'm sometimes asked, 'Excuse me, sir, but will this be on the exam?'

---

* Canada's animal symbol is a beaver, chosen because a beaver works hard—never mind the environmental disruption. It's fascinating the critter-symbols different nations choose. Russia's is a bear. England's a bulldog. Many nations have chosen the eagle, an aggressive raptor that sits atop both trees *and* the parasitic food chain, from which it swoops to harvest structure from their prey's flesh.

When walking back to my office after one of these tag-on lectures, a student briskly overtook me from astern. He said, 'You know, I've always wondered about the purpose of life. Now I think I know.' Whoa! I thought, what am I in for? Although many of us might have asked these questions in our youth, most of us have given up that search, turning to more tangible things like families, fortune and fun. In class that day, I had introduced Grobstein's definition of life with a slight modification*—which the student now repeated but changing the single word 'a' to 'the': *'Life. . . constitutes* the *spreading center of order in an otherwise increasingly less ordered universe.'*

For us more aged, jaded folks, it might be a bit much to import this as a purpose for our lives. But for this student, it was enough. He saw himself as one part of Earth's web of life, whose purpose was to be the spreading center of order in an otherwise increasingly disordered universe. So he hoped to pick a career that would, as directly as possible, allow him to assist the spreading—in such fields as aerospace, fuelcell technologies or hydrogen systems.

I shuddered. For in some way he was right. During the latter half of the 21st century, these will be the technologies that speed Earthlet spores on their way to colonize the universe. But first, these same technologies must be employed to help put Earth's house in order.

During the spring of 1996, my wife and I sailed our eleven-meter (35 feet) sailboat, *Starkindred*, from our home in Victoria, on Canada's west coast, to the Tahitian isles of the South Pacific. It's a great-circle passage of some 4,000 nautical miles—a little farther than the great-circle route from Victoria to London.†

Those night watches aboard *Starkindred*, during which I dreamily pondered entropy, life and our planet, were magical. The black ocean shushed past, a wake of phosphorescence marked our southerly course as, night after night, we

---

* See chapter 22 'Entropy and Living Planets.'

† A great-circle route is the shortest distance over the Earth's surface between two points on the surface. If the distances are significant, it can be much shorter than a rhumb-line course between the same two points, which crosses all meridians at the same angle—or, if you like, a straight line on a Mercator projection map.

gradually sunk the North Star. Here we were, slip-sliding along this profoundly important interface between Earth's two great conveyors of entropy and energy: her oceans and atmosphere.

From *Starkindred's* log: '1996 May 28, 2350h, 14° 42'N, 130° 09'W. Wind NE 18 knots, gusting 22. Broken clouds & waxing quarter Moon to starboard, clear sky and stars to port. Reefed mainsail and yankee. Full staysail. Boat speed 6.5 knots, surfing to 7.5. Waves 12-15 feet, long wavelength. Beethoven's Sixth booms into my head from the Sony Walkman.'

Here we are, more than a thousand nautical miles from the nearest land, sliding over a moonlit, silver-speckled, black, rolling ocean. Skittering wavelets atop majestic seas. Seas that have run a thousand miles effortlessly lift *Starkindred* to pass unperturbed on their journey across oceanic immensity. Spectacular night sailing. The second movement of the Sixth envelops my head, seems to fill the universe as I look to the masthead trilight and beyond to the stars. Although the symphony is named *Pastorale*, the interwoven motifs of the second movement seem an impressionist rendering of the braided waves and wavelets—all in wondrous empathy with ocean passages in small sailboats.

Then, across the ocean, comes the largest pod of Pacific white-sided dolphins I could have imagined. They've come to play. More than 50 animals. Cutting across the bow. Slaloming under the stern. Jumping. The phosphorescent wakes of dolphins and *Starkindred* intertwine. We're surrounded by living, silver torpedoes, filled with the joy of life, intimately connecting the ocean and sky with understanding I cannot hope to appreciate. They stayed with us for more than an hour—through midnight—and so joined yesterday with tomorrow.

I'm struck, again, with the reality that every corner of our planet is *alive*. Listening to its own kind of music. Partaking of its own kind of joy. *Waiting*. Waiting to spread throughout the universe—waiting to seed oases of structure throughout an otherwise increasingly disordered universe.

Part Eight

HARVESTING ENERGY SOURCES

27. THE GREAT ENEMY OF TRUTH

In which Jack Kennedy's observations to a Yale commencement echo across the decades.

'The great enemy of truth is very often not the lie—deliberate, contrived and dishonest—but the myth—persistent, persuasive and unrealistic.' This was John F. Kennedy speaking to the Yale graduating class on June 11, 1962.

As we set out on the next legs of our voyage, we should armor ourselves with this convocation wisdom. For there is no part of our energy system that suffers more from unrealistic myths and flawed convictions than the link we call energy (exergy) sources.

It seems to me there are three ways we acquire knowledge—by accretion, integration or substitution. Depending on which route we've traveled to knowledge, new information sometimes requires that we pay an emotional price for the insight, understanding and comprehension we've found. So let's try to anticipate how the route influences the emotional price.

'Accretion' is the process of gluing additional information to the body of knowledge we already have. It can be fun. Usually the new information is accepted willingly and without stress. A single, easily understandable factoid is added to our brain's immense memory capacity—disturbing no embedded knowledge, demanding little intellectual effort, requiring no emotional price. Everyone has their own examples—a new football scoring record would qualify.

'Integration' is the process by which we come to appreciate a new and better way to link together the knowledge we already have, to place bits and pieces within a clarifying pattern. Since integration requires refitting many data chunks, to test how they fall within a new template, the process often requires a little more thought, more reflection. Still, although the effort may be greater, the resulting synthesis can be joyous compensation. We seldom need to pay an emotional price. Indeed, we often get an emotional high. An integration experience for a student mechanical engineer could occur when, in a moment of revelation, she suddenly realizes that the vector calculus equations she's been applying to fluid mechanics can be wonderfully applied to the electromagnetic field theory that her electrical engineering roommate is studying, so that, together, they understand both fluid and electromagnetics a lot better.

'Substitution' is different. New, discordant, white-knuckle information crashes in upon long-held beliefs, upon our convictions. We first scurry about looking for counter-arguments, for rationales that will undermine the new information, allow it to be set aside. But if all defensive arguments fail, and if we're willing to push aside pre-existing flawed perceptions, then substituting new knowledge can be a painful process. None but the courageous will tread here. A childhood substitution experience is when you find, a few days before Christmas, in your parents' cupboard, the train set you'd asked for from the North Pole—and you suddenly realize there really isn't a Santa Claus. Or even tougher, because we're older with beliefs more solidified, might be when we come face to face with implacable flaws in political or religious ideologies we've held as long as we can remember.

There is another related issue. Although most of *Smelling Land* will continue to be about patterns, not numbers, sometimes we'll need a little numeracy—often when testing whether we might need to substitute new knowledge for old.

In everyday language we commonly speak of something being 'a lot more' or 'a lot less' than something else. In everyday language this usually works. But when understanding requires numerate comparisons, then answers like 'a lot' can be exasperating. What is 'a lot?' Twice? Ten times, a thousand times? To compare energy sources we'll often require numbers—or at least a range within which the numbers lie. Ranges are sometimes all we can expect because tight numbers are only possible when applied to specific sites, technologies and systems. Site-specificity is not what we're hunting. Understanding is.

On occasion, understanding requires a numerate answer, but there can be the converse—when people ask for hard numbers, but intellectual honesty demands that no such numbers be given, because the uncertainties are too large. A common but tedious example is when people want to know what the price of oil will be one year from now—no matter when 'now' is.

A different man, in different times, speaking a different language and living in a different country left us with much the same idea as John F. Kennedy, when Nietzsche wrote, *Überzeugungen sind gefährlichere Feinde der Wahrheit als Lügen*—convictions are more dangerous enemies of truth than lies.

During our voyage so far we've gained some knowledge by accretion, more by integration. But up to now, except perhaps for some places in the climate disruption chapters, little required substitution. Accretion and integration will continue throughout our discussion, but as we close in on exergy sources, many readers will encounter substitution rough spots. If you're one of these, and if you wish to avoid this choppy sailing, you can take shore leave to relax with friends where you can lift a few pints, and clink mugs while toasting things to which you all agree.

But I hope you'll clip your tether to the jackline, climb into your foul-weather gear and hang on.

28. FOSSIL SOURCES: A LOT WILL BE LEFT IN THE GROUND ❦

In which we discuss reserves, resources and the mythology that, sooner or later, our fossil sources will be gone. Then Weary Willie helps us understand what's happening.

Civilization's use of coal began more than 2,000 years ago but, compared with wood and dung, it remained a minor fuel up to the early Industrial Revolution. Then as expanding European cities depleted renewable firewood, they increasingly turned to coal. This is why, between 1840 and 1920, Europeans underwent the transition from wood (then the dominant renewable source) to coal (the prototype exhaustible source).

It was a time of extraordinary, parallel and synergistic transitions, not just from wood to coal, but from agricultural to industrial economies, from rural to urban populations and from muscle to steam engines. Steam engines powered locomotives that pulled trains that carried iron ore and coal to steel mills. In turn, the mills manufactured steel to build more railways and locomotives that hauled more coal and iron ore—to build not only more locomotives, but also ships that carried coal around the world where it could fuel more ships.

Transitions are never stress-free. It has been reported that burning coal was, for a time, a capital crime in parts of continental Europe. The reason was witch-huntery. Fumes from coal burning were sulphurous, the odors of Mephistopheles and hell. So people who burned coal were judged the devil's allies and some of these people were tied to a stake and roasted—using wood, I presume. For a time, England also declared burning coal a capital crime. But in England the transgression wasn't Satanism; it was toxic air. Pragmatic, those English!*

---

* The English events are covered in *Particles in Our Air: Concentrations and Health Effects*, edited by R. Wilson and J. Spengler (Cambridge, MA: Harvard University Press, 1996). *Air Pollution Vol. 1*, edited by Arthur C. Stern (San Diego, CA: Academic Press, 1976), covers some of the same history,

But our purpose is not to review more evidence of humankind's disquieting history. Our purpose is to get a feel for the fossil resources and reserves that remain in the ground and what this means for our evolving energy system.

We should start by first clarifying how 'resources' and 'reserves' differ. Following most conventions, such as the United Nations report, *World Energy Assessment: Energy and the Challenge of Sustainability,*[39] I'll define the 'total resource base' to mean the total of some commodity in the ground, like coal or oil. 'Reserves' are that portion of the total resource base that can be economically recovered at today's selling prices, using today's technologies and under today's legislation. 'Resources' are the remainder of the total resource base after subtracting reserves—they consist of material in the ground that is not economically recoverable under today's conditions.* Most of the report uses the year 2000 as a reference date, which is the date I'm using in these chapters on reserves and resources.

Figure 28.1 illustrates the relation between reserves and resources. The presentation was taken, with minor format changes, from the UN's report *World Energy Assessment* and encapsulates how the total resource base is partitioned and, by implication, how changing prices and technological advances move the internal boundaries between reserves and resources. This type of representation is commonly used by international and domestic organizations such as the U.S. Geological Survey, the World Energy Congress and the United Nations. The figure shows that resources are sometimes further subdivided into measured, indicated, inferred, hypothetical or speculative categories.

---

but speaks only of torture, not execution. Should you be interested in the often appalling history of coal, I recommend *Coal: A Human History,* by Barbara Freese.[40]

* Some people use the word 'resources' to mean what I've called the 'total resources base.' By this convention, reserves become a *component* of resources—rather than separate from resources, which is the convention used here.

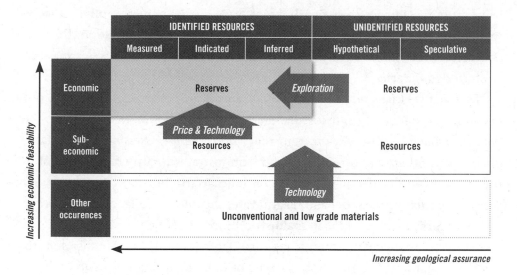

Figure 28.1 Schematic showing how resources and reserves are classified
*Prices and technologies can move resources in and out of reserves—as can exploration.
(Adapted from: World Energy Assessment: Energy and the Challenges of Sustainability,
Part III, Chapter 5, Energy Resources, 2000.)*[39]

The boundary between reserves and resources shifts when prices, technology or legislation change. For example, as prices increase while other factors remain the same, the reserve envelope expands to include some of the material previously counted as resources. The reverse happens if prices drop.

Price-change drivers can be fleeting or sustained. In the short term, geopolitics or a cold winter can trigger price flares. The time lag required for developing reserves means short-term price flares have little effect in moving resources into reserves. In the mid-to longer-term, technological advances can lower costs. Technological improvements include better technologies for finding and harvesting oil—such as improved geological analyses, advanced drilling techniques and superior processes for refining low-grade ores.

Figure 28.1 doesn't show the effect of legislation, but that's easily imagined. Legislation is the child of politics which, in turn, is the child of culture. In the United States, legislation such as oil-depletion allowances is well known. But legislation can impact in many different ways. For example, it might specify which public lands or sea bottoms are or aren't open for exploration. In some countries, legislation rides on the whim of kings or religious leaders—in western democracies, it's more likely to ride on the clout of special interests.

Now that we know what is meant by resources and reserves, let's turn to our question: how much of our fossil sources remain? For this we'll examine the data of Table 28.1, also drawn from the *World Energy Assessment*.* The table gives *global* reserves and resources and demonstrates that there is still a lot in the ground. Yet the idea of depletion continues to trouble many people. And it should, because *regional* depletion aggravates the international conflicts rooted in—or aggravated by—global maldistribution. In North America and the European Economic Community, regional depletion impacts oil most, natural gas next, coal last. For this reason, most of our remaining discussion will be about oil.

Crude oil is, effectively, the *sole source* of free-range transportation fuels. This contrasts with coal and natural gas, the fossil fuels commonly used to power stationary services (electricity generation, for example). Stationary services can always be powered by non-fossil energy sources. But today, free-range transportation cannot.†

---

* On some issues there is always disagreement. 'What's left in the ground' is surely such an issue. Those who disagree with the *World Energy Assessment*. . .include Colin Campbell and Jean Laherrère who wrote 'The End of Cheap Oil' for *Scientific American* in 1998, and Kenneth Deffeyes in *Hubbard's Peak: the Impending World Oil Shortage* (Princeton, NJ: Princeton University Press, 2001). To me it hardly matters. After a lot of thought and reading, I'll go with the UN report but, either way, we must get off fossil fuels.

† I'm speaking of 'conventional' crude. This contrasts with 'unconventional' oil resources unconventionally harvested. These include the Athabasca oil sands in Canada and the Orinoco basin reserves/resources in Venezuela. To harvest the oil sands (mixtures of bitumen, sand and water) we must first separate the bitumen from the sand and water, then process the bitumen into 'synthetic crude.' Unconventional doesn't mean unrecoverable. But they are more difficult and environmentally challenging to harvest—in part because so much material must be handled and demands for water are huge.

Type	Consumption		Reserves	Resources[a]	Resource base[b]	Additional occurrences
	1860-1998	1998				
Oil						
Conventional	4,854	132.7	6,004	6,071	12,074	
Unconventional	285	9.2	5,108	15,240	20,348	45,000
Natural Gas						
Conventional	2,346	80.2	5,454	11,113	15,567	
Unconventional	33	4.2	9,424	23,814	33,238	930,000
Coal	5,990	92.2	20,666	179,000	199,666	n.a.
Total	13,508	319.3	46,655	235,238	281,893	975,000

a. Reserves to be discovered or resources to be developed as reserves.
b. The sum of reserves and resources.

Table 28.1 Estimates of fossil energy in the ground.

(Adapted from: World Energy Assessment: Energy and the Challenges of Sustainability, Part III, Chapter 5, Energy Resources, 2000) 1 Exajoule = 6.12 x 109 barrels of oil.

Question: How quickly will we deplete our oil reserves if we continue using oil at the rate we were in 2000? Table 28.1 shows that, if we consider only conventional reserves and assume no expansion of these reserves by new discoveries, price increases, technological improvements or oil-friendly legislation, they'll be gone in about 42 years—about 78 years if we include unconventional resources. If we include the total resource base, our remaining time climbs to about 230 years. If we include what geologists call 'additional occurrences,' the time frame expands to almost 550 years.*

---

* If you want to follow this arithmetic, using the data of Table 28.1 we can start with the 1998 rate of consumption: 132.7 EJ/y (conventional) + 9.2 EJ/y (unconventional) for a total = 141.9 EJ/y. If we divide *only* the conventional reserves by the total (conventional and unconventional) annual oil consumption, we get a ratio = 600EJ/141.9 EJ/y = 42.3 years. Dividing conventional and unconventional reserves by total annual consumption, we get a ratio = (6004 + 5108)/141.9 = 78.3 years. Dividing resource base by total annual consumption, we get = (12,074 + 20,348)/141.9 = 228 years. Finally, if we divide the sum of the resource base plus additional occurrences (but *not* including coal from which we could also manufacture oil) by today's total annual consumption, we get: = (12.074 + 20,348 + 45,000)/141.9 = 546 years.

But this doesn't account for the reality that, so far, we have always consumed more oil each decade than during the preceding one, which, if nothing else were happening, would bring depletion closer. But neither do they account for new discoveries and technological advances that have steadily expanded resources to push depletion further away.

In Figure 28.2, the relative strength of these countervailing realities is given by the yearly ratio of 'identified reserves' *divided by* the 'annual consumption.' The numerator can be given in barrels of oil in the ground, and the denominator in barrels of oil consumed per year. The ratio comes out in units of 'years'—years until we'll run out. I find it remarkable that today the ratio is much the same as it was in 1900. In 1900, civilization had identified conventional oil reserves that, if used at the 1900 rate, would have lasted a little more than 40 years. In 2000, a century later, the conventional oil reserves used at the 2000 rate would also last a little more than 40 years.

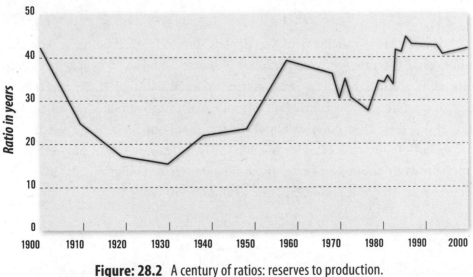

Figure: 28.2 A century of ratios: reserves to production.
*(Adapted from: World Energy Assessment: Energy and
the Challenges of Sustainability, Part III, Chapter 5, Energy Resources, 2000)*

The data of Figure 28.2 considers only *conventional* reserves because, in 1900, conventional reserves were all we understood. Today we know that we also have large non-conventional reserves, and we are harvesting them. So it's probably fair to say that the *de facto* ratio of 'annual consumption' to 'what's left in the ground' has never been higher and, therefore, the number of years until we could 'run out' has never been as long as it is now.

Because higher prices move resources into reserves, we might think the constant ratio of reserves to consumption has been the result of steadily rising prices. Wrong! In constant-purchasing-power US dollars,* the price of oil has stayed pretty much the same, albeit with short-term hiccups.† This is shown by Figure 28.3. Because we're using the year 2000 as the reference time frame, the early 21st century price flares, triggered in part by the Bush II administration's Middle East misadventures, do not appear in that chart.

If you still believe we could run out of conventional and unconventional oil, remember we can always manufacture synthetic oil from coal, as Germany did during World War II. Coal is just hydrogen-deficient oil. But as a source of petroleum-based fuels, all crude oil is hydrogen-deficient.‡ Coal is just more hydrogen-deficient.

We can encapsulate some broad truths as follows:

- Most countries that use oil don't have enough.
- Most countries that have oil don't need as much.
- This means most of the world depends upon the rest of the world.

In 2000, the United States imported 56% of its oil, a quarter of that from the Middle East. At the same time, Western Europe imported 60%, almost half from the Middle East. And 82% of Japan's oil came through the Straits of Hormuz. Today, worldwide fossil fuel interdependence grows year by year.

---

* The monetary currency of the international oil trade is US$.

† The price-flares of mid-2006 would, if sustained (they had already collapsed late in the year), greatly expand reserves in the longer term but, as we've noted, time-lags required for developing reserves, means short-term price flares have little effect on moving resources into reserves.

‡ Chapter 34 'Tethers and Transition Tactics' deals with the consequences of hydrogen deficient crude oil and especially low-grade ores like the oil sands. And it explains what we can do about it while simultaneously reducing CO_2 emissions.

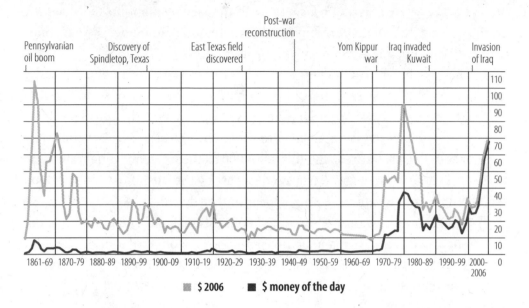

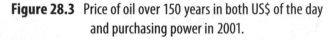

Figure 28.3 Price of oil over 150 years in both US$ of the day
and purchasing power in 2001.

(Source: British Petroleum's Statistical Review of World Energy, June 2002)[41]

'Running out has the advantage of being easy to understand but the disadvantage of being wrong.'

When Cesare Marchetti spoke these words in 1979, six years after the OPEC oil embargo, it was a deeply contrarian statement. It still is! But in the context of the International Symposium on Hydrogen Air Transportation where Marchetti was speaking, it was more than contrarian, it made many in the audience angry, because the symposium's rationale was to find a new aircraft fuel before we ran out of the old fuel. People dislike having their rationale punctured.

Yet speaking from both historical evidence and logic, Marchetti was right. Globally, there is little meaning to 'running out' of fossil fuels. There are two reasons for this. First, we can't consume all our fossil resources because, long before they'd be gone, the resulting environmental disruption would have destroyed our economies absolutely, and thereby removed any need for fossil fuels. Second, I believe, the maldistribution of fossil fuels in Earth's crust, stirred together with political and religious fundamentalism, has become *the* toxic brew that feeds so many of today's geopolitical conflicts and, with but a few pauses, can only get worse until military conflicts devastate us all. We aren't threatened by near-term global fossil fuel depletion. But if our appetite for fossil fuels persists, irreversible climate volatility will be our fate—if it isn't already.

So let's ask: if we burn only today's identified fossil fuel *reserves*, what will be the impact on atmospheric CO_2?

In 1997, Dr. Hans-Holger Rogner of the International Atomic Energy Agency studied this question.[42] His assumptions for the study were:

- No further discoveries;
- No further improvements in technologies for finding and extracting oil;
- Prices at or below the equivalent of US$20/barrel.*

If all the reserves meeting these criteria are consumed, then atmospheric CO_2 will rise to more than 250% of pre-industrial age levels—that is to more than 700 parts per million by volume (ppmv). To place this in context, remember that for more than the last 600,000 years the *maximum* atmospheric CO_2 the world has experienced was 310 ppmv.

Civilization *may* be able to survive the more than 450 ppmv CO_2 that, unavoidably, we are now destined to reach, without falling over the metastability lip into an entirely new climate. But should atmospheric CO_2 rise to more than 700 ppmv, we will experience a global environmental catastrophe unlike any other in recorded history. Even those in history/myth, like the Great Flood. We can take no comfort in hoping that prices will rise to cap fossil fuel use before CO_2-induced catastrophe. Rising prices won't.

I wrote that last sentence the day oil prices rose to above US$66 per barrel— then the highest in history. I thought of what I'd already drafted for Chapter

---

* The word 'equivalent' means at price and quantity ratios (between gas, oil and coal) equivalent to these ratios in 1997, when the paper was written.

35 'OK! Now Tell Me about Cost'—where I'd written that light Arabian crude could be pumped for less than US$2 per barrel.

With these high oil prices on the morning news, I felt a jab of self-doubt, wondered if I'd overstated the low costs of extracting Saudi crude. So I called my friend Ben Ball. Ball has enjoyed a remarkable career in the energy business that includes teaching petroleum management at MIT's Center for Advanced Engineering Study, serving as vice-president of Gulf Oil, and now acting as a much-in-demand consultant. I reached Ball at his home in Sugar Land, Texas, and asked, 'Do you think it's too aggressive for me to write "light Arabian crude can be pumped for US$2 per barrel?" '

'No!' he said, 'But maybe too conservative. You could've said ' "between one and two dollars." ' Then, with his wonderful Texas drawl and engaging humor, he went on to explain some fundamentals—which I jotted down as fast as my pencil could scratch.

'Awl* prices don't follow the supply-demand price curves taught in Economics 101. In traditional economics, the high-cost suppliers begin producing when price exceeds some threshold, and go out of business when the price drops below a different, lower threshold. But in the oil business, at least in the short-to medium-term, it's exactly the opposite. The high-cost producers are pumping full blast all the time. You don't hear about Texas producers increasing or decreasing supply. It's the one-dollar per barrel producers who turn their spigots on and off.

'Price flares have nothing to do with conventional economic fundamentals. It's whim. Anyone who thinks they can predict the price of oil for next year is either blowing smoke or terribly naïve. I tell folks that the best you can do when forecasting oil prices is to say that they will be somewhere between $10 and $100 per barrel. Oil price predictions are just WAGs.' I assumed a WAG was a professional oilman's term for something I hadn't yet learned. So I asked—to learn it was a 'Wild Ass Guess.'

Ball continued. 'There's something else at play. Over the longer term of eight to ten years, we sometimes speak of 'price attenuation'—because it takes that

---

* When talking to a Texan we must remember it isn't oil, it's 'awl.'

long to bring on a new oil field. For example, if we have prices in the range of US$25-US$35, producers will go out looking for oil even under 10,000 feet of water. That's why we often speak of a seven-year half-life—because it takes seven years for half the high-cost oil supply to respond to a price change. And it's important to understand that production costs are overwhelmingly the cost of exploration and development.'

Finally, Ball directed me to that day's New York Mercantile Exchange (NYMEX). This is a trading house that allows us to buy or sell futures in commodities like crude oil ('puts' and 'calls' if you're a pro). Like any trading market, price is a balance between folks who believe price will go up and those who believe it'll go down. That day, the NYMEX futures price for the year 2011 was more than US$20 *below* the trading price of the moment. Ben laughed. 'Anyone who believes we're running out should look at these 2011 oil futures and invest every last penny. If they don't, they're either stupid or lying.'

So what were my conclusions on this day oil hit US$66 per barrel?

First, as we should all know, it's difficult to keep our heads screwed on when the morning headlines tells us the sky is falling. Second, remember the fundamentals! As long as Arabian crude can be pumped for a few dollars a barrel, the price cannot be understood with the neat logic of Economics 101. Moreover, when we realize that the cost of oil is largely the cost of exploration, it's sobering to remember that the Saudis have a country of sand floating on oil. Since the day I phoned Ball, the price of a barrel of oil has gone down, then up, then down, then up and down again.

It even peaked above US$100. Of course, since I spoke with Ben, the value of the US$ against world currencies has dropped between 12 and 16%. Moreover, if you look at the 2006 $ curve of figure 28.3 you'll see that we've been there before—back in 1979. For another context, in the year 2007, ExxonMobil posted the highest profits, ever, by a US corporation. Methinks the price we're paying at the pump has a little more to do with corporate profits and greed than 'running out.'

For a short while, when oil reached its highest prices, American newspapers and some congressmen enthusiastically spoke of vast reserves of oil 'right next door' in Alberta, Canada. Then, beneath a full page of this kind of report, I read a small item in the bottom-right corner. A Saudi ambassador, speaking about the Alberta 'find,' had reminded a reporter that Saudi Arabian oil can be pumped for 'between US$1 and US$3 per barrel.'

So Ben Ball was right. We have it straight from the camel's mouth.

Now let's look to the future. Fossil *resources* we'll always have because, as you now know, if we get anywhere near consuming all of them, we-be-done-gone. But eventually, one way or another, two realities will reduce our fossil *reserves*. First, the cost to civilization of burning massive quantities of fossil fuels will finally be seen as not only unsustainable, but ruinous. Second, the benefits of hydrogen technologies will become increasingly apparent. Fossil-fueled technologies will lose their grip on energy systems, making fossil sources increasingly irrelevant (something difficult to visualize today). Civilization will begin using less and less oil.

Thinking about this, my thoughts tumble to Emmet Kelly. Emmet created and played perhaps the world's most famous sadsack clown, Weary Willy. Of Willy's many routines, his best may have been the heartbreaking, yet joyfully simple performance, 'Sweeping up the Spotlight.' From high in the circus tent, a spotlight controller directed a wide beam of light to the floor enveloping Willy. Then Willy, using his broom, began sweeping the spotlight's edges inwards. The spotlight controller saw what Willy was up to, and gradually shrank the size of the illuminated circle on the floor until, with one last swish of the broom, the spotlight—and Willy vanished. It was a magnificent, happy-sad routine that reminded a damp-eyed audience of Kelly's genius.*

To me, the entire floor of the big tent can be a metaphor for resources. The area illuminated by the spotlight, a metaphor for reserves. Prices, technologies and demand determine the spotlight's diameter. But, in the end, Willy's broom—a metaphor for environmental intrusion, geopolitical instabilities and hydrogen technologies—reduces the size of the reserves until they vanish. The floor of the big tent will still be there. But the spotlight will have vanished.

---

* If you'd enjoy watching Kelly's routine, you can get a video or DVD of *The Greatest Show on Earth*. It's a dated movie, but it can be fun watching this Hollywood snapshot of the Ringling Bros. and Barnum & Bailey Circuses.

29. RENEWABLES AND CONVENTIONAL *WISH*DOM

In which the public's well intentioned but unrealistic hopes for renewable sources cause us to ask: What is their real promise? Where do they contribute today? How might they contribute tomorrow? How will hydrogen help?

Whales are a renewable energy source. At least they were until we learned that, when looking for oil, it was both possible and much cheaper to send young men a few miles inland to drive drills into the ground rather than half-way around the world in dangerous wind ships to drive harpoons into whales. What's more, we were beginning to run out of renewable whales.

Yet today, for many people the imprimatur 'renewable' bestows an unquestioned goodness that pre-empts any need for further thought or examination. What surprises me about this genuflection to *all* renewable energy sources is its contrast to the glaring problems with many renewable other things—like fish and freshwater. Harvesting fish and freshwater have brought about some of the most critical environmental stresses caused by humankind. For that matter, so has harvesting forests. Could this muddy thinking be another artifact of language—of unexamined words shaping actions or, at least, *wish*doms?

Renewable energy certainly has a role, but why has our passion for renewables become so indiscriminate? Certainly, there are places and times when renewable exergy sources make a lot of sense. I'm not anti-renewable, just pro-rationality.

The oil-supply shocks of the mid 1970s launched our devotion to renewables. Devotion matured into conventional *wish*dom during the 1980s. Industrial nations had awakened to the possibility of depletion of inexpensive, light, sweet crude oil under *their* ground. It wasn't a big step to believe that eventually all Earth's fossil fuels would be gone. To a lesser degree, but using the same logic, some worried we would also run out of uranium, the exergy source for nuclear power. These concerns were founded on an undeniable fact: both fossil and fissile (uranium) sources are finite and, in principle, exhaustible.

So people decided we should use sources that are continuously replenished. If we use renewables, it was reasoned, we'll never run out. The renewable mantra was born. Strangely, without examination, the motivation that launched the inexhaustible mantra soon broadened to include the idea that harvesting renewable energy was inherently environmentally friendly.

As with many ideas that appear sensible at first glance, most of us didn't look beyond first glance. Nor did we look *back* to how denuding the forests of Europe had brought environmental damage and a firewood crisis that curtailed the growth of cities, forcing people to use coal. We didn't remind ourselves that over-harvesting renewable water accelerated the collapse of great civilizations like Mesopotamia, or that similar short-sightedness apparently destroyed Mayan civilizations, the Easter Island peoples and others.*

Because hindsight was not used to aid foresight, the renewable religion gathered converts. It was encouraged by government pamphlets showing photos of windmills and families picnicking by streams, and proselytized by a growing industry consisting of self-appointed saviors of the planet.

Sometimes, in order to better understand how people think, I've asked, 'Why do you think renewable energy, like sunlight and wind, is desirable?' More than once, the answer has been: 'Because sunlight and wind are free!' *What?* Please tell me which energy sources *aren't* free. Surely, oil and uranium in the ground are free. So is coal. The cost comes in the equipment that *harvests* the source—transforming it into a currency—and then in the cost of the technologies that deliver the services, in the wages of people operating the systems, and so on. In fact, per unit energy, it costs more to harvest wind and sunlight than to harvest coal or uranium.

---

* Many books speak to the lack of environmental foresight that destroyed civilizations—and the Garden of Eden (thought to have been located in Mesopotamia). I find Ronald Wright's *A Short History of Progress*[20] to be one of the most remarkable. Wright's telling of the self-destructive history of Easter Island peoples begins on p. 57. A more well known book that covers similar themes is Jared Diamond's *Collapse*.[43]

None of this is to say renewables shouldn't contribute. Of course they should. In appropriate venues, renewable energy has historically been *the* source of both clean energy and substantial economic growth, as shown by the hydraulic power of Niagara Falls. Renewable wind lifted* our sailboat, *Starkindred*, across the Pacific, and during the day, photovoltaic solar panels harvested renewable sunlight to recharge batteries that, in turn, powered our running lights and refrigerator. Anchored in South Pacific atolls, we found that if we supplied the ice cubes, other sailors would supply the rum.

In isolated settings, renewables can serve not only practically but also lyrically—witness a dreamy plume of wood smoke meandering upwards from a cabin nestled on a snowy wooded hillside. Of course, 'isolated' is key. In Denver Colorado, after a short revival of firewood home-heating, the urban air quality became so intolerable that wood-fire heating was restricted. Today, Denver air quality is flagged by color code—a 'red day' forbids wood fires. And in many cities, building codes now outlaw wood-burning fireplaces.

I'll recount an experience that speaks to the triumph of denial over understanding. As part of a course in Advanced Energy Systems, my students must complete a self-defined project. Three students proposed examining opportunities for renewable energy sources in Canada. I thought the topic fine but, looking closer, noticed they had excluded hydraulic power. When I asked why, it became clear they didn't consider hydraulic renewable because, they argued, it was already established, produced a lot of electricity, was run by large utilities and was environmentally intrusive. I was stunned. By excluding hydraulic power, these students had ignored the essential adjective in their project's title. All because they didn't *like* naming hydraulic power 'renewable'— in spite of the fact that it's continuously renewed by rain.

Later, when digging into this unsettling experience, I found this kind of fuzzy thinking wasn't confined to a few well-meaning students. It's much more

---

* Non-sailors sometimes think the wind 'pushes' sailboats along. Sometimes it does. But much more often—especially if the sailors are competent—the sails act the same way as wings on an airplane and 'lift' the sailboat over the ocean. But this is the fussiness of a sailor and fluid dynamicist speaking.

pervasive, infecting public-beholden bureaucracies. The 1995 edition of the *Electric Power Annual* removed hydroelectric power from the renewable category. I understand it's still excluded. Later, in 1999, the US Department of Energy disqualified hydroelectricity from the Renewable Portfolio Standard.

Officialdom has become so addicted to 'renewables as the panacea' that, when our largest and most successful renewable brings negative environmental impacts, it is banished from the renewable family. The truth, of course, is that any renewable source will come with environmental intrusion, especially if it makes an important contribution to our exergy needs. Only the little guys, harvesting minuscule exergy, won't. But if they grow up, watch out!

Let's be sure we have a common understanding of what we mean by a renewable exergy source. Some might argue that because oil, gas and coal are continuously laid down in Earth's mantle they are renewable. It's a cute argument. But the issue comes down to comparing deposition time scales with harvesting time scales. None of us needs reminding that, today, the rate at which we're extracting our fossil resources is many times faster than the rate at which they are being deposited.

To focus on the main players and avoid nitpicking, I'll define renewables as those sources that can make important contributions to civilization's exergy needs, while being harvested at rates well below Nature's replenishing rates. We'll first group all renewable sources within three broad categories and then select, from these, those that meet our criteria.

1. Solar-sourced energy/exergy:
   - sunlight itself
   - hydraulic power
   - wind
   - biomass
   - ocean waves and currents
   - ocean thermal and saline gradients.

2. Energy/exergy from Earth's core:
   - fluid geothermal flows
   - lithosphere hot rocks.

3. Energy/exergy from Earth-Moon interaction:
  • tidal currents
  • tidal rise and fall.

Of these three groups, solar-sourced flows dominate—overwhelmingly. They deliver exergy at rates more than 10,000 times greater than flows from the Earth's core, and 100,000 times greater than Earth-Moon interactions. So, although there may be appropriate site-specific applications for renewables like geothermal and tidal energy, they can never contribute at anywhere near the magnitude needed to power civilization's energy system. What's more, were we to try to harvest these minor sources at rates that could make an important contribution, we'd grossly violate the overarching criterion for environmental gentility set out earlier: to intrude as little as possible upon Nature's flows and equilibria.*

So to focus on the main chance, we'll concentrate on solar-sourced exergy. And within this group of six options, we'll concentrate on those that have had, are having, or are likely to have the greatest potential. These are hydraulic power, direct sunlight, wind and biomass. These are the renewables that will contribute most to the coming Hydrogen Age.

To place renewable sources in context, we need to think about the services they must power if they are to make a significant contribution. And to do that we must return to the idea of exergy. Services can be divided into low exergy-grade heat services and high exergy-grade work services.

Any energy source can provide low exergy-grade services such as space heating. Sunlight for example, can directly warm your living room (especially

---

* There are only a few places around the world where we can intrude on ocean wave-patterns, currents or thermal gradients to build the typically massive structures required for harvesting exergy. Almost always these are along coastlines with many competing needs. I once visited the site of one of the more interesting, the Hawaii Natural Energy Institute's OTEC project on the west coast of the Island of Hawaii. Alone, the energy product could never be economical. However, several co-produced services and products nudged the scheme toward cost-effectiveness. One was spirulina production. A second application used cold water, drawn from the deep ocean and run through pipes positioned a few centimeters above gardens. These deep-ocean-cooled pipes condensed water from the air and then dripped that water on plants on this, the dry, leeward side of the Island.

if you have large windows), or your swimming pool (especially if the water is circulated outside your pool though black PVC pipes). Some sources, wind for instance, must first manufacture an intermediate currency—usually electricity.

But to make a significant contribution, exergy sources must be able to support high exergy-grade services—most importantly, information and transportation. Information services require electricity. In contrast, most transportation services require chemical fuels—although fixed-route transportation can use electricity. Either way, it's the high exergy-grade services that are the challenge.

Any worthwhile exergy source, therefore, must deliver either electricity or chemical currencies. Of these, with the single exception of biomass, renewable sources now only produce electricity. This leaves about two-thirds of the energy market—the chemical commodities and fuels market—inaccessible to almost all renewables. Toward the end of this chapter, we'll touch on how things will improve in a fully developed Hydrogen Age. But today, if we want renewable sources to produce high exergy-grade currencies, they are trapped within the electricity ghetto. That is why this chapter is really a discussion of the prospects for renewable-sourced *electricity*. If we want renewables like wind, solar and tides to contribute to the much larger transportation market, we must await the advent of the Hydrogen Age.

All renewable sources share two generic features.

First, Nature delivers her renewable energy diffusely, spreads it over large areas, rarely shoots out her energy and exergy in concentrated zaps. Good thing, too! We'd be discomfited if all Nature's exergy came in lightning bolts and tornadoes. So if we hope to harvest enough renewable exergy to make an important contribution, we'll require immense surface areas. And one way or another, the area required for renewable energy always encroaches on competing uses for land and oceans. In the unique case of hydraulic power, we're lucky. Nature mitigates her diffuse exergy supply by doing most of the gathering for us, using her brooks, streams, lakes and rivers.

Second, as we've said, we must never harvest renewables at rates that approach Nature's replenishing rate. Today, our problems with renewable forests, fish and freshwater are our punishment for having neglected this obvious rule. When we run the numbers on renewables, especially if we believe they can meet all our exergy needs, we come smack up against Nature's replenishing rate.

Today, hydraulic power is by far the world's most important renewable exergy source and, when available, usually the lowest cost. In British Columbia, hydraulic power generates 78% of the electricity. In Canada, it generates 60%. In the United States, 10%. In New Zealand, 62%, Norway, 100%.* Like any source, hydraulic has both positive and negative attributes—as you consider the following list, you can decide which is which for your region:

1. Compared with many renewables, hydraulic power has the great advantage that its energy can be *stored*—as hydraulic potential behind dams, usually in reservoirs upstream of the generators.

2. Depending upon the setting, hydraulic power often requires that large land areas be flooded. (In northern Quebec, the James Bay hydraulic power development floods an area larger than Switzerland, providing electricity not only for Quebec but also for Ontario and several northeastern American states even, sometimes, Times Square on New Year's Eve.)

3. The infrastructure for hydraulic electricity generation often delivers by-products like irrigation and flood control. It can also produce electrolytic hydrogen that, these days, is frequently used to make ammonia for fertilizer.† (The Three Gorges development on China's Yangtze River will deliver 18,000 MW of electricity. It will also provide irrigation, stabilize agriculture and break the centuries-old pattern of annual floods that repeatedly destroyed farmlands and drowned people. Its construction

---

* A characteristic of many renewables when they are used to generate electricity is this: the ratio of installed generation capacity (called the 'nameplate capacity') to the electricity generated (by the same equipment) is usually surprisingly low. In contrast, hydraulic power is one of the few renewables where the ratio can be close to unity, because hydraulic power is not hobbled by intermittency and is also, most often, the least-cost source of electricity. So utilities will keep it producing whenever possible.

† Besides fertilizer, ammonia is an ingredient of conventional explosives. During World War I, a dam for making ammonia for explosives was built on the Tennessee River at Muscle Shoals. A short history of the Muscle Shoals development and the political arm-wrestling it fomented—which in part, led to the establishment of the Tennessee Valley Authority—is outlined in John Kenneth Galbraith's entertaining and informative book *Name Dropping*, pps. 34-36.[44]

required displacing several million Chinese from their traditional homes and farms—some to modern, clean cities higher up the bank. The Yangtze development is controversial, particularly in our often sanctimonious, judgmental West.*)

4. Although hydraulic power is renewable, the infrastructures aren't. It's not just that the immense concrete structures, penstocks, flumes and turbines won't last forever. It's also impossible to build dams without causing silting and other environmental intrusions, most of which grow with time. (Speaking again of the Yangtze, during a spectacularly scenic trip through the Three Gorges in 2001, I often stuck my hand into the fast-flowing water to test its clarity. Each time, with the waterline at my wrist, I couldn't see my fingertips. The tonnage of silt flowing by was beyond my comprehension. Chinese engineers told me the structures had been designed to avoid silting problems. I have my doubts.)

Hydraulic power will be an important renewable energy source for many years. Moreover, large untapped reserves remain in developing countries, though it will always come with trade-offs—many of them environmental. So, as with every decision, we must balance the trade-offs—in this case, the environmental intrusion of flooding, silting and destroying wildlife habitats against the benefits of economic strength and CO_2 avoidance.

Before dealing with the second and third players, direct sunlight and wind, let's start with a parable. Imagine you run a company, Green Widgets Inc., employing several hundred people. Imagine your business requires that you provide widgets the instant customers want them. Imagine you have two groups of employees, the 'reliables' and the 'whimsicals.' Reliables always come to work on time, work steadily, are willing to produce day or night, rarely call in sick and, when they do, schedule their sick leaves to coincide with low

---

* Environmentalists successfully lobbied the US Export-Import Bank to deny funding for China's 18,000 MW Three Gorges Project. (It proceeded in spite of this.) To groups such as Friends of the Earth, it seemed of little concern that if Three Gorges were not built, the electricity would be generated by coal, the worst of the CO_2 emitters.

widget-demand times. Whimsicals work well sometimes, but you never know when they'll show up—it's often when you don't need them. And when you could really use their help they've often decided it's picnic time. Moreover, each whimsical requires many times more floor area, for lunchrooms, lounges and workspace, than do the reliables.

As manager of Green Widgets, with which group of employees are you happiest? Before you answer 'the reliables, of course,' beware!

What if the whimsicals are the darlings of the media, of political correctness and, therefore, of the legislators? To encourage hiring whimsicals, those legislators have ruled that you *must* employ the whimsicals any time they show up, even if that means you will need to lay off reliables. Whimsical boosterism is so pervasive that, to be seen as a good corporate citizen, you've hired several— had them photographed, smiling, as they enter your operations on a day they've decided to work. You've posted the photos on your website, but have not posted photographs of laid-off reliables heading out the back door.

As you have guessed, the reliables are a metaphor for conventional electricity generation plants, like fossil-or nuclear-fired thermal stations and large hydraulic installations. The whimsicals are a metaphor for wind-powered turbines and solar-powered photovoltaics.

We'll now examine the special features of both windmills and solar panels when used for electricity production. Windmills power alternators that can directly manufacture AC electricity, which must, in turn, be synchronized with the AC frequencies of the electricity network. In contrast, photovoltaic panels produce DC electricity, which must be inverted to AC before it can also be synchronized and fed to the distribution grids. Here are some observations:

1. The 'nameplate' capacity on a solar panel or wind turbine is the device's output under ideal circumstances. For wind, this means at optimum wind speeds when the windmill axis is perpendicular to wind direction.* For solar, 'nameplate' capacity is met only when the panel is orientated perpendicular to the incoming sunlight from a cloudless sky, at cool

---

* Vertical axes, sometimes-called eggbeater windmills, operate independent of wind direction. However, these designs have not been competitive with horizontal-axis machines.

temperatures, and without even the smallest shadow falling on a corner of the panel.* Too often, when popular articles claim wind or solar can provide enough electricity to support, say, 1,000 homes, the claim is based upon continuous output at their nameplate capacity. This is terribly misleading. Solar and wind generators rarely operate at their ideal outputs—in fact, they usually operate at less than 50%. Frequently they don't output at all. *(At Green Widgets Inc., you'd risk indictment if you were to issue a business prospectus claiming your whimsicals produce as much as your reliables.)*

2. Trustworthy electricity supply is not merely desirable, it can be a life-death requirement. So if solar or wind *alone* feed the electricity grid, we must have, on standby, conventional 'spinning reserves'—ready to take over if the wind goes calm or a rain cloud covers the Sun. 'Spinning reserves' are power plants that, although delivering no power at a given moment, are kept whirring so they can be called upon, instantaneously, to match a sudden electricity demand. Plants with a spinning reserve don't need to be brought up to speed. One example is a hydraulic turbine with just enough water flowing through to keep it spinning while not producing electricity. Another example is a steam-driven turbine with just enough steam to keep it spinning under no load. *(At Green Widgets Inc. when whimsicals are on the job, reliables must be kept standing by in the parking lot so they can instantaneously go to work if the whimsicals down tools for a pickup football game.)*

3. We've observed that *all* renewable energy sources require a lot of land to harvest comparatively little exergy. We've also noted that, in the special case of hydraulic power Nature uses her streams and rivers to do most of the collection for us. Not so for wind and sunlight. A useful rule of thumb

---

* Increased temperature reduces the performance of photovoltaic panels; therefore, when in the tropics, higher insolation is subject to a countervailing disadvantage imposed by higher temperatures. For most conventional photovoltaic (PV) panels, a small shadow covering as little as 1% of the panel will shut down the whole panel. Fortunately, this shortcoming can be overcome with a different arrangement of diodes interconnecting the crystals. In some applications this can be very important, as when PV panels are installed on sailboats where rigging and sails are always, one way or another, casting shadows on the panels—something I've learned from bitter experience, in spite of solar-panel salesmen's promises it would not happen.

is that the rate of electricity *delivery* from renewable sources can approach 0.1 W/m²—but it's usually less. But the rate of electricity *use* in urban areas is about 1.5 to 3W/m².* So if we hope to use renewable sources to power our cities, we must dedicate land areas to harvesting renewables that are at least fifteen times greater than the area of the cities. This, in a world with ever increasing urbanization and encroachment on forests and wetlands—the lungs and kidneys of Earth. And we also need land for agriculture and recreation. And we should leave a little space for critters too. For comparison, the rate of electricity delivery from fossil-fired or nuclear electric utilities is typically about 3,000W/m² or higher.† This means conventional electricity generation requires less than 1/1,000th of the area of the cities they serve, while renewables typically require fifteen times more area than the area they serve. *(Heads-up for Green Widgets: if you decide to expand your whimsical workforce, you will need more than 10,000 times the floor area per employee.)*

4. It's often argued that, by using energy storage, we can mitigate the intermittency that hobbles solar-and wind-derived electricity. In principle, the idea is correct. But then we must add the capital requirements and the land area needed for the storage technology. Moreover, we must account for round-trip efficiencies of both putting the electricity into storage and taking it out again. These are rarely above 80% and commonly not better than 50%.‡

---

* The density of total energy use—electricity plus all other forms—is between 5-10W/m².

† This typical footprint of a generation station includes land for parking lots, office buildings and security perimeters. But it does not include an estimate of the area occupied by the mines for, say, coal, gas or uranium.

‡ Conventional storage batteries are only feasible for small capacities. What is small? Well, batteries store enough energy for starting our cars. And they are used to provide sufficient backup power for our telephone system to last, typically, two days. Defense establishments and hospitals usually have standby generators, with batteries to bridge the time until generators are brought up to speed. But it's unrealistic to imagine battery capacities that could keep, say, the northeastern US operating during a week of calm winds and gray skies. Hydrogen proponents suggest electrolyzers, hydrogen storage and fuelcells could provide this storage role. Such an exergy conversion-storage train might combine to give, *at best*, a round-trip efficiency of about 50%, obtained from a 90% electrolyzer efficiency times a 55% fuelcell efficiency. Of course, if we hope to use such a sophisticated storage complex, we must account for the capital cost of the two exergy-conversion technologies plus hydrogen storage infrastructure.

Before leaving solar, I point out these paragraphs speak only to photovoltaic generation. Other technologies are under development that, if successful, will not *directly* convert sunlight into high exergy-grade currencies. Rather they will redirect incoming sunlight using mirrors that concentrate the heat so it can be used to power heat engines. The trick is similar to when, as kids, we used magnifying glasses to set dry leaves afire. These solar concentration schemes, which go by names like 'power towers' or 'solar troughs', have been under development for a number of years, at places such as Sandia National Laboratories. In principle, the heat can be stored to mitigate solar intermittency. Other technologies hope to use chemical conversion processes that employ sunlight to split water, producing hydrogen. But so far, none of these approaches seem to give a serious challenge to photovoltaics.

I wish all challengers well. But we should not forget those fundamental constraints that will never vanish.

From the time our ancestors began two-legged walking until the days of early steam engines, biomass was our staple exergy source. It still is in many parts of the world. Globally, about 11% of civilization's energy comes from biomass. In China, it's 19%, in India 42% and in many of the poorest countries, up to 90%.[42]

One way or another, all biomass was harvested from sunlight—either by photosynthetic life, or by animals that ate photosynthetic life. It follows that the two main divisions of biomass energy are vegetation biomass and animal biomass. In addition to the obvious use of vegetation biomass, like wood, developing nations use dung for heating, cooking and sometimes as a feedstock for biogas generation. It makes sense for developing nations to continue using their traditional fuels as long as it serves them well. Yet unless development patterns take a U-turn, when these nations become wealthier their use of biomass energy will decline. And I see no U-turns in the offing.

In industrialized nations, where biomass exergy has become almost trivial, it is remarkable that now it's one of our conventional wishdoms. So let's take a closer look—not so much at its traditional role in developing nations—but rather at how it might contribute in industrialized nations.

Vegetation can be sub-divided into woody biomass (such as trees, shrubs, bamboo, palms), non-woody biomass (sugar cane, cotton, cassava, grass, bananas, plantains, swamp plants), processed-waste biomass (cereal husks and cobs, pulp mill black liquor, municipal waste) and processed fuels (charcoal, alcohols and vegetable oils).

Again, we'll begin with some general observations, a mixture of positive and negative:

1. Unlike wind or sunlight, biomass is not constrained by intermittency.

2. Unlike wind, sunlight or hydraulic power, biomass products are inherently chemical currencies, which are important for transportation services.

3. In addition to requiring a lot of land as do all renewables, biomass also requires a lot of water. (Hydraulic power also requires water, but this is flow-through water. In contrast, biomass absorbs water and agriculture evaporates it. Therefore, significant biomass fuel production will further stress our diminishing freshwater reservoirs. As we entered the 21st century, one-third of the world's population lived with moderate to high freshwater exhaustion. This could rise to two-thirds by 2025.)[42]

4. Biomass plantations require fossil fuels for manufacturing fertilizers and pesticides, for running harvesting equipment and biomass processing. Depending upon the crop, the input of fossil energy ranges between 4% and 25% of the biomass output. Moreover, if the crop must be moved more than about one hundred kilometers for processing, transportation imposes additional fossil fuel demands.* Those who claim carbonneutrality for these fuels either gravely misunderstand what's happening or they're lying.

5. When fermentation is used to manufacture liquid fuels, like ethanol from grains, considerable CO_2 is dumped into the atmosphere. (This is a large-scale version of the fermentation that produces bubbly CO_2 in our beer or, if more to your taste, champagne.)

---

* These data and some of the arguments were taken from the tables and discussion on pps. 56–162 of *World Energy Assessments*. Much of this information seems to have been provided by Dr. José Goldemberg, the driver of the largely failed Brazilian commitment to fuel all their cars with ethanol. So I doubt the numbers are biased against biomass sources.

6. In some tropical countries, large tracts of deforested, degraded lands might benefit from well-managed bioenergy plantations.

7. Genetically modified crops could increase bioenergy yield by up to a factor of two.

8. In developing countries, burning biomass fuels has been identified as the cause of most energy-related health problems. The most common are respiratory failures from particulate matter and carbon monoxide poisoning.

Biomass will continue to contribute in poorer, underdeveloped countries. And it might return to make a minor contribution in rich, developed countries, especially with the tax incentives so favored by farm lobbies and agribusiness. However, contrary to their proponents' claims, biofuels are anything but carbon-neutral.

We must remember that we're looking for ways that renewable energy can make hydrogen—because hydrogen is the only chemical fuel (currency) that can be used without emitting CO_2.

In 1979, the Soviet Union invaded Afghanistan ostensibly to help the Afghan government protect their country from attacks by Mujaheddin Islamic fundamentalists. The Afghan government had been building schools and libraries and encouraging the education of women. But the Mujaheddin didn't like this and so they attacked and killed Soviet and Afghan soldiers from their mountain lairs. Yet because they were killing Russian soldiers, we in the West took to calling them 'freedom fighters.'*

Witnessing a conflict between the Soviets and Afghan freedom fighters, it was easy for the United States to know which side it was on.† To show their

---

* Forefathers of the guys who took on the Twin Towers and now are killing American, Canadian and other coalition troops from countries who believe—or claim to believe—they can bring democracy and a measure of women's liberation to sad, troubled Afghanistan.

† The US supported the resistance fighters with money, military intelligence and weapons, like stinger missiles that were more than 70%-effective in taking out USSR aircraft. For a comprehensive history of the tangled web of US, Saudi and USSR involvement, read Steve Coll's *Ghost Wars: The secret history of the CIA, Afghanistan, and Bin Laden, from the Soviet Invasion to September 10, 2001.*[45]

displeasure with the USSR, the United States led a group of western nations in boycotting the 1980 Moscow Olympics. Yet boycotting the Olympics wasn't enough. Congress also cut off American wheat shipments to the USSR.

Oops! Cutting off wheat shipments angered the American farm lobby. Where would farmers sell their crops? Congress responded with a brilliant two-birds-with-a-single-stone solution. They introduced legislation providing subsidies and tax exemptions that encouraged farmers to grow crops for ethanol production. Voila, this gave farmers something to grow and simultaneously helped solve the nation's 'energy crisis.' Thus the Soviet invasion of Afghanistan begot 'gasohol'—gasoline-alcohol fuel blends.

We seemed to forget that oil was used to power the farm equipment, natural gas to make the fertilizer and water to irrigate the crops. Moreover, the fermentation process in ethanol factories belched CO_2. What's more, patriotic consumers were tricked: the volumetric exergy density of alcohol is about two-thirds that of gasoline, so people were getting less exergy from a gallon of gasohol than from a gallon of gasoline. This has continued until today.

When I first drafted this chapter in late 2002, my wife and I frequently enjoyed watching the *Lehrer News Hour* on PBS during dinner. Over and over again, the news was preceded by a vision of cornfields waving in the wind, over which a voice solemnly intoned, 'Can you imagine a world when we're not diminishing our resources, we're growing them? Ethanol, cleaner burning fuel, made from corn,' beneath which a byline identified the sponsor: 'ADM: the Nature of Things to Come.' When the 2007s rolled around, misrepresenting ethanol's advantages and hiding its disadvantages became a staple of lobbyists.

Back in the 1970s, the Gods of Energy Planning Foolishness had never had it so good. Recently they've had it even better. Today, if we substitute realism for wishdom, the prospects for renewable sources are gloomy. But happily, at least for wind and sunlight, the situation will improve over the second half of the 21st century and will continue to improve thereafter. Why?

For the answer, we must look ahead to a fully developed Hydricity Age when hydrogen will be the dominant exergy carrier, dominant because I see no reason why the ratio of chemical and electronic currencies shouldn't be about the same as today.* And that means hydrogen-powered services will require about twice as much exergy as electricity-powered services. Hydrogen pipelines will become the trunk exergy networks across continents, and perhaps even between some continents. In Chapter 42 'A Round Trip to the Deep Future,' I'll develop these ideas.

In this deep future, whenever renewable wind and sunlight are available, they will produce hydrogen and feed it to the grid where it will be stored. Hydrogen pipelines crisscrossing the continents will be immense storage sponges. Moreover, just as today large tanks store oil, in the future hydrogen-storage tanks will dot the nation to supplement the (by comparison) lower-capacity pipeline storage. The intermittency limitations that hobble many renewables today will be gone.

The trouble is, I expect we'll need to wait more than a half-century before hydrogen pipelines begin replacing electricity grids†—although we might start earlier by converting some of today's natural gas pipelines.‡ Until then, we should build hydrogen systems as quickly as possible.

The coming Hydrogen Age will remove a major barrier to using renewables. But not all renewables. Biomass will remain yesterday's fuel. And we should leave humpback whales undisturbed, roaming the oceans, enjoying their families, singing their songs.

---

* Chemical (protonic) and electronic currencies were explained in Chapter 19 'Can Something Better Come Along?'

† Since *before* World War II, hydrogen pipelines have run through the Ruhr Valley in Germany. France also has an extensive network of hydrogen pipelines.

‡ Provided they have been constructed of hydrogen-compatible materials—or can be retrofitted with liners to avoid hydrogen embrittlement.

30. FISSILE SOURCES: 'RUNNING OUT' WON'T BE A PROBLEM 🌶

In which we look at our fissile fuel reserves, because they're our main chance for a carbon-free future.

In Chapter 28 '*Fossil* Sources—A Lot Will be Left in the Ground,' we realized that depletion won't cap fossil fuel use, but the risk to international peace should. And the global environment will. Then in Chapter 29 'Renewables and Conventional *Wish*dom,' we found that while renewable energy sources will contribute, they cannot provide all the exergy we'll need.

This means nuclear power is the only carbon-free exergy source that can give us a chance of producing the *massive* quantities of hydricity we'll need, at least for the next several centuries. So we must turn our attention to uranium, the principal fissile fuel of nuclear-produced exergy.

Today, all nuclear power is derived from fissile material. An atom of fissile material is capable of being fissioned—split into fragments—when struck by a neutron.* After splitting, the fragments have less total mass than the original atom. It is this *mass reduction* that releases energy according to Einstein's famous relationship, $E = mc^2$. Here 'm' is the mass exchanged for energy and 'c' is the speed of light. The energy released is what we've named nuclear energy.

To understand which materials are fissile we'll need to return to the idea of isotopes, which we introduced in Chapter 13 'Bubble, Bubble, Signs of Trouble.' Every element in the periodic table—like hydrogen, oxygen, or einsteinium—actually consists of a *family* of isotopes of that element. Each member of a family has the same number of protons (positively charged subatomic particles) in its

---

* 'Capable of being' does not mean, 'will be.' A neutron striking the nucleus of a uranium-235 atom has a better chance of splitting the nucleus if the neutron is traveling at modest speeds, which we call 'thermal' speeds. But neutrons flying away from a fission event are traveling very, *very* fast—far above thermal speeds. The role of the 'moderator' in a nuclear power reactor is to slow these fast neutrons to the thermal speeds more likely to achieve the next fission. Today's moderators are light water (typically in US reactor designs), heavy water (in Canadian designs) and graphite (as in the designs of the former USSR, like the Chernobyl reactor).

nucleus. It is the number of protons that defines the chemistry of an element. But while all isotopes of a single element have the same number of protons, each isotope is distinguished by its number of neutrons. If you want to consider this further, turn to Appendix B-3.

All isotopes of the metal uranium have 92 protons. Naturally occurring uranium contains two isotopes. About 99.3% of the metal is the isotope uranium-238 (^{238}U) which has 146 neutrons in addition to its 92 protons. The remaining 0.7% is uranium-235 (^{235}U) which has 143 neutrons, and it is *only* this uranium-235 isotope that is fissile.

'Enriched' uranium is uranium that has been processed to have a higher fraction of uranium-235. Reactor-grade nuclear fuels are typically enriched— some to contain as much as 5% uranium-235. Bomb-grade uranium is enriched to well above 90% uranium-235, a difficult and very expensive task. The fact that bomb-grade uranium is entirely different than reactor-grade uranium is just one of many reasons a power reactor could never explode like a nuclear bomb.*

In contrast to fossil resources that have been accumulating over the ages, the fissile isotope uranium-235 has been self-depleting (through the process of radioactive decay) since our planet was born. When our world was new, more than a third of its uranium stock was uranium-235. Today, it's down to 0.7%. So taking the *very* long view, the uranium-235 will ultimately be gone whether or not we use it to power our energy system.† Still we needn't worry. Naturally occurring uranium will contain about 0.7% uranium-235 for many times longer than the 10,000 years or so we believe civilization has existed.

A second difference from fossil resources is that the high neutron fluxes (flows of neutrons) inside a nuclear reactor 'breed' additional fissile material from the uranium-238 isotope. Although uranium-238 itself is not fissile, it is fertile. ('Fertile' means capable of being a feedstock from which we can breed more fissile material.) When uranium-238 absorbs one of the neutrons flying

---

* The meaning of enriched and depleted uranium is explained further in Appendix B-4.

† Some 1.8 billion years ago, natural processes established a reactor running on uranium that, in the world of the day, contained about 3% ^{235}U. It started spontaneously and ran for many thousands of years in Oklo, in Gabon, Africa.

about inside the reactor it begins a process that ultimately leads to the fissile plutonium isotope, plutonium-239. Because some breeding occurs in power reactors, by the time spent fuel is removed from the reactor, about half the energy has come from the original fissile uranium-235, and the other half from such bred fissile isotopes as plutonium-239.

This contrasts with burning a fossil fuel. Burning coal doesn't produce more coal.

Third, instead of being stored or buried, spent nuclear fuel has the potential to be reprocessed (outside the reactor) to breed many times more fissile material, which can then be repeatedly recycled through conventional nuclear reactors. Today's once-through nuclear power plants only use a small fraction of the fuel's ultimate potential. Uranium can be used to fuel advanced 'fast-neutron' or 'breeder' reactors.* The so called 'spent fuel' from today's reactors isn't spent at all. Today's governments, at least in North America, ignore this reality and push ahead with plans to bury spent fuel when many times the exergy that was extracted by the once-through reactors is still there, waiting to be claimed. (Appendix C-3 has more information.)

Fourth, *even* when used in once-through reactors, a kilogram of natural uranium provides about 14,000 times as much exergy as a kilogram of oil. If reprocessing were used, uranium could yield more than a million times more energy than the same mass of oil. Earlier chapters pointed out that environmental intrusion is always, one way or another, rooted in material. So the astonishing difference between the mass-to-energy ratios of fossil and fissile sources should lead us to expect that fissile sources are comparatively environmentally benign. As they are.†

To determine how much fissile material is left in the ground, let's again turn to the United Nations report, *World Energy Assessment: Energy and the*

---

* The adjective 'once-through' means a reactor that does not recycle its spent fuel. 'Breeder' refers to a reactor that can breed even *more* fissile fuel than was put in. The adjectives 'thermal neutron' and 'fast neutron' (often abbreviated simply to 'thermal' and 'fast') are applied, respectively, to reactors that use moderators to slow the fission-product neutrons (thermal reactors), or operate without moderators (fast reactors) that leave the neutrons whizzing about 'fast.'

† Compared with coal mining, the overburdens and tailings from uranium mining are small. Still, early-days uranium-mining procedures have, in some places, left a legacy of toxic tailings.

Challenge of Sustainability[42], which classifies uranium reserves as 'reasonably assured deposits recoverable at no more than US$80 per kilogram.' By this criterion we have 2.3 million tonnes. If the recoverable price were increased to between US$80 and US$130 per kilogram, the reserves would increase by up to another 0.9 million tonnes. The World Energy Assessment suggests these known reserves are more than enough to meet the needs of existing and planned *conventional* nuclear power plants throughout the 21st century. If we reprocess the spent fuel from once-through reactors, the ratio of reserves to consumption jumps to more than double.

In addition, we now reprocess the bomb-grade uranium and plutonium from decommissioned weapons, turning it into peaceful power plant fuel —a beautiful example of 'beating swords into ploughshares'

'Thorium, another constituent of Earth's crust, is not fissile but it is *fertile*. Thorium can absorb a neutron that starts a process that kicks it up to fissile uranium-233.* If we add thorium to our fissile reserves, the reserves to consumption ratio takes another leap, probably tripling our fissile reserves.

Large quantities of both uranium and thorium are also stored in unconventional deposits, as in uranium-containing phosphates in sedimentary deposits. And there is the prospect of harvesting uranium from seawater although, using today's technologies, the process is very expensive.

For all these reasons, those who claim that the use of nuclear power is limited by Earth's remaining fissile materials are even more mistaken than those who argue the greatest danger to our energy supply is that we'll run out of fossil fuels.

To those who still worry about what we'll do after fissile material is gone, I'd say: First, let's get through the next 500 years. I'm sure we will find another energy source within the next 500 years. We've already got some options on the horizon, like controlled nuclear fusion. Let's have faith in the ability of the next ten-to-fifteen generations of people.

---

* If you're wondering why I didn't include this isotope when we spoke of uranium having two isotopes, uranium-235 and uranium-238, it's because the uranium-233 isotope does not occur naturally.

Producing nuclear-derived energy currencies is dominated by the capital cost of the power plant rather than the cost of fissile sources, which represent only a small fraction of the cost. After capital cost, fueling, operating and maintenance add to the cost of producing currencies. Today, the cost of the uranium seldom accounts for more than 2% of the cost of nuclear-derived electricity. Therefore, if all else stays the same, doubling the price of uranium would increase the cost of nuclear-sourced electricity by only about 2%.

Unsurprisingly, the cost of uranium is not the same as the cost of nuclear fuel fabricated from uranium. Depending upon the reactor type, fabricated fuel costs range from 3–20% of the total cost of nuclear-generated electricity. But the price that determines the uranium reserves/resources boundary is set by the price of the ore, not by the cost of fabricating the fuel. For comparison, the price of coal typically constitutes between 30–60% of the cost of coal-sourced electricity. Natural gas prices are an even higher fraction of the cost of natural gas-sourced electricity.

To put the disparity between the respective costs of coal and uranium in context, we need to know the comparative capital, operating and maintenance costs of coal and nuclear power plants.

When the next-generation nuclear power plants are built over the next decade, they will cost, about 20–25% more than a new-design coal-fired plant. Moreover, in this comparison, I've not included the cost of CO_2 capture and sequestration, which will drive coal-fired plant capital costs over the top.

For coal, the operating and maintenance costs are about the same or slightly higher than for nuclear plants. So on balance, it's reasonable to assume that the difference in the price between fossil and fissile-generated electricity (and hydrogen) will primarily reflect the difference in fuel costs—where nuclear wins hands-down.

It seems we can draw several conclusions:

- The price of *fissile*-derived electricity and hydrogen will always be comparatively insulated from uranium (or thorium) price flares.

- In contrast, the price of *fossil*-derived electricity and hydrogen will always be very sensitive to coal, oil or natural gas price flares.

- While significant increases in uranium prices will impact hydricity costs only moderately, they will substantially increase uranium reserves because the higher prices will transform resources into economically recoverable reserves.

- The long-term price of nuclear-derived energy currencies will be much more stable than the price of fossil-derived currencies, like gasoline or diesel. (The price-at-the-pump wiggle-waggles will be gone, and so too will the political flares that result.)

- Business investors abhor uncertainty. So the comparative price stability of nuclear-derived currencies will be an attractive platform for energy system investment. People and corporations developing business plans for energy-intensive industries will have confidence that their planning is founded on relatively stable prices and an assured supply of hydrogen and electricity.

For the next several hundred years, nuclear fission is our only realistic chance to provide all the carbon-free exergy we need, so it's nice to know we won't be running out of fissile reserves. Still one problem remains. Many people are afraid of nuclear power. But I'm not. I'll tell you why in the next chapter 'You've Got to Be Carefully Taught—Know Nukes.'

31. You've Got to Be Carefully Taught*—Know Nukes

In which, reflecting on the troubled, misunderstood history of nuclear power, we ask: If we were to 'know nukes' would it free us to enjoy the environmental benefits of Nature's most ubiquitous energy source?

In 1999, less than three months before we entered the new millennium, the headline 'Radiating Terror' shrieked from the front page of the *Guardian Weekly*—in red ink. The story continued with sub-headlines, 'Japan Nuclear Blast Triggers Crisis' and 'Hundreds of Thousands at Risk from Radiation in Worst Accident Since Chernobyl.' We learn that this was the Asian country's 'worst nuclear accident' and that President Clinton had offered 'all possible assistance to Japan's Prime Minister Keizo Obuchi.'

Citing the International Atomic Energy Agency, the story reported that of the three workers closest to the 'flash,' two were almost certainly going to die and a third might survive although it was 'touch and go.' The worker who received the largest radiation dose died just before Christmas. The second worker survived to see the 21st century, but died shortly thereafter. The third lived. Throughout the story, 'nuclear blast' language reinforced the too-frequent confusion between nuclear weaponry and nuclear power.

Remember, I'm not quoting the *National Enquirer*. I'm quoting the *Guardian Weekly*, one of the world's most respected newspapers. Published halfway around the world from Tokaimura where the accident happened, the Guardian carries copy from two other respected papers, the *Washington Post* and *Le Monde*. With news and opinion drawn from among the best newspapers of three nations, the *Guardian* is normally expected to provide broad-based, balanced views. Most of the 'nuclear blast' story was taken from the Washington Post, although I suspect the headlines were pure *Guardian*.

---

* Song title from the Richard Rodgers and Oscar Hammerstein musical, *South Pacific*.

The emphasis journalists give their stories depends on the distance between where the story happened and where the readers live. As distance increases, interest decreases. So for the Tokaimura story to reverberate around the globe the way it did, the media had decided to treat the news as very big indeed. By the criterion of media attention, let's compare Tokaimura's 'big' nuclear power plant mishap with other energy accident stories.

Less than a year before Tokaimura, my daily local newspaper, Victoria's *Times Colonist*, ran a brief story about an accident that had happened the previous day. It had happened a mere 40 crow-flying kilometers (25 miles) due east of Victoria outside the charming Washington town of Anacortes that borders the San Juan Islands archipelago. Folks from Anacortes sail to Victoria for a flower-potted harbor of old-world gardens and grace. We often sail to Anacortes for a wee bit of stateside culture and less expensive whiskey. In short, we're friends and neighbors.

The Anacortes accident happened in an oil refinery. But unlike Tokaimura, three people weren't expected to die. Eight people had been killed.*

Because Anacortes is a neighbor I was surprised the accident was not bigger news in the *Times Colonist*. I looked in Canada's national newspaper, the *Globe and Mail*. Nothing there. The *Washington Post* carried a brief piece on an inside page, but I found nothing in the *New York Times*. Nothing appeared in the *Guardian Weekly*.

The message was clear. Once you're out of the immediate neighborhood where workers are killed, families broken and wives widowed, eight people dying in an oil refinery explosion is not news. But can you imagine the media frenzy if these men had been killed in a nuclear power plant? Accidents are always flukes. Still, there are many more flukes that injure and kill people in the fossil energy industry than in the nuclear energy industry.

During the months between the Anacortes and Tokaimura accidents, another fatal fossil-energy accident struck northwestern Washington. It happened on June 10, 1999, near Bellingham. A pipeline carrying gasoline leaked and an

---

* The accident occurred in late November 1998. Initial reports claimed five had died. Later it was learned 'six union members and two supervisors' had been killed.

explosion resulted. The accident killed two young boys who were playing along the banks of the Whatcom Creek, and a teenager who was fly-fishing. The explosion made no international news, little national news, and triggered only a brief piece in the *Times Colonist*.

It was clear that I didn't need to look around the world to find fossil fuel accidents that far outweighed the Tokaimura accident—even Whatcom Creek had more deaths. I only needed to look next door. But we're used to people being killed in fossil energy accidents. It's not big news.

On February 15, 2005, the very day I wrote the 'It's not big news' sentence, a brief announcement appeared on page 13 in the *Globe and Mail*: 'Blast Kills Hundreds in Chinese Coal Mine.' More than 203 miners had been killed in Fuxin, Liaoning Province. More were still trapped and many injured. Do you recall hearing of this coal-mining accident? I doubt it. The story went on to report that, that year, at least 5,000 people had been killed in mining accidents in China alone*.

Thousands of people killed in coal mines each year are not newsworthy. Yet on each anniversary of the Three Mile Island accident—the most severe US nuclear power accident ever (although no one died)—news stations march out their 'We Should Be Quaking about Nuclear Power' stories.

These last few paragraphs help to illustrate the safety of nuclear energy compared with fossil energy, and the irresponsibility of a media that so grossly distorts the true balance. And we haven't yet considered the safety record of renewable energies. I'll say something about that next.

In 2000, I traveled to New Zealand and China. On New Zealand's South Island I visited the Manapouri power project, an impressive hydraulic installation where four generators were built *inside* a mountain. On a wall in the gallery

---

* In January 2008, the Chinese press were proud to announce that the 3,600 lives lost in coal mining fatalities were 20% lower than the year before. The recent record is: 7,200 deaths in 2003; 6,027 deaths in 2004; 5,986 deaths in 2005; 5,986 deaths in 2005; 4,746 deaths in 2006. So coal mining is getting less dangerous in China. In the US, over the same four years, 2003 through 2006, coal mining deaths were 30, 28, 22 and 47.

from which visitors could look down upon the generators, a plaque had been mounted in memory of the sixteen workers killed during construction of this renewable energy installation. Yet the total output from Manapouri is only 585 MW—much less than the typical output from a *single* reactor in a nuclear generating station. Traveling on to China I visited the Three Gorges hydraulic project on the Yangtze River where, the day before I arrived at the construction site, several workers had been killed. The loss of life had been mounting steadily since the project began.

In July 2007, a 6.8 magnitude earthquake struck almost directly beneath one of the world's largest nuclear plants in Kashiwazaki, Japan. This is the kind of event that nuclear-phobic people like to say will cause meltdowns, massive radio-toxic leaks and horrendous loss of life. But it turns out Kashiwazaki served as an unplanned experiment that showed how resistant a modern nuclear plant can be. The plant continues to operate and no significant radio-toxic leaks occurred. In contrast, in the area immediately surrounding the power plant, hundreds were injured and at least nine died.

The Tokaimura nuclear accident, the Washington State accidents, China's coal mining accidents, the Manapouri and Three Gorges renewable energy accidents, and the Kashiwazaki earthquake got me smelling land. I knew Japan generates some 30% of its electricity from nuclear power. So if Tokaimura was Japan's worst nuclear power accident,* could it mean that, compared to others, nuclear might be one of the world's safest energy sources, perhaps the safest? I looked up the numbers.

According to the OECD publications, before the September 1999 accident, Japan's nuclear electricity production had totaled 480 GW*y (gigawatt-year) and, as far as I was able to determine, the only fatalities were those at Tokaimura. Therefore, the ratio of Japanese fatalities to the amount of electricity delivered

---

* The Tokaimura accident was not, strictly speaking a nuclear *power* accident. Rather it occurred when processing nuclear fuels.

comes out to be 0.004 persons killed per GW*y.* The worldwide nuclear fatality record is 0.0084 killed per GW*y.

Lives have been lost in generating electricity from nuclear power. But what about other sources? What's stunning is the fact that, worldwide, natural gas generation—which has the best safety record of all non-nuclear electricity generation technologies—kills ten times more people per GW*y than does nuclear generation. Coal-fired electricity generation kills more than a hundred times more people per GW*y. By the way, the nuclear safety record I just quoted *includes* the world's worst nuclear power accident in Chernobyl, Ukraine—which involved a design that could never have been approved in any OECD country.†

Then it struck me: If the Japanese had been using any other source to generate this much electricity, it would have cost the lives of many more than the two who died at Tokaimura, 'Japan's worst nuclear power accident.'

Reflecting on these truths pulls me back to a childhood memory—lying on the living room floor, listening to the music from *South Pacific*, enveloped in 'Some Enchanted Evening' and 'Bali Ha'i.' But now, my head swirling from thoughts of the media's infatuation with distorting the environmental and safety record of nuclear power, I especially remember 'You've Got to Be Carefully Taught'. With his prescient lyrics, Oscar Hammerstein II illustrated how we can be taught to

---

* Some readers may stumble over energy measured in GW*y. So let's review the difference between power and energy. The *rate* at which energy is delivered is 'power,' commonly measured in watts. If we multiply the *rate* of energy delivery by the *time* it's delivered, we get the total energy delivered—commonly in watt-hours, or W*h. But a W*h is a tiny amount of energy when we're talking about national energy use. So units of GW*y are commonly used. A GW*y is the energy used if one billion watts were delivered continuously for one year—or, if you like, the energy that would be used by 10 million, 100-watt light bulbs left burning for one year.

† These numbers were drawn from *Electricity Generation Systems and Sustainability Interdisciplinary Evaluation*, a study prepared by Switzerland's prestigious Paul Scherrer Institute. Most of this report discusses risk in terms of loss-of-life expectancy (LLE) for the *full* energy system chain, including both operation and construction accidents as well as emissions and decommissioning. In my view, LLE is probably a better way to judge the comparative safety/risk of various energy options. But this chapter began with the media's response to the Tokaimura *accident*, so I decided to express the safety records in terms of loss-of-life per GW*y. Readers looking for further information, and data applied to different countries, should peruse the Paul Scherrer Institute's website.

hate some people. The media's selective response to energy accidents reminds me that we have been taught to hate some technologies. Topping that list is nuclear power. I can't resist. Here are the lyrics:

> *You've got to be taught to hate and fear,*
> *You've got to be taught from year to year,*
> *It's got to be drummed in your dear little ear,*
> *You've got to be carefully taught.*
>
> *You've got to be taught to be afraid,*
> *Of people whose eyes are oddly made,*
> *And people whose skin is a different shade,*
> *You've got to be carefully taught.*
>
> *You've got to be taught before it's too late,*
> *Before you are six, or seven or eight,*
> *To hate all the people your relatives hate,*
> *You've got to be carefully taught,*
>
> *You've got to be carefully taught.*

The musical *South Pacific* was one of the few good things to emerge from the 1940s war in the Pacific. A different legacy was nuclear bombs. Which has led to a different tragedy: most people don't distinguish well between nuclear power and nuclear weapons. The media—I expect more in ignorance and laziness than purposeful dishonesty—repeatedly blur the difference.

Here is a test: when you next read a newspaper headline that speaks of something bad about nuclear, see if you can tell from the headline whether the article is about nuclear weapons or nuclear power—or, for that matter, about nuclear medicine. I've been applying this test for years and, most often, I can't tell until I've read deeper into the piece.

In contrast, when reading a headline about the problems with oil, I don't expect the article to be about napalm which, of course, is an oil. Yet the fuel of nuclear power is much more different from the ingredients of a nuclear bomb than are the ingredients of gasoline different from napalm.

All this suggests we should understand some fundamentals about nuclear power. A good place to start is to examine the causes of public concern, which seem rooted in three fears:

1. How to deal with the waste products (spent fuel).
2. The risk that a nuclear power plant accident could kill many people.
3. The possibility that bad guys might acquire spent fuel and use it to make bombs.

Underlying all three is the fear of radiation. So let's first talk about radiation: what it is, where it comes from and what it does. Then, with a better understanding of radiation, we'll return to the three fears.

Nuclear radiation means radiation that originates from the *nucleus* of an atom. It's different from the radiation released from the energy given up by decelerating electrons when they fall into lower energy levels to produce X-rays, northern lights and the illumination from a fluorescent light. The adjective 'nuclear' tells us the radiation's source. In contrast, the adjective 'ionizing' tells us what the radiation is capable of doing. So if we want to speak about the health effects of radiation, whether or not it's ionizing radiation is more relevant than whether or not it's nuclear radiation.

Ionizing radiation has enough energy to ionize molecules—the building blocks of material, which include your building blocks. When ionizing radiation strikes living biological material it can either stimulate positive biological responses, have little effect, or (when there is too much) cause biological damage, like cancer.

The statement 'ionizing radiation can stimulate positive biological responses' may astonish. Certainly I was astounded when I first learned of this reality. We've all been carefully taught that no good can come from ionizing radiation, that the smallest dab, however slight, will increase the likelihood of cancer. But for those who have gathered and examined the extensive epidemiological data, it's become evident that low doses of radiation—but still well above normal background levels—can bring measurable biological benefits.*

To most people, including me, the acronym ABS means 'antilock braking system.' For others it means 'American Bonsai Society.' But to researchers who studied the after-effects of Hiroshima and Nagasaki, it means 'atomic bomb survivors.' One finding from these epidemiological studies (of some 30,000 survivors monitored for more than half a century) has been that, if people weren't directly killed by the explosion—or if they didn't die shortly thereafter from extremely high radiation doses—the mean lifespan of the ABS cohort was significantly *longer* than the mean lifespan of the general Japanese population. This healthier, longer-living ABS cohort received between twice and twenty times the radiation dose received by Japanese living far from Hiroshima and Nagasaki.

By now, many studies have confirmed a beneficial relationship between ionizing radiation and health—called the 'hormesis' relationship. These studies show that if you enjoy exposure levels up to about twenty times normal background radiation levels, you're likely to live a longer, healthier life than do 'normal-exposure' folks. The University of Pittsburgh's Dr. Bernard Cohen analyzed lung-cancer rates against average radon† concentration for over 1,600 American counties representing more than 90% of the US population. His results provide some of most persuasive data in support of the hormesis response to ionizing radiation.[46] Many other studies confirm the phenomenon. And recently Ed Hiserodt published his book, *Under Exposed—What If Radiation Is Actually Good for You?*[47]

---

* Of course, I'm not speaking of medical use of radiation targeted at killing malignant growths— which is an entirely different issue.

† Radon is a gas that results from the atomic decay of radium. It is everywhere—because trace radium is everywhere. (One of the highest concentrations is in the stone used to build Manhattan's Grand Central Station.) Most radiation from radon decay consists of alpha particles, which have the highest 'weighting factor' for biological intrusion of all ionizing radiation types. So the worldwide ubiquity of radon, plus the fact that its background level can be easily determined, makes radon tracking ideal for determining the biological effects of ionizing radiation.

To accept this extensive data and information probably requires substitution learning, a concept we introduced in Chapter 27 'The Great Enemy of Truth.' But if we're willing to put ourselves through the stress of substitution, we might come to a surprising conclusion and it's this: if a nuclear power plant accident released so much radioactive material that public exposure reached as high as twenty times normal (but not greater), the exposed population would probably live longer and be healthier. The results of Chernobyl bear out this phenomenon.* And except for Chernobyl, there has never been a nuclear power plant accident that exposed the public to anywhere near this level of radiation.

The other famous accident occurred at the Three Mile Island generation station in Pennsylvania. Three Mile Island *might* have exposed the public to ionizing radiation but, if so, at levels only infinitesimally above normal background radiation—not by multiples of the background radiation. As frequently pointed out, reporters coming from around the US to cover the Three Mile Island story received more ionizing radiation during their high-altitude flights than they received at the plant's site. I suppose we could claim the accident bequeathed the community a very small health benefit but, because the increased dose was so minute, neither the radiation nor the hormesis effects were measurable. On the other hand, the Three Mile Island-hyping media certainly brought on anxiety-induced health problems in some residents.

After the first edition of *Smelling Land* was published, I learned of an extraordinary, accidental experiment[48].

About 25 years ago in Taiwan, during the manufacture of recycled steel, the 'melt' was accidentally contaminated with cobalt-60—a radioactive isotope of cobalt with a half-life of 5.3 years. The steel became rebar and found its way

---

* If you want to dig further, a good place to start is with the extremely comprehensive report of The Chernobyl Forum (2003–2005), *Chernobyl's Legacy: Health, Environmental and Socio-Economic Impacts*. This report was produced by an impressive group of international agencies that included the IAEA, WHO, UNDP and UNEP. WHO (the World Health Organization) contributed most of the health study results. Their 2006 report, 'Health Effects of the Chernobyl Accident and Special Health Care Programmes' may be obtained by accessing bookorders@who.int.

into the walls of more than 1,700 apartments—home to about 10,000 people. The average radiation dose received by persons living in these radiotoxic apartments was found to be 400 mSv, or about 100 times higher than normal background levels.

Later, when this accidental contamination was discovered, an epidemiological study was launched to establish how many more of the exposed residents had died from cancer and had had children with congenital deformations in comparison with a similar number in the unexposed population.

The results were stunningly unexpected: in the irradiated group, the incidence of cancer was reduced to three percent compared with the general population; congenital birth malformations had fallen to seven percent. Said another way, cancer incidence was 30 times lower, birth defects fifteen times lower.

Yet when I tell my colleagues about these results, rather than finding it cheering that we may have discovered a 'magic bullet' for cancer reduction, their resistance to the idea is often striking—emphasizing the pain of acquiring knowledge by 'substitution.'

Undoubtedly, we'll need more evidence of this strong hormesis confirmation. Still, in many ways we already have it, because the Taiwan accidental experiment confirms the long-standing data from the ABS cohort, the seminal work of Bernard Cohen and others.

The determined resistance to changing the so-called 'linear no-lower threshold' (LNT) model for describing the health effects of low-dose radiation reminds me of the scorn first heaped on two Australian physicians, Robin Warren and Barry Marshall, who postulated that peptic ulcers are caused by bacteria. Later, in 2005 they received the Nobel Prize in Physiology or Medicine for their work.

Will insistence on holding with the LNT model continue because it's so comfortably consistent with fears we've been so 'carefully taught?' Can we imagine the outrage if the public learned that these 'carefully taught' regulators were holding back the pursuit of information that might prevent 97 out of every 100 cancer deaths?

That said, excessive radiation *is* dangerous. So what is excessive? The average annual dose received by an American is 3.6 mSv.* Epidemiological studies show that the range of radiation exposure that brings health benefits lies from this normal exposure level to somewhere between 30-40 mSv. At least until the Taiwan accidental experiment, increased cancers were thought likely at doses above 400 mSv, and a person receiving a quick dose of about 4,000 mSv has a 50% chance of dying.

If I quickly ate 10,000 aspirins, I'd be in trouble. Yet we know small, regular doses of aspirin can be good for us. I swallow one with my orange juice during breakfast.† It seems we can expect the same kind of effect from ionizing radiation—a little more than normal is probably good for you, but a thousand times more could do you in. This got me thinking that an annual, full-body medium-dose ionizing radiation zap might be good preventative medicine. I'm ready to sign up.‡ My physician chuckles at the idea but has never followed through, probably mindful of legal risks. And I'm confident my health insurance won't cover it.

The two Tokaimura workers probably took the equivalent of more than 10,000 aspirins. People a few blocks away likely gained health benefits. So much for the *Guardian Weekly's* claim: 'Hundreds of thousands at risk.'

Now that we understand something about ionizing radiation, let's turn to the three issues that seem to lie at the root of public concern about nuclear power.

For the public's first concern, what to do with the spent fuel, the short answer is: bury it! This we can do very safely, using well understood and proven technologies. An even better answer is to reprocess the spent fuel, and use it

---

* Albert B. Reynolds, *Bluebells and Nuclear Energy*.[49] There are many sources for this kind of data. Reynolds' book is one of the easiest to read and understand. Data on background radiation levels is given in Table 2.4.

† In our litigious world I should add, 'before you follow my one aspirin-per-day example, you should check with your family physician.'

‡ These epidemiological studies are based on very large populations and therefore are robust for average impact on large populations. Individual response can deviate from the average response just as, for a few people, one aspirin a day may be too many—which is why some physicians, to be conservative, recommend only a baby aspirin a day.

to generate ten to twenty times more electricity, and then bury what remains. This strategy would simultaneously reduce the spent fuel we must ultimately bury and reduce its toxicity. Before we further discuss these answers, we should clarify a few things about spent fuel.

Spent fuel from a nuclear reactor contains four things:

- Some unused fuel, such as left-over uranium-235 and fissile plutonium-239 that has been 'bred' within the reactor;

- A lot of uranium-238 which, although not fissile itself, can be 'bred' to plutonium-239 that is fissile;

- Fission fragments—the larger bits and pieces that flew off when nuclei of uranium-235 or plutonium-239 were fissioned (fission fragments, which have the highest radio-toxicities of the spent-fuel ingredients, include relatively short-lived radioactive isotopes of elements like xenon and strontium); and,

- Some longer-lived actinides—radioactive elements that have atomic weights greater than uranium, but because they are long-lived, they contribute little to spent-fuel radioactivity.

Spent fuel *doesn't* contain the smaller bits, such as alpha and beta particles and neutrons. They are long gone. With this short primer on spent fuel, we should encapsulate some realities about the radio-toxic fission fragments:

- All fission fragments exhibit half-lives: the time it takes for half the material to be gone. These half-lives range from less than a minute to over 10,000 years.

- For perspective, we should remember that the half-life of many conventional industrial waste toxins—mercury and lead, for instance— are infinite. They hang around forever.

- As a rule of thumb, if a fission fragment is highly radioactive, it's using up its kick quickly and, therefore, has a short half-life. High radioactivity means the stuff will go away quickly. Low radioactivity means the stuff stays around much longer. So those who would claim the horrors of spent fuel cannot claim it will both stay around for hundreds of thousands of years and all the while emit high radiation. Can't have it both ways.

I've been speaking about spent fuel, not about the total 'nuclear waste' which, in nations like the United States and Russia, overwhelmingly originates from weaponry. Decommissioning weapons is a very different matter and it's an issue about which I cannot speak with any knowledge. But for nuclear power, straightforward technologies for the long-term, safe storage of spent fuel are proven and ready to deploy. It's the politics that has twisted our knickers. Politics—underpinned by persistent, disingenuous claims that we have no way to deal with spent fuel. These claims come from those who appear not overburdened with either knowledge or honesty.

Let's return to my earlier suggestion that we could substantially reduce the amount of spent fuel we need to bury. Today's 'once through' nuclear power strategy means the spent fuel typically contains considerably more than 90% of its original energy. Indeed, it's possible to recycle spent fuel to generate at least ten times more power. Moreover, if we were to follow a policy of reprocessing and reusing spent fuel—recycling at its best—we would reduce to about one-tenth the spent fuel we must ultimately bury. This could be the gain if we used breeder reactors. The gains from using only today's thermal reactors would be less.

Today, one drawback to a recycling strategy is that the price of reprocessed fuel is somewhat higher than virgin fuel. Yet this is mitigated because fuel cost is a comparatively small fraction of the cost of nuclear-derived electricity and, as well, we would have less spent fuel to ultimately bury. So the reduced disposal costs might reduce total life-cycle costs. In the United States, the Carter Administration forbade reprocessing, presumably because it might give bad guys a chance to steal the fuel. Later, in 1981, the Reagan Administration lifted the ban. However, by then potential suppliers were gun-shy about making the investments required for an industry that could be shut down by presidential whim.

In contrast, the French, British, Russians and Indians regularly reprocess their spent fuel. Today, the Japanese send their spent fuel to Europe for reprocessing. For another wrinkle on this business, I found it interesting that the *New York Times* of October 5, 2004, carried an article on how the US was shipping weapons-grade plutonium to France for reprocessing into civilian power plant fuel.*

Yet getting more energy and less waste from nuclear fuels can get even better. Earlier I spoke of what we can do, today, with conventional reactors. If we move beyond present reactors to next-generation *breeder* reactors, and use not only spent fuel but also the 'depleted uranium' (the by-product of enrichment processes and sometimes used for bullets), we could, in principle, get up to 100 times more exergy for each shovelful of uranium taken from the ground. We'd then have only about one-hundredth the spent fuel to bury.†

There is another point. Despite what we've been carefully taught, dangerous radioactivity from spent fuel *does not* last for tens of thousands of years. Rather, depending on the type of reactor, within 400 to 1,000 years spent-fuel toxicity returns to about what it was when the uranium was in the ground. For context, the CO_2 residence time in the upper atmosphere is roughly 300-400 years, and more than 1,000 years in the deep oceans. So the times required for fossil and nuclear waste products to return to their respective background level toxicities are the same order of magnitude.

Although hang-about times for fossil and nuclear wastes are similar, there are several important differences. The most important may be the sheer magnitude of waste material.

Let's compare the mass of spent fuel from nuclear power plants to the mass of CO_2 from coal-fired power plants, for equal output power. For a precise comparison, we'd need to know the type of nuclear and coal-fired power plants. But no matter which of today's designs we select, for the same electricity product the answers will come out in this range:

- if the nuclear plant generates a single kilogram of spent fuel,

---

* During the same week, President George W. Bush proudly proclaimed that the US would never listen to France when determining US foreign policy. I reflected that if the US had listened rather than disparaged, it might have avoided two tragic quagmires in a row, first Vietnam and then Iraq.

† Appendix B-4 explains enriched and depleted uranium.

- the coal-fired plant will emit between 50,000 and 350,000 kg of CO_2 —plus other nasties.

If in the future we adopt breeder reactors, the mass ratio of spent fuel from coal-fired power plants to spent fuel from nuclear power plants jumps to around five million to one.

In this narrative, I've only identified CO_2 as spent fuel from fossil-fired stations. There is also the water vapor and nasty things like heavy metals and ash. So the huge differences in the amounts of total waste materials from nuclear and coal-fired stations are even larger than these numbers indicate.

Because the mass of nuclear waste is so small, it can be carefully and safely stored for astonishingly long time frames. In contrast, the massive volume of fossil waste means that, practically, it must be pitched, unconstrained, into the mixing bowl of our atmospheric and oceanic global commons.* So it's not just the difference in amounts that is of vital importance. It's the *certainty* that continued dumping of spent fossil fuel effluent into the atmosphere will bring global disaster[†].

The second public concern is that an accident might cause a nuclear power plant to have a nuclear explosion. It's absolutely, physically impossible for a nuclear power plant to explode like a nuclear bomb. This is not just because nuclear power plants are probably the safest of 20th century technologies, although they are.[‡] Rather, as we've said several times, it's because the fuel for a power plant is entirely different than the fuel for a bomb.

---

* As we've noted, some claim that we can sequester CO_2 emissions underground, say, in depleted natural gas fields. Analysis shows that this might work in limited circumstances, when, for example, natural gas is used to produce hydrogen in stationary settings *and* when there is nearby underground storage *and* if there isn't too much CO_2 to store. So while the concept may serve old technologies fighting back, it doesn't really serve the long-term environment. Sequestration was first introduced in Chapter 17 'Hydrogen: The Case for Inevitability' and will be discussed further in Chapter 32 'Harvesting Hydrogen.'

† Carbon dioxide is by far Earth's most critical geopathogen. Spent nuclear fuel is simply not a geopathogen

‡ Modern commercial aircraft are also designed with exceptional concern for safety, although they can't be made nearly as safe as a nuclear power plant. Toward the lower end of public safety we

One way to appreciate this reality is to compare nuclear power plant fuels with dynamite. A stick of dynamite can explode. But fine grains of dynamite, diluted several hundred times by mixing uniformly throughout a large bucket of sand, cannot. Reactor-grade uranium is not only diluted because it contains a much smaller fraction of fissile uranium-235, it is further diluted because in reactors it's chemically bound to oxygen as a metal oxide (UO_2). It's not a pure metal (U) as it is in a bomb.

Even the best reporters sometimes give (probably unintentionally) misleading information on the difference between nuclear fuels and bomb-grade nuclear material. In an otherwise excellent essay in the March 18, 2006 issue of the *Globe and Mail*, Paul William Roberts, author of *A War Against Truth: An Intimate Account of the Invasion of Iraq*, wrote 'Of course, most nuclear reactors require enriched uranium fuel, which also can be used to make a nuclear weapon.' He's correct that enriched uranium is used in most power reactors, although not all.* What he neglected to mention is that the *degree* of enrichment is entirely different. The enriched uranium used in nuclear power reactors contains *at most* 5% uranium-235 *and* it's further diluted because it's in the form of a *metal oxide*. In sharp contrast, the *pure metal* uranium used in a bomb contains well above 90% uranium-235.

Accidents can happen at nuclear power plants, but they can't be nuclear explosions. There might be fires, steam explosions or although less likely—the release of radioactive materials. To date, only the Chernobyl accident exposed the surrounding population to radioactivity in concentrations capable of health damage. In that case, the release was only possible because Chernobyl-style reactors have no containment vessels. Such designs have never been licensed in any Western country—or, after Chernobyl, anywhere in the world. What aggravated the Chernobyl accident—but did not cause it—is that the Chernobyl reactor employed a graphite moderator that caught fire and thereby helped spread radioactive materials in the fire's plume. And among the horrors

---

might find domestic water supply, especially in smaller towns and rural communities and especially in less-developed nations.

* Canada's existing CANDU reactors use unenriched uranium (although Canada's advanced ACR-1000 units employ fuels that are slightly enriched to between 1.5 and 2% uranium-235).

Chernobyl visited upon firemen was that the roof had been tarred and the firefighter's boots became stuck in the molten tar. Almost all Western reactors use either ordinary or heavy water moderators.* Water can't catch fire.

Finally, after the infamy of September 11, 2001, it is reasonable to ask: what would happen if a jumbo jet made a direct hit on a nuclear power plant? Answer: all the people in the aircraft would die. Some workers in the power plant might be killed. But, for any western power plant, structural analysis has shown it would be impossible for the impact to break through the more than a meter thick, reinforced-concrete containment vessel and to so severely damage the reactor core that either a core meltdown or a release of radioactive material could result. A wide-body jet, aimed with pinpoint accuracy to strike the most vulnerable part of a western nuclear power plant would, almost certainly, not result in a single fatality in the surrounding communities. Terrorists have many easier, headline-grabbing targets—as we're often reminded.

Now we'll look at the third public concern: what if bad guys steal nuclear material to make their own bombs? This time I can't make a declarative statement like, 'a nuclear power plant can't blow up in a nuclear explosion' or 'the technology for safe spent fuel disposal is well-known and proven.'

It's conceivable that bad guys could steal material from a nuclear power plant and use it as feedstock to manufacture a crude bomb. The question is: why would they try? As we know, the ratios of both plutonium-239 and uranium-235 to the other stuff in a nuclear power plant (or its spent fuel) are far different than the ratios needed for a bomb. So aside from the difficulty of performing the theft, if they successfully nicked the stuff it would be extremely difficult and risky to handle (spent fuel gives much more than the healthy dose we spoke of earlier). They would require unusually sophisticated, expensive technologies for separation and enrichment in order to approach bomb-grade material for even a crude, unreliable bomb.†

---

* The world's first reactor, CP-1, that was built beneath the Chicago stadium in 1941 used a graphite moderator. I'm honored to have a small piece of that graphite sitting on the coffee table in my living room.

† I've chosen the phrase bomb-grade rather than weapon-grade. Weapon-grade terminology is more common in the industry. My choice is based upon the fact that the fuel for a nuclear-powered

Why do I say this? One reason is because it would be easier for terrorists or rogue nations to use a simple reactor to breed plutonium from unenriched uranium. Such a dedicated, plutonium-producing reactor is much easier to design and build than the sophisticated power reactors—or even the enrichment technologies needed to turn stolen power plant material into bomb-grade material. Then, even if they get enough uranium-235, or plutonium-239, the bad guys would still need to make the nuclear bomb—a non-trivial task. Contrary to something else we're often told, it would probably be easier to build a Boeing 747 in a (rather large) backyard than a nuclear bomb—and a lot safer. There are many easier ways to kill lots of people.

Finally, like it or not, we cannot put the nuclear bomb genie back in its bottle. Just as our forefathers couldn't stuff the crossbow or gunpowder genies back in theirs. The danger of terrorist-deployed nuclear weapons cannot be dismissed. But it's foolish to think that the peaceful use of nuclear power plants significantly increases terrorist opportunities.

Substituting new knowledge for long-held beliefs is the supreme challenge. However, as proven by both analysis and experience, nuclear power plants are the safest and most environmentally non-intrusive of exergy sources. They can be used to manufacture electricity today, and both hydrogen and electricity tomorrow. Nuclear is the only exergy source that has any chance to be deployed fast enough, and to deliver enough exergy to roll back CO_2 emissions while, simultaneously, supporting all the services civilization requires. Moreover, nuclear delivers its energy reliably and requires the smallest land 'footprint' among exergy sources—an important environmental consideration with the growing demands for space that encroach upon our remaining wilderness.

Naturally, as with any technology, precautions must be taken. But I fear thoughtful precaution has been trumped by hyperbole. We must re-examine what we've been carefully taught.

---

aircraft carrier or submarine is far from bomb-grade, but it can still push military ships along quite nicely.

Nuclear energy is a masterpiece of an inside-in technology—a concept we'll discuss in Chapter 37 'Chasing Locomotives.' In the present chapter we never really looked inside this inside-in technology. So my observations may have smacked of superficiality or, worse, that I wanted them accepted with a dose of faith. For those who would like to know nukes better, I recommend *Bluebells and Nuclear Energy*[49]—a book that gives an excellent and lay-accessible description of how nuclear power plants work.

Nuclear power is much too important an issue to leave to the shrillies. So if you are truly concerned about civilization's future, please take time to be sure you know nukes—and then take time to explain what you know to your friends. I expect you'll be surprised by the positive reception you'll get.

Part Nine

WORKING WITH HYDROGEN

32. HARVESTING HYDROGEN

In which we review some of the many ways we can produce hydrogen.
Some clean, some not so clean. Some for today, some for later.

People often forget that to harvest hydrogen we need both material from which to mine the hydrogen, and exergy to do the mining.

We must repeatedly remind ourselves of that truth, because it's so different from all the other currencies. For example, crude oil delivers both the material and the exergy of gasoline. And electricity can hardly be said to even contain material, unless we mean those electrons flashing by. So while throughout *Smelling Land* we've repeatedly said that any exergy source can be used to make hydrogen, we didn't say all exergy sources can provide the material from which we get the hydrogen. Sunlight and nuclear power can be used to mine hydrogen, but they obviously can't be the material from which the hydrogen is mined.

Water is always the prime hydrogen ore. In the future, water will be the sole material source. But no matter what process we use to harvest hydrogen, water always provides at least half the hydrogen atoms. I'll use a few simple equations to demonstrate why—starting with the steam methane reformation process which dominates hydrogen production today.

The overall process for harvesting hydrogen using steam-methane reforming (SMR) is shown in this equation:

$$CH_4 + 2H_2O + \text{energy } (\textit{emits CO}_2) \rightarrow 4H_2 + CO_2 \qquad 32.1$$

If we count hydrogen atoms on each side of equation 32.1, we find that of the eight atoms of hydrogen that end up in the four molecules of hydrogen, four atoms came from the two water molecules ($2H_2O$) and four from the single methane molecule (CH_4). So half came from water and half from methane.

But if we produce hydrogen any other way, from coal for example—as has been done for more than a century to make coal gas (earlier called town gas)— more than half the hydrogen atoms come from water. A reasonable, albeit simplistic, chemical formula for coal is CH.* If I use this formula to represent coal, the overall process equation comes out like this:

$$2CH + 4H_2O + \text{energy } (\textit{emits a lot of } CO_2) \rightarrow 5H_2 + 2CO_2 \qquad 32.2$$

Counting hydrogen atoms in the coal equation, we find that, this time, 80% of the hydrogen atoms came from water, and only 20% from coal. If you have now become excited about a higher percentage of hydrogen coming from water, beware. Compared with the first equation, this time we have two, rather than one molecule of CO_2 heading off into the great beyond.

Now comes the brighter future. The next equation shows that when we use non-fossil energy sources, say sunlight or nuclear, all the hydrogen comes from water:

$$2H_2O + \text{energy (emits no carbon)} \rightarrow 2H_2 + O_2 \qquad 32.3$$

How about that!

Before we describe individual harvesting technologies, I should say something about the input exergy sources, because they will always be important to both systemic and environmental performance. I've written equations 32.1 through 32.3 so they can tell can us more than just where the hydrogen atoms come from. I've also added, in italics, the typical emissions from the exogenous† exergy source used to mine the hydrogen. The exogenous exergy used today in the SMR processes normally comes from burning additional natural gas. For coal, the exogenous exergy usually comes from burning additional coal. The exogenous exergy for both processes could come from a clean source, such as modern high-temperature nuclear reactors, but this requires institutional partnerships that have not yet occurred—except perhaps at the

----

* This is simplistic as a chemical formula for coal (in part because coal can have different H/C ratios-more typically between 0.7-0.9—but it also contains other unpleasantries such as heavy metals.) Still using an H/C ratio of unity is not over—simplistic for demonstrating the typical amounts of hydrogen that come from coal compared with the amounts from water.

† I'm using 'exogenous' to mean energy from outside the basic process. The methane that is 'split' during the SMR process brings most of the exergy to the party.

discussion level. When we produce hydrogen by electrolysis the process, itself, is environmentally benign. However, if the electricity used for electrolysis comes from a coal-fired generating station, we can't be smug about the system's environmental gentility.

Electrolysis is a good starting-point for thinking about the different, clean ways we have to produce hydrogen. At some point during teenagehood, your high school chemistry teacher probably used small electrolysis 'factories' to split water into hydrogen and oxygen. And, unless your classmates were particularly well-behaved (or unimaginative), one of your classmates would have dreamed up some prank to recombine the H_2 and O_2 with a pop. This would have brought squeals from some and (feigned) innocent surprise from others. Teachers used electrolysis to show that water is composed of hydrogen and oxygen—at least that was their objective then. Now some high schools also use it to show the essential features of the coming Hydrogen Age. I've been told some even have tiny fuelcells powering little model cars running around a track. The demonstration can illustrate the beauty of this systemically closed fuel cycle. We know all non-carbon exergy sources can be used produce electricity. And because we can always convert electricity to hydrogen using electrolyzers, if we want cleanly produced hydrogen, electrolysis will, at least for a while, be the clean technology to beat. Electrolyzers use direct current (DC) electricity. So if we take the electricity from the electricity network, the grid's AC must be changed to DC.

This is a good time to remind ourselves that electrolyzers and fuelcells are reciprocal electrochemical technologies. Electrolyzers accept inputs of water and DC electricity to give outputs of H_2, O_2 and (normally) heat—with, of course, a little exergy destruction along the way.* Most fuelcells accept inputs of H_2 and O_2 (typically carried in with air) to produce DC electricity, water and heat.†

---

* High-temperature electrolysis can be designed to use heat *inputs*.

† To find out more about the prospects of building reversible fuelcells and electrolyzers, you can go to the Canadian Hydrogen Association's website under the *Smelling Land* entry where you will find an article titled 'Reversible Fuelcells and Electrolyzers.'

Normally we think of electrolyzers as producing hydrogen. But they simultaneously produce oxygen, often a marketable by-product.* Moreover, electrolyzers can be designed to produce a second valuable product, heavy water (D_2O).† Heavy water is the moderator in Canadian nuclear power plants, a feature that allows these reactors to be fueled by natural, rather than enriched uranium.

Co-production will characterize the future: Harvesting technologies that co-produce more than one currency or commodity. The most obvious co-production scenario is to harvest both hydrogen (and oxygen) *and* electricity from the same energy source at the same site. Let's discuss how this will work using two categories of exergy sources—renewables and nuclear.

Most modern electrolyzers can adjust instantaneously to variable rates of input electrical power. So using electrolyzers gives us another way to exploit the synergies of hydricity systems. When electricity is made, it must be used; when hydrogen is made, it can be stored. As we discussed in Chapter 29 'Renewables and Conventional *Wish*dom,' intermittency and unpredictability are constraints on many renewable energy sources, such as wind and solar. However, co-producing hydrogen and electricity will mitigate these limitations. When electricity demand is high the renewable source can deliver electricity. But when the electricity demand falls, the output from the renewable can manufacture hydrogen.

Next-generation nuclear power stations provide one of the most promising prospects for co-production of electricity and hydrogen. Most nuclear power plants operate best when supplying continuous power. That is why they

---

* If you ever have the opportunity, visit the museum at Peenemünde where Wernher von Braun and his team developed rockets during World War II. You'll find the displays located in the one remaining turbine building, which produced electricity used for electrolysis. But the product was not hydrogen, as you would expect knowing that hydrogen is one of today's rocket propellants. Rather, the key product was oxygen used to oxidize the rocket's alcohol fuels. Oxygen is also the prime product of electrolysis in submarines, where it's used to keep the submariners breathing.

† Many people have seen the historical, but error-ridden movie, *The Heroes of the Telemark*, about Norwegian saboteurs who attacked a German-operated heavy-water plant near Rjukan on a Norwegian fjord during World War II. The plant did exist, and was producing heavy water using electricity generated at the nearby Vemork hydroelectric station.

are normally dedicated to 'base load.' But there is another reason for steady operation: Nuclear plants have comparatively high capital costs, offset by low operating costs, so keeping these plants running at close to full power makes good economic sense.

For these reasons, I think the ability to co-produce hydrogen and electricity should be included in the design of all new nuclear power stations. A technical trick to help is this: It should be possible to mount both DC and AC generators on the same turbine output shaft. The AC generator will directly feed the electricity grid. The DC generator will feed electricity to electrolyzers that produce hydrogen. For intermittent renewables, the advantage of electrolysis is that it can follow *input* power changes almost instantaneously. For nuclear power plants, the flip-side is that electrolysis can follow load (demand) changes quickly, allowing the generating station to produce hydrogen as the demand for electricity drops—say in the middle of the night. The advantage of having both DC and AC generators on a common output shaft is that there will be no need to rectify AC electricity to DC. Rectification reduces overall efficiencies and requires expensive AC-DC conversion equipment. With twinned AC and DC generators, the nuclear power plant can directly feed electrolysis without the AC-DC step.*

We'll find many more opportunities for co-producing hydrogen and electricity. It's a fertile field for innovative business people and imaginative engineers. It will build on several co-production systems that, for decades, have already been co-producing more than just electricity, hydrogen and oxygen.

One example is the chloralkali process which uses input DC electricity to manufacture chlorine gas (Cl_2), hydrogen (H_2) and sodium hydroxide (NaOH). Common names for sodium hydroxide are 'lye' and 'caustic soda.' Worldwide, today's annual production of sodium hydroxide is some 50 million tons. Of the many uses for caustic soda, manufacturing soap is one and whitewashing your barn another. An interesting footnote to the chloralkali process is that Dupont developed the Nafion membrane to separate chlorine from the caustic

---

* The feasibility of this scheme will depend on an analysis of such things as capital costs, the price for 'off-peak' electricity, and the estimated load fractions to hydrogen and electricity.

soda during chloralkali manufacturing. The Nafion membrane was later used in most of the first-generation polymer electrolyte membrane (PEM) fuelcells.

Now let's look towards other clean ways to produce hydrogen. As we know, Nature uses photosynthesis to harvest hydrogen directly from water. Some engineers and scientists hope to mimic Nature, by harvesting hydrogen from water but *without* the electricity-to-hydrogen step of electrolysis. One promising avenue is to use the heat from high-temperature nuclear power plants to split water directly in thermochemical cycles. These processes are not problem-free, in part because they often require prickly chemicals, like hydrogen bromide and sulfuric acid. Still, I have little doubt that one or more practical processes will ultimately be successful. Other people hope to develop chemical cycles that can directly convert the exergy of sunlight into hydrogen and pure oxygen—again, without first manufacturing electricity for electrolysis.

Now let's talk about the hydrogen's purity. The requirements for purity depend upon how the hydrogen will be used. Today, one of the largest hydrogen uses is in refining and upgrading 'heavy' petroleum ores, like the oil sands in Canada and Venezuela. These refining processes do not require very high purity hydrogen. Another major use of hydrogen is in the production of ammonia (NH_3) for fertilizer, where again purity requirements are not demanding.

In other applications, as in the production of float glass or microelectronic chips, high-purity hydrogen is essential. And high-purity hydrogen is important for hydrogenating food stuffs, and for most (but not all) fuelcells. Reasonably high-purity hydrogen is needed if the hydrogen is to be liquefied. (The hydrogen used for rocketry at the Kennedy Space Center is produced by the SMR process in Louisiana, then purified and liquefied before being loaded aboard large cryotank trucks or railcars and shipped to the spaceport.)

Thinking about the different hydrogen purity requirements, my mind jumps to *Animal Farm*.[50] 'All animals are equal but some are more equal than others,' applies to hydrogen no less than to the four-legged community that commandeered the farm.

Large established industries always fight back. The fossil source industry is fighting back. Consider the coal lobby. Their arguments dwell upon the strategic need for a secure supply of domestic energy—arguing it's coal. They will do their best to tighten up coal processing and claim they can deliver 'clean coal.'

Cleaner coal, maybe! Clean coal, not a chance!

Those who hope to return to producing hydrogen from coal by using cleaner coal processes must capture and sequester the CO_2 emissions—and hold them captured for at least several thousand years in such places as depleted natural gas fields or underground salt domes.* Thus, if we want to keep using coal, we must keep identifying suitable CO_2 sequestration sites not too far from the coal-to-hydrogen manufacturing. This will be difficult. (To get a feel for the sheer magnitude of CO_2 that would need to be sequestered, look back to Chapter 31 'You've Got to Be Carefully Taught—Know Nukes' for a comparison of the mass of spent fuel produced by coal versus the mass produced by nuclear.)

Another approach to sequestering might help. We might be able to use chemical processes that trap the effluent CO_2 as carbonate rocks. In some ways, this would mimic how Nature traps carbon dioxide in the oceans to form oyster shells and coral, and to form the white cliffs of Dover—or the chalk formations through which Canadian and German soldiers carved tunnels near Vimy during World War I, or in which the French carved caves for storing champagne.

There is another issue. When we use coal to make hydrogen, there are *two* streams of carbon dioxide emitted. The first can be a comparatively pure stream—from the basic coal-to-hydrogen process set out in equation 32.2. The second comes from combusting the coal burned to provide the exogenous energy. In this second stream, the CO_2 is diluted by the nitrogen that constitutes 78% of the incoming combustion air. Before we can capture the carbon dioxide from the second stream for later sequestration, we must separate it from

---

* The production of hydrogen and liquid fuels from coal is an established technology—used, for example, by Germany during World War II. It's the plan to capture and sequester the CO_2 emissions that is new.

the rest of (what engineers call) the flue gases. All of this will be technically, exergetically and economically demanding.*

In Chapter 10 'From Oil Lamps to Light Bulbs: Pathways to Environmental Gentility,' we set out the core choice when we want to clean the place up: Do we add collectors to catch pollutants? Or do we change the process to stop making pollutants? Carbon sequestration is archetypical of adding collectors. Collectors are the weapon of choice when old technologies fight back. And their substantial war chests can sustain the battle for a long time—all the while claiming clean sustainability. Some claims might persuade for several decades. None for centuries.

---

* Richard Doctor and John Molburg offer a useful discussion in Chapter 6 'Clean Hydrogen from Coal with CO_2 Capture and Sequestration,' published in *The Hydrogen Energy Transition*.[51]

33. But Doesn't Hydrogen Explode?

In which we examine what happened during the disasters of 5/6 and 9/11, to conclude that hydrogen may be the safest fuel of all.

Dr. John Bockris, one of the founders of the International Association for Hydrogen Energy, had just completed a lecture about the benefits of hydrogen when a member of the audience challenged: 'But doesn't hydrogen explode?' Controlling his irritation, Bockris replied, 'Of course it can. That's how fuels work!'

Still, the questioner had raised the important issue of safety. Whenever we introduce innovation, especially technological innovation, we must evaluate safety with great care. And a good platform from which to discuss the safety of hydrogen is to review two disasters.

The first happened on the grotty evening of May 6, 1937, when the airship LZ—129 *Hindenburg* burst into flames over Lakehurst, New Jersey. The news media immediately jumped on hydrogen as guilty of causing the accident. Some, in ignorance, still do. But more recent research has exonerated hydrogen, showing it probably had nothing to do with *causing* the accident.

The second occurred during the morning of September 11, 2001, when two hijacked commercial aircraft were flown, at speed, into Manhattan's World Trade Center. You probably think this attack on civilization had nothing to do with hydrogen. And you'd be right. Except that, *if* the aircraft had been hydrogen-fueled, the loss of life would probably have been more than halved and, almost certainly, both buildings would still be standing.

As the first chair of the US Department of Energy's Hydrogen Safety Review Panel, Dr. Addison Bain came to this responsibility from a uniquely relevant background. After earning his engineering degree, he spent four decades with the US Air Force and NASA, continuously involved with America's aerospace adventures. Bain was responsible for the acquisition of rocket propellants—primarily liquefied hydrogen and liquefied oxygen. (In a few circumstances solid propellants are used, such as by the space shuttle's booster-rockets during the first 140 seconds after launch.) Bain's multi-year experience with propellants persuaded him that liquid oxygen is the dangerous ingredient of rocketry—and that hydrogen is the safest of chemical fuels.

Bain also doubted that hydrogen caused the *Hindenburg* accident. So when he had a late-career opportunity to pursue a doctorate, he proposed sleuthing out the true reasons for the *Hindenburg's* demise as his research topic. His research supervisor acquiesced, with the caveat, 'OK, but you better have a backup topic.'

Bain traveled to Friedrichshafen, Germany, where the *Hindenburg* had been built, scoured old reports and design manuals, interviewed witnesses and survivors, analyzed films and photographs. In the end, his forensic engineering paid off.

Based on Bain's research, the National Air and Space Museum in Washington re-explained their *Hindenburg* exhibit. PBS produced a documentary on his findings as part of their *Secrets of the Dead* series. As a result, almost everyone in the aerospace and hydrogen communities knows that hydrogen, which had been used for buoyancy rather than as a fuel on the *Hindenburg*, did not cause the airship's demise. She fell to the ground burning because, Bain argues, the outer fabric had been doped (painted) with a flammable mixture of aluminum powder flakes and fine iron-oxide. The paint had been used to reflect sunlight to prevent the gas cells within the dirigible from overheating. As NASA's former propellant engineer, Bain realized these doping ingredients were similar to those used in the space shuttle's solid-rocket boosters, which provide more than 80% of the takeoff thrust during the first few minutes after launch. (Following that brief high-thrust send off, H_2-O_2 rockets push the shuttle on its journey.)

Bain's story demonstrates two things: How a top engineer, whose career working with hydrogen helped send men to the Moon, came to view hydrogen as the safest of fuels; and how hydrogen, used as a buoyant lifting gas, did not cause the *Hindenburg* accident.*,†

So how safe is hydrogen as a fuel?

I consider it preposterous to assign a single number, like IQ, to quantify a person's breadth of intellectual abilities. Similarly, no single parameter can rank the safety or danger of different fuels. Fuels have a mosaic of properties, each of which sets out a different aspect of the fuel's safety. I can't itemize all the properties that govern hydrogen's safety—and certainly can't recount the many experiments and accident anecdotes from which insight is born. So I'll simply set out some of hydrogen's most distinguishing features:

- We know hydrogen is Nature's lightest element. So if gaseous hydrogen escapes outdoors, its extreme buoyancy quickly takes it into the sky.

- If liquefied hydrogen spills, before it can ignite some of it must first boil-off to become a gas and then the gas must mingle with enough air to reach a combustible mixture. But gaseous hydrogen rises so quickly that it's difficult to accumulate a sufficiently large hydrogen-air mixture to sustain a substantial fire. This is one reason hydrogen is probably the safest fuel for transportation applications.

- Hydrogen is both colorless and odorless. This can be a danger since neither your eye nor nose can detect its presence. Outdoors this is not particularly troubling—because the hydrogen quickly goes up and away. However, if hydrogen were to leak into an enclosed space—say from a parked vehicle in

---

* Bain's career and how it led to his *Hindenburg* research is recounted in his book *The Freedom Element, Living with Hydrogen.*[52] It's a book filled with entertaining anecdotes on his work at the dawn of space voyaging, spiced with easy to understand technical ideas.

† Bain's analysis of the cause of the accident is now widely accepted in the aerospace community. However, readers interested in a counter view on the accident—and a critique of Bain's 'paint was the cause' conclusion—should obtain the paper by Dessler, A. J. et al, 'The Hindenburg Fire: Hydrogen or Incendiary Paint?' *Buoyant Flight*, Vol. 52, Jan/Feb and April/May 2005. Dessler introduced several interesting ideas: How the fabric and dirigible frame would have acted similar to the 'mantle' in a gas lamp, allowing the 'invisible' hydrogen flame to be easily seen. Dessler points out that although the aluminum powder flakes and fine iron oxide were ingredients of both the *Hindenburg* fabric—doping and solid rockets, they existed in different proportions and configurations. Dessler attacks the doping as the cause postulate, yet doesn't propose an alternate cause, nor does he challenge other aspects of Bain's observations on hydrogen's safety.

an underground garage—then, to provide warning, hydrogen sensors must be located at strategic locations in the garage.

- When burning, hydrogen is sometimes called the 'invisible flame.' That's because the flame's emissivity (the radiative energy the fire sends out to its surroundings) is much lower than the emissivity from burning conventional fuels.* In some cases, hydrogen's low emissivity might be a danger. For example, if you approach a hydrogen stove with a burner left on, you might not see the flame.† In contrast, you can see a propane or natural gas flame—because it is the carbon content within these fuels that radiates.‡ But in most accident scenarios—especially transportation accidents—hydrogen's low emissivity is a safety feature. People often die from, or are severely burned by, the radiative heat from intense fires, even when a long way from the inferno. If an accident involves a hydrogen fire, the danger of radiative scorching is much reduced.

- Hydrogen has wide flammability limits (from 4-75% volume in air), so very-lean to very-rich hydrogen-air mixtures can ignite. In practice, this wide flammability range has been an advantage. Any spilled hydrogen either ignites before significant quantities build up, or the hydrogen dissipates into the sky before a fire starts.

- The detonation (explosive) limits for hydrogen-air mixtures are much narrower (18.3-59% volume in air) than the flammability limits. This, combined with very low ignition energies, means that an accidental mixture

---

* Although you wouldn't be foolish enough to stand *inside* a bonfire, it's sometimes difficult to get close enough to toast your marshmallows. That's because the bonfire's flames have high emissivity: They radiate a lot of their energy into the surroundings. Low emissivity is the opposite: As long as you're not *in* the fire, you'd feel comparatively comfortable standing close by. Hydrogen is one of the rare fuels with a very low flame-emissivity.

† To reduce this danger, some people have proposed adding to hydrogen trace compounds with high flame-emissivity. Other impurities, like the mercaptons added to natural gas, could be used as odorants. However, a disadvantage to these approaches is that very pure hydrogen is required for many applications, such as providing the fuel for most fuelcells.

‡ We can often 'see' something is hot even when there isn't a flame. For example, the heat from your living room radiator changes the air's refractive index and buoyancy, which allows you to see the air turbulence waving above the radiator.

of hydrogen and air is likely to begin burning and be up and away before it reaches explosive mixtures.

These last two properties were confirmed during a traffic accident shortly after midnight in August 1972, near Tallahassee, Florida, when teenagers ran a red light. They collided with a tractor trailer hauling 605,000 liters (16,000 US gallons) of LH_2. Oops! The tractor trailer jackknifed and slid down the pavement on its side. One of the diesel saddle tanks ruptured to spill diesel that ran through the ditch, ignited and, in turn, set fire to hydrogen that was venting from the LH_2 tank's pressure relief valve. The venting hydrogen reportedly burned 'like a blow-torch' until it was gone. But there was no explosion.

When hydrogen is used to fuel aircraft or space vehicles, it must be carried on board as LH_2—which means at a temperature of about 20 Kelvin (-253°C or -424°F). In addition to rockets, before we're deep into the Hydrogen Age, I expect LH_2 will fuel many surface vehicles. At 20 K, the cryofuel is very, *very* cold. So it will be obvious that if you spill LH_2 on your hand, you'll have a nasty case of freezer burn. A less obvious but potentially more important safety issue is this: If an LH_2-handling technology, say a refueling nozzle, is not properly insulated, the outer surface can become cold enough to liquefy some of the surrounding air. Now we have liquid air (a mixture of liquid oxygen and nitrogen) dripping onto the floor. If the floor is a flammable material the result could be fires or explosions.*

Zooming to aircraft fuels: Jet A is used in commercial aircraft and JP-4 in military aircraft. Both fuels are members of the kerosene family. In the 1970s, researchers at Wright-Patterson Air Force Base shot armor-piercing incendiary and 'fragment-simulating' bullets into Styrofoam-lined aluminum containers filled with LH_2 and JP-4, to test which would be the safest aircraft fuel when

---

* In 1984, Toronto hosted the 5th World Hydrogen Energy Conference. Our Institute for Hydrogen Systems at the University of Toronto 'garaged' visiting hydrogen-fueled vehicles from Canada and around the world. The fuel had to be drained before the vehicles could be moved to indoor storage. In one case, we unwittingly used an uninsulated copper tube to remove LH_2 from a car. The copper became cold enough that liquefied air began collecting on the copper tube. The team quickly caught the error. A more serious accident occurred at the California Rocketdyne test facility, when larger LH_2 flows were involved, and liquid air accumulated and dripped on the asphalt floor.

in combat. They concluded LH$_2$ was a 'more forgiving' fuel than JP-4. The incendiary weapons ignited the LH$_2$ but the hydrogen never exploded. The hydrogen fire was reported 'less severe' and 'expired more quickly' than the comparable JP-4 fire. Conclusion? In combat (and by extension civilian accidents), LH$_2$ would be a safer aircraft fuel than JP-4 (or Jet A).

We're now ready to review what happened when terrorists hijacked two Boeing 767s and flew them into the World Trade Center. The towers collapsed because tons of Jet A, disgorged from shattered tanks, had burned through the towers' steel backbones, destroying their structural integrity. The heat softened the steel allowing the weight of the top floors to begin collapsing. Once they began to fall, their momentum built to pile-drive the lower floors into the ground.

Although this collapse mechanism is well-known, another piece of supporting evidence comes from recalling the sequence of events. About sixteen minutes *after* the first Boeing 767 hit the North Tower, the second aircraft struck the South Tower—but many floors lower than where the first aircraft had crashed into its target. Yet the South Tower collapsed first—some 30 minutes *before* the North tower came down. In the South Tower, it was the greater weight of the extra floors *above* the fire that allowed them to fall first. In contrast, because there was less weight in the fewer floors above the impact point in the North Tower, the burning Jet A needed more time to weaken steel columns before they softened enough to allow the upper floors to sledgehammer the lower ones.

What if it had been LH$_2$-fueled aircraft that had flown into the World Trade Center?

The outcome would have been different. Obviously, there would still have been enormous damage, but I doubt the towers would have collapsed. Spilled liquid hydrogen can't burn until gaseous hydrogen has boiled-off. The boiled-off hydrogen would have rapidly floated *upwards* to mix with air and become a flammable mixture. Almost no burning liquid hydrogen would have run down through the buildings' steel backbones. Moreover, as the hydrogen-air fireball floated into the sky, very little radiative heat damage would have occurred *outside* the fireball. Structural damage, fire and loss of life would have been confined to the floors the aircraft struck and, of course, to the aircraft and its passengers—with perhaps some damage on higher floors as burning hydrogen-air mixtures traveled upward. But the lower floors would have been largely unscathed. Many more people would be still alive. The twin towers would probably still be standing.*

The safety of any fuel depends on many factors—the fuel's fundamental properties, the technologies in which it is used, and the circumstances of the accident. The work of hundreds of researchers, of which Bain is just one, indicates that, when used properly, hydrogen is probably the safest fuel of all.

By the way, in spite of Bockris' correct response, 'that's how fuels work,' his answer needs a little qualification when hydrogen is used in fuelcells. Unlike today's internal combustion engines, when used in fuelcells, hydrogen doesn't explode. Instead, fuelcells oxidize hydrogen, using peaceful electrochemical reactions—analogous, in fact, to reactions your digestive track uses to oxidize food.

I'm indebted to Addison Bain for clipping this gem from the August 2002 issue of the *Orlando Sentinel:*

> Today, of course, thanks to the educational efforts of the bottled-water industry, we consumers are terrified of our tap water, because we know that it contains some of the most deadly substances known to man: chemicals. To cite one example: Bottled-water industry researchers recently stated that virtually every sample of tap water they tested contained large quantities of hydrogen, which is a type of atom believed to have caused the *Hindenburg* dirigible disaster. This is why millions of consumers now prefer bottled water.

---

* Although I've checked the arguments of this paragraph with top liquid hydrogen and structural engineers, readers should be aware that the ideas in this paragraph are purely mine. The logic has not been sanctified, I expect not even considered, by any official body.

All the same, I'll be staying with tap water—although, if reading the *Sentinel*, perhaps diluted with whiskey. Plus an ice cube.

34. TETHERS AND TRANSITION TACTICS

In which we set out actions that can rapidly reduce CO_2 emissions before we get to the fully developed Hydrogen Age.

If we could quickly jump to the Hydrogen Age, we'd have our best chance to reduce CO_2 emissions soon enough to avoid toppling over the metastability lip into climate catastrophe. But the transition to fully-developed hydricity systems will take many more than a few years. So in the interim we must use complementary tactics to reduce carbon dioxide quickly. We need short-term tactics embedded within a longer-term strategy that will take us to the Hydrogen Age. When we choose shorter-term acceleration tactics that can quickly take big chunks from our CO_2 emissions they must be:

- Comparatively easy to achieve,
- Possible *without* requiring significant changes to the entire energy system chain, and
- Able to achieve *significant* CO_2 reductions within one or two decades.

The second point needs explanation. The most expensive components of our energy system lie in the 'service technologies' link—the aircraft, road vehicles, trains, ships, information systems *and* all their supporting infrastructures. Therefore, if we hope to cut CO_2 emissions quickly, our bridging tactics must not significantly disturb that link of our system chain—at least not too quickly or by too much. In turn, this means the 'services' and 'currencies' links will hardly be disturbed. And like it or not, this also means that we'll continue using hydrocarbon fuels for longer than some might wish.

At first blush, this reality might seem to mean that we can't do much. But with a closer look, things turn brighter. There are actions that can meet our three criteria for rapid and significant CO_2 reductions—while steadily pushing hydricity systems ahead. So we must zoom to the 'sources' and 'harvesting technologies' links of our five-link energy system chain, to see how we can use cleaner ways to produce the currencies we use today.

Figure 34.1 The five-link energy system chain highlighting
the two right-most links.

There is no fundamental barrier, either technical or economic, that prevents us from supplying *all* our electricity from non-CO_2 emitting exergy sources. So if we are realistic about the magnitude of the climate-disruption dangers, then our only responsible path for new electricity generation is to never again build, anywhere in the world, another coal-fired station.* All new electricity capacity must use sustainable, non-carbon sources. The leading candidates are nuclear, hydraulic and sometimes wind power—but any other non-carbon source that can deliver electricity with comparable reliability and cost should be deployed whenever appropriate.

After the moratorium on new coal-fired electricity generation is in place, we should move to a moratorium on new natural-gas generation which, although emitting much less CO_2 than coal-fired generation, is still a carbon dioxide contributor. As existing coal-fired plants age, they should be retired as rapidly as economics allows and replaced with new non-fossil utilities. In most cases this means nuclear.

Deploying new non-carbon generating capacity—while decommissioning older fossil-fueled plants—will bring large CO_2 reductions. And it can be done, just as fast as we can build new and decommission old. No new technologies are needed. No economic uncertainties exist. And the approach meets the second criterion, because the energy system chain between currencies and services will be undisturbed.

---

* Coal proponents will object, saying 'clean coal' (by which they mean coal-fired electricity generation with CO_2 sequestration) must be an option. If there were any realistic probability that *all* the CO_2 from the complete coal-fired cycle could be securely sequestered for thousands of years, I'd say: OK! But there is no such possibility. As I remarked earlier, 'cleaner coal' is possible. But 'clean coal' is an oxymoron.

The *production* of hydrocarbon fuels generates between a tenth and a third of the CO_2 emissions from our fossil-fuel system—depending upon the grade of the fossil reserves and the refining processes used. Carbon dioxide doesn't only come from our cars and trucks, trains and airplanes, but also from the processes used to produce fuels for our cars, trucks and so on.

Yet we're lucky: Nature laid down her fossil reserves so that, by using hydrogen as a tether—as a kind of marriage broker that can unite fossil and non-fossil sources in the production of fuels—we can substantially reduce CO_2 emissions during the refining processes. A simple explanation goes like this:

1. The hydrogen-to-carbon (H/C) ratios of fuels like gasoline and diesel lie between about 2.1/1 and 2.3/1. This means that these fuels contain *more* than two atoms of hydrogen for every atom of carbon.

2. But the H/C ratios in our hydrocarbon resources contain less than two atoms of hydrogen for every carbon atom. In conventional crude, the H/C ratio ranges between 1.8/1 and 1.9/1. For the heavy oils of Athabasca (Canada), Orinoco (Venezuela) and Olenek (Siberia), it's between 1.4/1 and 1.6/1. For most coals (which some believe we'll use in the future to make liquid fuels) it lies between 0.7/1 and 0.9/1.

3. When used as sources for conventional liquid transportation fuels, this means that all fossil reserves have too much carbon, too little hydrogen.*

4. To achieve the correct ratios for fuels, one way or another, all today's refining processes reject carbon, mostly as CO_2.

5. Yet it's possible to change the processes so that, rather than *reject carbon*, we *add hydrogen*. If the added hydrogen is harvested from water using a non-carbon exergy source, we can massively reduce the CO_2 emitted during the harvesting link of our energy system.†

---

* Natural gas has a H/C ratio of 4/1, but it's never used as a sole feedstock for conventional fuels.

† This approach is coherent with the principle of changing the process rather than adding a collector, a concept which, frankly, I enjoyed introducing in Chapter 10, 'From Oil Lamps to Lightbulbs: Pathways to Environmental Gentility' (in this case, carbon sequestration technologies).

By changing the process we will have both decreased the amount of crude oil taken from the ground and decreased the CO_2 emitted as we make the fuels.

To visualize what's happening, here is a simplified explanation: If we want ten liters of gasoline we can take, say, thirteen liters of reserves from the ground and then correct its H/C ratio by pitching out carbon until we have ten liters of gasoline. Alternatively, we can take about nine liters of reserves from the ground and correct its hydrogen-to-carbon ratio by adding hydrogen, and we'll still get our ten liters of gasoline. The second route leaves four more liters of reserves in the ground. And it eliminates the CO_2 emissions these four liters would otherwise have bequeathed to our global commons.

From Chapter 28, 'Fossil Sources: A Lot Will Be Left in the Ground,' we know the world won't run out of fossil resources. But for those of us living in North America, the Far East and Europe we have certainly been mopping up our light (higher H/C ratio) reserves. And in our attempt to claw back to self-sufficiency, we're digging up what remains under *our* ground. And what remains is the heavier (lower H/C ratio) stuff. So as time progresses the average H/C ratio of our own reserves will become lower and lower, requiring that we reject more and more CO_2—or, if we do it right, add more and more non-fossil derived H_2.

Like the non-fossil electricity generation tactic, the tether-hydrogen tactic also meets the second criterion, because the energy system chain between currencies and services is undisturbed. This will reduce the requirements for rapid and massive infrastructure investments, and allow many of us to continue fossil fueling a little longer—although we'll be watching more and more H_2-fueled vehicles overtake us in the passing lane.

We've set out the first principles of how we can use hydrogen as a tether to link exergy sources. Experience tells us that first principles often hide in-practice difficulties. Yet what's surprising about *this* principle is that the implementation can be straightforward, and the results even better than we might expect. That is why I'll now spend a little time reviewing the origins of the approach, which can be more broadly described as integrated energy systems. Then we'll examine how it can be, and is being, used today.

I was introduced to the concept of integrated energy systems by people at the International Institute for Applied Systems Analysis (IIASA) in Vienna, who were collaborating with Wolf Häfele and his colleagues at Kernforschungsanlage (KFA) in Jülich, Germany. I paraphrase their view of integrated energy systems as:

Industrial linkages that use exergy and material inputs from multiple-exergy sources to manufacture multiple-exergy currencies.

In some ways, these principles have been in regular use for many years. It's common in places like Germany's Ruhr Valley and the US Gulf Coast to use the by-product of one industry as the feedstock of another. But the IIASA and KFA teams extended the concepts and objectives well beyond then current industrial practice. They pointed out the many advantages of including inputs from non-fossil sources, such as high-temperature nuclear reactors, and increasing the menu of output currencies to include methanol, hydrogen and electricity.[54] Still the combined forces of comparatively inexpensive energy,* the public's misguided fear of nuclear power and industry's inattention to environmental impact has meant that, over the intervening years, adopting these principles has not moved much beyond the most obvious applications.[53] Moreover, when the theme is followed today, the multiple inputs come only from *within* the fossil fuel industry to the *exclusion* of non-fossil inputs.

Many exergy-producing regions of the world can employ tether-hydrogen. Each has different site-specific advantages and disadvantages. But to illustrate principles, I'll sketch out what is happening, and what could happen, as we harvest North America's largest reserves of oil—the Athabasca oil sands in northern Alberta.

Oil sands are a gucky mixture of bitumen, sand and water. Some say they contain more oil than does Saudi Arabia. In Athabasca today, hydrogen is the

---

* 'Comparatively inexpensive' does not mean tested against some arbitrary price. Rather, it means energy price's effect on the structure of society. Energy prices by the criterion that energy has been repeatedly substituted for capital, labor, and most important, technical and institutional innovation.

tether that joins natural gas and bitumen to produce synthetic crude oil—which, in the trade, is called syncrude. As we remember from earlier chapters, when natural gas supplies the hydrogen by using conventional SMR processes, the basic process emits a lot of CO_2. In addition, when the external heat needed to drive the reformation process comes from burning more natural gas, still more CO_2 emissions float into the air.

We must also remember that the product of the oil sands industry is syncrude. It is not gasoline or diesel. To get these energy currencies, the syncrude must be shipped to conventional refineries that produce the diesel, gasoline, and other petroleum products—and more CO_2.

In 1987 Canada's Advisory Group on Hydrogen Opportunities completed its report *Hydrogen: National Mission for Canada*.[55] To help the advisory group's deliberations on this issue, Dr. E. J. Wiggins of the Alberta Oil Sands Technology and Research Authority did a study showing that, when producing 735TJ/d (about 17,000 tons per day*) of syncrude, the emissions of CO_2 could vary from 44,600 tons per day down to (in principle) zero tons per day—depending upon the process used to achieve the syncrude's required H/C ratio. Today, the numbers might change a bit. But not by much. Counting atoms is a rather straightforward game. And the H/C ratios of coal, bitumen, natural gas and hydrogen do not go wandering about.

To give a few more specifics from Wiggins study:

- When coal is used to achieve the required H/C ratio, the CO_2 emitted is 44,600 tons per day—*more than two and a half times* the 17,000 tons per day of synthetic crude produced.

- When natural gas is used, the CO_2 emitted is 20,900 tons per day—still more than the tonnage of synthetic crude.

- When nuclear generated H_2 and O_2 is used (the oxygen is used to produce more hydrogen by partial oxidation) the CO_2 emitted drops to 6,400 tons per day—little more than a third of the tonnage of the synthetic crude product.

---

* Conventional tons: 2,000 lbm (pounds mass).

Hydrogen: National Mission for Canada also observed that there is also a case when CO_2 emissions can be reduced to essentially zero, which is when all the added hydrogen comes from electrolysis. The pitch, light ends, and off-gasses are converted to liquids and electrolytic hydrogen supplies the fuel-fired heaters. The by-product oxygen is either vented to the atmosphere or used for other industrial processing.

But there's more, which is why I earlier wrote 'the results can be even better than first principles suggest.' Upgrading bitumen to the correct H/C ratio is the second step in manufacturing syncrude. The first step is getting the bitumen, sand, clay and water to the bitumen upgraders. Today that is done mostly by huge draglines, conveyors and trucks. What's more, massive quantities of water are used. If you ever have a chance to see these operations, take it. You'll stand in awe.

But now newer, cleaner techniques are being developed for getting the material to the upgraders. Working with various oil sands industries, Atomic Energy of Canada Ltd has designed processes that use nuclear generated steam from next-generation Advanced CANDU Reactors (ACR) for a process called Steam Assisted Gravity Drainage (SAGD).

While the initial application for nuclear steam is the SAGD process, the quality of the steam produced will open up codicil opportunities. Steam leaving an ACR reactor comes out at pressures of 6MPa (about 870psi). But the pressures needed for SAGD range between 2 to 3MPa (290–435psi). Therefore it will be possible to skim off the high pressure steam and run it through turbines that generate electricity *before* it leaves the turbine at the correct pressure for the SAGD application.

And it still hasn't ended. A major difficulty with expanding the oil sand operations is the extensive and growing demand for water. However, the low exergy-grade (low temperature) tail-end steam from this process is well suited to multiple-step distillation of dirt-laden water. This can clean up the water for recycling. There can be still more benefits, but I think you now have a sense of things.* These processes *can* work.

Alberta with its Athabasca reserves and similar regions around the world are faced with both opportunity and danger. The opportunity is to grow wealthy. But over the next decade or so, it will be dangerous for these regions to enjoy the wealth flowing in from harvesting heavy oils and bitumen—while thinking themselves immunized from the intensifying global pressures to reduce CO_2 emissions as climate disruption moves ever higher on political agendas.

Therefore, it seems to me that an appropriate business interruption insurance would be to put in place technologies, such as using nuclear power for both harvesting and for tether-hydrogen upgrading. Reasonable people will know we cannot wean ourselves from fossil-derived liquid fuels for several more decades—even after climate catastrophes begin to hit us in the face. They will accept that we must use our fossil resources a little while longer. But the public will be infuriated by producers who continue to use today's wasteful, CO_2-belching processes for harvesting heavy oils. Producers that have deployed low carbon-emission integrated processes will be a few steps ahead of the game—and they'll have a bulwark against legislation that might shut them down. They will also have technologies and expertise valuable to those who are harvesting similar low-grade fossil sources, but who lacked earlier foresight and are then threatened with being shut down.

The idea of tether-hydrogen processes is distinguished from 'neat' hydrogen systems. I'm using 'neat' as we might when speaking of whiskey. Neat whiskey is imbibed straight, not diluted with soda or, worse, a soft drink. Neat hydrogen applications will surely shape our culture, because they will involve things like fuelcell cars and trucks, liquid-hydrogen fueled airplanes and the unimaginables we've suggested will come along later. Tether-hydrogen applications are less likely to influence culture—although they will surely help clean the place up.

---

* Most information on these processes has not been released into the public domain, but one Internet link is www.strategywest.com/downloads/chemeng20030814.pdf.

And while tether-hydrogen processes are not glamorous, over the next 30 years they will employ more hydrogen and have a greater impact reducing CO_2 emissions than neat hydrogen systems. Tether hydrogen and neat hydrogen are our two paths to a fully developed Hydrogen Age. The first can be fairly fast and meaty. The second will take longer but, in the end, will be better.

Both are essential.

35. Ok! Now Tell Me About Cost 🌶

In which a journalist's challenge sets us off on the winding path to cost and price—and then back to cost.

By now I should expect it. Still, I'm often caught off-guard.

It happens when a reporter comes to talk about hydrogen. I try to steer our conversation to the 'why hydrogen' rationale, then to the idea that people want energy services not energy, then to how the *system* is evolving. Soon the interviewer's face betrays his coming trump: 'But my readers want to know how much hydrogen will cost!' The craving for a number has pushed aside interest in how things work. Hard has driven out soft.

Nevertheless, cost and price do have a role in systemic evolution. And that's what this chapter is about.* Let's draw our first example from personal transportation, which requires two kinds of purchases—the vehicle and its fuel.

We'll begin with fuel. Regardless of what advertisements claim, we know there is little difference between the gasoline we get from, say, BP or Shell. That's why oil companies try to lure us with scratch-and-win cards, or addict us with PetroPoints. For most of us, it's the price per liter (or gallon) that dominates where we buy our fuel. If one service station's prices are even half-a-cent higher, the station across the street gets the business.

This contrasts with buying a car. If price were the sole criterion, we'd all buy the cheapest jalopy we could find. We don't. Rather, once we've determined our budgetary snack bracket, we select our car using many criteria—starting with whether we want new or pre-owned. Then we consider carrying capacity (for both people and things), and if we want off-road capability, leather seats, a disk player, or electronic maps integrated with GPS. Will the car fit in the garage or will it be parked on the street? What image will it project to friends, business clients and associates? Sometimes we consider practical things—like how far

---

* To some people, economics means cost and price. But economics is a more comprehensive discipline within which cost and price are merely inputs.

we drive per year and, therefore, the importance of the vehicle's fuel efficiency. Will the vehicle also be used as the family sedan, a taxi, a pick-up truck, or all the above? These are some of the criteria we use to judge how well each candidate car (the service technology) will deliver personal transportation (the service)—and whether the service is tangible (delivering kids), or intangible (the joy of driving with the top down).

When we focus on the fuel, the only issue is price. But with the car, the decision is a many-featured thing. This is so obvious it nudges trite. Yet these simple ideas will help us shape expectations for future energy systems.

I'll use 'price' to mean what we pay for something, and 'cost' to mean what it costs the supplier to deliver it. We often like to believe that, in a free enterprise system, the relationship between the two is both straightforward and tightly linked—thinking the system will always introduce another supplier if cost and price stray too far apart. But the relation between production cost and selling price is far from straightforward.

For an example, let's go to the heart of today's energy system, to oil, to compare price and cost. Arabian light sweet crude can be pumped for less than US$2 per barrel—a cost unlikely to change much over the foreseeable future. But the price for this Arabian crude, landed in Europe or North America, typically ranges between US$20 and US$60 per barrel (bbl). To stay in business, North American domestic crude cannot be priced higher than the price of competing Arabian crude.

Herein lies a strategic reality: If North Americans want to further develop the immense Athabasca oil sands of northern Alberta, the plants must be able to produce syncrude at costs that don't exceed the price of landed Arabian crude. Today, syncrude from oil sands can be produced for less than US$30/bbl. But imagine if North America produced so much syncrude that it began backing out significant Arabian imports. It would then be an easy matter for Middle East oil-producing states to lower their price well below $20/bbl, or below $15/bbl—how about $5/bbl? It will always be easy for OPEC (Organization of the Petroleum Exporting Countries) to undercut the price of North American-produced crude oil.

Conclusion: As long as cost alone is the determinant it will never be possible for North America, or Europe, or Japan, to become oil independent. If a country is *determined* to achieve oil independence, it must have the *political* will to set a minimum price, say US$40/bbl, and hold it there. This would protect indigenous-sourced synthetic crude by giving it a firm price floor—eliminating any risk of being undercut by imported crude. Yet this kind of political will has never existed and I see no such will on the horizon. So oil exporting nations can easily adjust their price to prevent any prospect of oil independence in Western nations—thereby keeping the United States and many other industrialized nations sucking oil through a 3.2 kilometer wide teat called the Strait of Hormuz.

Before leaving cost and price, we observe that we have price flares on a frequent but unpredictable basis. Higher oil prices are never caused by higher production costs. Hurricane Katrina, which smashed oil-drilling platforms in the Gulf of Mexico, was blamed for the disruption of the oil supply and therefore caused higher oil prices. Oil prices surely flared. But soon after, both Shell and ExxonMobil recorded their highest profits—ever! During the same time frame, the Canadian oil-producing province of Alberta announced unprecedented profits. So this oil *price* spike cannot have been triggered by the *cost* of production. Rather, it was triggered by producers seeing an opportunity to get even richer.

We began by speaking of cars and gasoline, then of crude oil. Cars are *technologies*; gasoline and crude oil are *commodities*. Companies selling technologies usually have more flexibility to set prices than do companies selling commodities.

The commodity water plumbed into my home is used for drinking, to wash dishes and clothes, to flush toilets, give showers and baths, to water the lawn. If water prices increased so that I had to reduce consumption, lawn watering would be the first to go, drinking last. Water is sold to me at the price of the *least-valued service* for which the supplier can sell it—in my case, lawn watering. It's not sold at a price commensurate with its most essential service, keeping me alive.

The 'marginal utility' price of a commodity is the price associated with its least-valued use. If your success depends on selling a commodity, say natural gas, there are two ways to break out from this marginal utility trap.

The first strategy is to raise your selling price above the least-valued market and move up-market-value to the *next* least-valued. Trouble is, the next up the value hierarchy may be not much further up. And it would also mean selling less natural gas. The second strategy is to move up the energy system chain to services. This requires changing your focus from selling commodities to selling services.

Consider urban public transportation. If you run a natural gas utility you can sell your product directly to a transit authority so they can fuel their buses with natural gas. You'll be selling them a commodity. Or, you might cut a deal that expands your business into full-service public transportation. Now natural gas cost is buried within the overall cost of delivering the service. You'll be selling public transportation, not natural gas, and the price you can charge will be determined by the value (to the public and city governors) of cleaner urban public transit.

How does a meat packer allocate its costs to produce hamburger, sirloin, liver and tongue when they all come from the same cow? How does a refinery allocate the cost of its many petrochemical products when they all come from the same crude oil?

It's a fool's game for an outsider to try to determine the real cost of a single product from among the multiple-commodity outputs of the petrochemical industry. Production costs assigned to different petrochemical commodities reflect the accounting department's twiddling to meet overall corporate objectives more than they reflect the true costs of the input crude oil and manufacturing processes.

So as we continue this search for the cost of hydrogen especially when compared to the cost of gasoline or diesel—all I can do is use broad-brush comparisons of the costs attributable to input feedstocks and processing complexity.

The reporter is becoming impatient. He wants to move on from this tortuous discussion of cost and price. All he wants is a simple answer to a simple question: 'What will be the cost of a liter (or gallon) of hydrogen compared to a liter of gasoline?' Trapped, I answer: 'Less!' And I'm reminded of a television staple, a courtroom lawyer cutting off a witness trying to give a complete answer, because the lawyer demands a simple yes or no. In my case, the journalist was demanding a simple more or less. So my answer is 'less.' But correct as it is, it sticks in my craw.

Today, steam methane reforming (SMR) dominates hydrogen production. Other methods—for example, electrolysis—are only competitive when their product can match SMR prices. Therefore, to compare the cost of producing hydrogen with the cost of producing gasoline, we should consider SMR-derived hydrogen. To develop comparative costs, we'll start from the right-hand end of our five-link systemic chain and move up through the first three links from the energy source to the harvesting technology and on to the currency.

For SMR hydrogen, the exergy feedstock is natural gas and the harvesting technology is steam methane reforming. For gasoline, the exergy feedstock is crude oil and the harvesting technology is oil refining. The comparative costs to produce hydrogen or gasoline are determined by their comparative feedstock prices multiplied (in some way) by the costs of building and operating the relevant harvesting technologies.

Natural gas has traditionally been priced less than crude oil (per unit of exergy delivered). And the SMR process for hydrogen production is simpler and cheaper than the refining process used to make gasoline. Since both the feedstock and processing costs are typically lower for hydrogen production than for gasoline production, hydrogen *costs less to produce* than gasoline. This means cheaper per unit exergy, which is what counts because it's exergy you're buying.

But the reporter isn't thinking this way. He's thinking volume not energy, let alone exergy. He wants to know what a liter (or gallon) of hydrogen will cost because, he says, his readers want things explained in quantities they're used to. And although it's exergy we buy, we can get useful information from volume.

When hydrogen leaves an SMR plant it does so as a gas, not a liquid. And speaking of liters of gaseous hydrogen is especially meaningless. But if a reporter demands an answer, we'll need to first choose both the pressure and temperature to know how much gaseous hydrogen will occupy one liter. No matter what the pressure and temperature, a liter of gaseous hydrogen contains far less exergy than a liter of liquid gasoline. This means the cost of producing a liter of gaseous hydrogen is much, *much* less than the cost of producing a liter of gasoline.

Of course this comparison is silly, and unfair to gasoline. We should have, at least, compared gasoline to liquefied hydrogen.

Hydrogen is liquefied by cooling it to about 20K (-253°C, -458°F). The cryofuel liquid hydrogen (LH_2) contains about one-third the exergy-per-unit volume of gasoline or diesel. Therefore, even though the cost of liquefying hydrogen can increase its cost by as much as 50 to 80%, the cost of a liter of LH_2 will still be less than the cost of a liter of gasoline. So whether talking about gaseous or liquid, the cost of producing a liter of hydrogen will always be less than producing a liter of gasoline.

By the way, although liquid hydrogen contains less exergy per unit volume than gasoline, it has much more exergy per unit-mass. A kilogram of hydrogen contains almost three times as much exergy as a kilogram of fossil-age liquid fuel. This means you can expect a kilogram of hydrogen to cost much more than a kilogram of gasoline—as it should since it's carrying much more exergy. Another interesting tidbit: LH_2 contains even more exergy than energy—which can be a very important attribute, as we'll see in both Chapters 39 'Hydrogen On Board' and 40 'Contrails Against an Azure Sky.'

Few things are as simple as we'd like. But we can surely say that the cost of hydrogen will never be a major hurdle on our path to a brighter Hydrogen Age.

We've now discovered that the cost of hydrogen per unit volume is irrelevant. The cost of hydrogen per unit exergy moves closer to relevance. But what the reporter *should* be asking is: What will be the cost of H_2-fueled *services*—like the cost of driving to the cottage? The answer to this kind of question is set by the performance of the two technology links: Harvesting technologies, and service technologies.

It will also depend on techno-economic synergies and how we account for what economists call 'externalities.' To get to the *true* cost of a service we need to know the cost contribution from each of the two technology links, and then add the cost of externalities:

- *Harvesting Technologies:* When we produce hydrogen and electricity using co-generating technologies, we win both technological and economic synergies. This is true whether the harvesting technology is a conventional generating station, like a hydraulic or nuclear plant, or whether it's an intermittent technology like a windmill or solar panel. As we observed several times, when the generation capacity of a utility is not required for electricity, this capacity can be used for hydrogen production. Or, if a solar panel is used to produce hydrogen during the day, that hydrogen can be fed to a fuelcell to give electricity for late night television. Or stored to fuel the car in the morning.

- *Service Technologies:* Almost all H_2-fueled technologies are more exergy-efficient (exergy out divided by exergy in) than their fossil-fueled counterparts. Fuelcell surface vehicles enjoy efficiencies about two times better than today's spark-ignition engine vehicles. Diesel engines close the gap a little, and hybrids may close it a little more, but a H_2-fueled vehicle still stays on top. When we consider hydrogen-fueled aircraft, the light weight of LH_2-and its clean-burning properties (which reduce maintenance costs) will allow these aircraft to fly more efficiently, carry greater loads and have greater range. All these translate into cost reductions.

- *Externalities:* An external cost is what economists call a cost that must be paid by people 'external' to the transaction. These costs are not included in the price we *directly* pay for a service, but they are real costs that somebody must pay. Often it's all of us. Today, externalities include the economic damage of acid rain and smog. Tomorrow, they will include the impossible-to-overestimate costs of climate catastrophe. Soon, legislation will force some external costs into the price of fossil-derived fuels—but never all. As these externalities are internalized they will further shift cost and price in favor of hydrogen, and against natural gas, oil and especially coal. It's easy to wrap our heads around the concept of externalities. It's also easy to expect vested interests will put up one hell of a fight to block legislation that will internalize these costs.

I'll close this chapter with two stories that illuminate the many aspects of external costs. About two decades ago, one of my students chose a thesis topic to examine the economic damage of poor air quality. Not unexpectedly, the student found that poor air quality increased societal costs, things like lost workdays due to respiratory ailments, increased medical expenses, increased home painting and other maintenance, increased commuter expenses due to accidents and delays in smog-bound traffic.

But one item I hadn't anticipated has stuck in my head. You might find it intriguing too. Particulate matter over cities causes these cities to stay dark longer in the morning and get dark earlier in the evening. Therefore, the lights are kept on longer in the morning and turned on earlier in the evening. The study was done at the University of Toronto, where I found it ironic that the electricity for lighting was then generated (at least in part) by the Lakeview coal-fired station located in the city's western suburbs. Like most North American cities, Toronto enjoys prevailing westerly winds. These winds carried the plume eastbound from Lakeview over the city—shutting out more sunlight, increasing the need for lighting, requiring more electricity from Lakeview, which pushed out yet more smoke. Isn't this a masterpiece of positive feedback aggravating the problem? (Fortunately, Lakeview was taken out of service in the early 2000s—and the stacks demolished June 12, 2006.)

Before abandoning external costs I leave you with the second example. That is the cost of maintaining a military presence in the Middle East to ensure oil keeps flowing from the Gulf of Arabia. American taxpayers pay most of this cost.* It may be one of the more easily determined of oil's external costs—though I expect, these days, it might be difficult to persuade someone to do

---

* With the massive run-up in national debt during the first seven years of the 21st century, it's more likely to be their grandchildren.

the calculations.* And how would they assign a price to the collateral human suffering and lives lost?

Cost and price, *in their fullest sense*—which must include the external costs of climate catastrophe—will always be the issue as we continue our voyage anticipating the future. We'll be looking beneath dollars, euros and yen to see how the energy system works. We'll also be looking outside, to see how these workings strike our moist blue planet.

That's where the real cost lies.

---

* Earlier attempts to determine the excess US military costs to keep this oil flowing can be found in *Fueling Global Warming: Federal Subsidies to Oil in the United States.*[56]

36. WE NEVER KNOW EVERYTHING, BUT. . .

In which we look at three currency waves—slave renewables, slave exhaustibles and liberated renewables—to find we live at the cusp of an unprecedented transition.

About the future, we never know everything. But we always know something. And for some things we can project the deep future better than tomorrow. This is especially relevant to our energy system when that 'something' is energy *currencies*. To remind ourselves where currencies fit, let's take another peek at our five-link energy system chain.

Figure 36.1 The five-link energy system chain highlighting currencies.

Looking at the chain, we remember that energy sources are harvested to make energy currencies, which are then used to energize service technologies. When planning for the future, it's prudent to distinguish among those things which are impossible to project and those that are pretty damn certain.

If we escape climate disruption, a century from now the staple currencies will be the hydricity pair. This knowledge is a robust tether to the deep future. It helps us leap the tangles of today's vested mantras, ill-found hopes and conventional wishdoms. In Chapter 43 'A Journey to the Deep Future,' we'll use this perspective on the importance of currencies to consider deep future infrastructures and sources.

To view the sweep of the past and the future, I find it helpful to group currencies in three categories—*slave-renewable* currencies, *slave-exhaustible* currencies and *liberated-renewable* currencies:

- *Slave Renewables* are currencies indentured to a single *renewable* source. Wind, firewood and dung are examples. We've spoken of how the material of primitive currencies is usually the same as its source. Firewood is the same material as a tree. But whether or not a slave currency is made of the same material as its source, it will always be tied to that source or group of sources. Dung is not the same 'stuff' as livestock, but it can only be harvested from livestock. Dung can't be harvested from hydraulic power.

- *Slave Exhaustibles* are currencies that can be harvested only from a single, *exhaustible* source. Natural gas, gasoline, diesel and coal are examples. If we speak of natural gas or coal out of context, we can't know whether we're referring to a currency or its source. In contrast, the names 'gasoline' and 'diesel' distinguish these currencies from their crude oil source.* Yet whether or not a slave exhaustible carries the same name as its parent, in every case it will always be a prisoner of the fossil fuel family. We can't make gasoline from wind.

- *Liberated Renewables* are currencies that can be harvested from *any* exergy source. And the currencies, themselves, are inherently renewable. They come down to the electricity-hydrogen pair, which I've called hydricity. Today the hydricity currencies are usually manufactured from unsustainable, exhaustible fossil sources like coal and natural gas—although sometimes renewable sources, predominantly hydraulic, contribute. Because any exergy source can be used to manufacture the liberated renewables, this means that all technologies that use hydricity can be powered from any source. Today different sources produce electricity in Bangladesh, Germany, Japan and the Congo. But the TV and light bulbs don't care.† So it will be with hydrogen and hydrogen technologies.

Now we're ready to look at the evolution of market share among slave-renewable currencies, slave-exhaustible currencies and liberated-renewable currencies. Figure 36.2 gives the picture over four centuries. Figure 36.3 expands the time frame to two millennia.

_____

* In extenuating circumstances, gasoline and diesel can be manufactured from coal, as it was in Germany during World War II.

† Providing, of course, the frequency and voltage are compatible—something easily done.

I assembled the data to the year 2000 using well-documented records of energy-source market share, combined with knowledge of how these sources produced either electricity or chemical currencies. Beyond 2000, the curves were developed using characteristic technology substitution times, such as those found by Marchetti and many others.* This means that the data up to the year 2000 can be taken as an accurate representation of history. Beyond 2000 the projections are founded on logic, but not data, and therefore may deviate, somewhat, from our future. It's unlikely, however, that the rate of change to hydricity will be slower than the historical rates of change found by people like Marchetti—especially because the dangers of anthropogenic CO_2 emissions will become a major driver of energy system evolution.

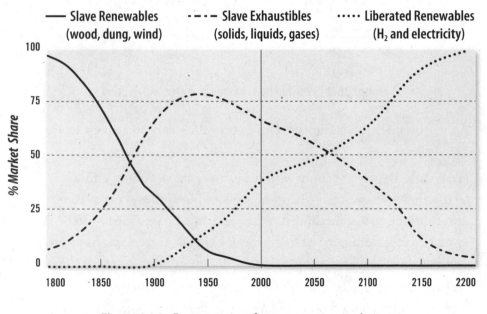

Figure 36.2 Four centuries of energy currency evolution

---

* Cesare Marchetti presented energy system market penetration rates in Chapter 8 of *Energy in a Finite World*.[57] Subsequently, penetration rates have been further studied by Marchetti and others, with no real change in the characteristic timelines set out in *Energy in a Finite World*.

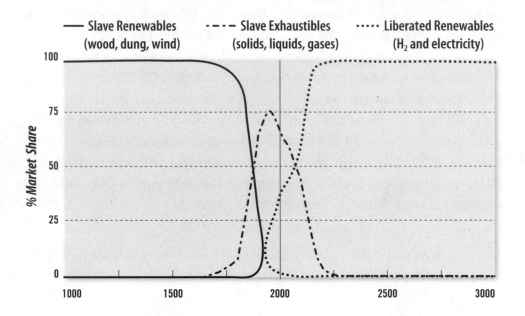

Figure 36.3 Two millennia of energy currency evolution

As you ponder the significance of figures 36.2 and 36.3, keep in mind the following:

- First, these figures are about *currencies*. They are not about either technologies or sources. When I began looking at these curves I'd often slip into thinking that something or other just didn't make sense—to be jarred awake by reminding myself that I was looking at currencies, not sources. Then the curves made sense again. You might experience much the same.

- Second, the 'liberated renewables' (electricity and hydrogen) do *not* include hydrogen used to manufacture conventional liquid currencies, such as gasoline and diesel. These carbon-based currencies remain slave exhaustibles. So the CO_2 emission reductions when using tether hydrogen will be larger than if we only consider the growth of liberated renewables.*

---

* Chapter 34 'Tether Hydrogen: Key to Transition Strategies' explains why.

The further we look ahead, the clearer is our knowledge of energy currencies, but the fuzzier our knowledge of technologies. In the deep future and forever after, hydrogen and electricity will sit plunk in the middle of our five-link chain. Of course, we've said that before. But by classifying currencies as slave renewables, slave exhaustibles or liberated renewables, we have another, valuable way of looking at our future. The more ways we have the better.

The figures indicate a slight fall-off in the hydricity growth rate over the next several decades. My rationale is this: Electricity growth during the 20th century was dominated by heavy-load applications—such as manufacturing, lighting and air-conditioning—but the 21st century's electricity growth is shifting to lower-load, higher-value services, like information technologies. When hydrogen-powered services begin to make substantial contributions, these will be in heavy-load transportation and material-intensive industries. Thus toward the second half of the 21st century, we can expect a rapid expansion of hydrogen services. This will re-energize hydricity growth and return the liberated renewables to their historical growth rates.

The projections of Figures 36.2 and 36.3 were based on a business-as-usual expansion of liberated renewables, which is equivalent to assuming hydricity markets will not be influenced by the climate challenge. Later, in chapter 45 'If Shocked Awake, How Fast Could We Get to a Hydrogen Age?' we'll start from a different premise—to project how fast hydricity can capture markets *if* we truly awake to the dangers of business-as-usual.

Figures 36.2 and 36.3 show we are now a little beyond the midpoint of the slave-exhaustible era, which lies between a forever past of slave renewables and a forever future of liberated renewables. In the fullness of time, the less than two centuries dominated by slave exhaustibles will seem but a blip.

In the sweep of energy system history, we live in interesting times.

Part Ten

INTERMEZZO

37. CHASING LOCOMOTIVES

In which locomotives remind us that technologies compete but fuels seldom do, and that technologies habitually evolve from inside-out to inside-in.

Jammed within a few decades surrounding the mid-20th century, diesel locomotives competed with steam locomotives for the honor of pulling trains. Diesel locomotives won. This was a competition between service technologies—diesel locomotives challenging steam locomotives. It was not a competition between fuels, not a competition between coal and oil. Yet the victor in the locomotive contest determined the victor in the fuel contest.* Coal lost. Oil won.

We like to believe that things like price and availability determine the outcome of economic competition, especially competition among fuels. But neither price nor supply determined whether coal or oil would fuel locomotives. Instead, the coal versus oil competition slipstreamed behind the locomotive competition. The winner in the coal versus oil contest was determined not on a fuels playing field, but rather on a locomotive playing field using locomotive competition rules. Fuels merely watched from the bleachers—rapt aficionados, critically interested in the outcome, but glued to their seats, unable to participate in a game determining their future.

Technologies compete. Fuels watch.

The 20th century also launched the Electricity Age, an age distinguished by the profound cultural impact of electric technologies. Swelling electricity production differentiates the 20th from previous centuries. But electricity

---

* Before steam locomotives were eclipsed by diesel, some steam locomotives were oil-fired for example, locomotives built by Baldwin Locomotive Co. for the Santa Fe Railway System. But later diesel locomotives fully pushed aside steam locomotives, whether fired by wood, coal or oil. And that reinforces the fact that, overwhelmingly, technologies compete rather than fuels. A steam engine is an *external* combustion process, so anything that can be burned externally (that is outside the pistons, not outside the locomotive) can energize a steam engine. A diesel engine is an *internal* combustion process, so the technology is very fussy about the fuel it consumes. Because the fuel fed to diesel engines is so engine-specific, the fuel itself came to be called 'diesel.' Diesel fuel was designed for the technology, not the other way around.

production was merely the response from the bleachers. Down on the playing field, the real action was competition among service technologies.

Beginning in the 1890s, service technologies powered by electricity competed with those that weren't—electric streetcars displaced horse-drawn streetcars, electric light bulbs beat out coal oil lamps. Then, about mid-20th century, the nature of the competition changed. Increasingly *all* the competitors were powered by electricity. Subways challenged streetcars for urban transportation, fluorescent lights fought with incandescent light bulbs for illumination markets and, in the century's final years, email knocked out fax.

Déjà vu is a phrase popularized by Yogi Berra when he spoke his famous redundancy: 'It's *déjà vu* all over again.' But sometimes it's exactly the right phrase. For just as the 20th century launched the Electricity Age, the 21st century will launch the Hydrogen Age. And while details of the hydrogen launch are unpredictable, the patterns of 21st century energy transitions will be classic *déjà vu*.

Repeating patterns of the 20th, during the present century our energy system will be shaped by competitions among service technologies rather than currencies. During the first decades, competition will be among technologies that are *hydrogen-fueled* and those that aren't. As the decades proceed, winners and losers will be selected from among competitors, *all of whom* will use hydrogen spiced, sometimes, with electrical technologies when they enter the fray.

Let's stay with our locomotive example. Before the mid-21st century, probably after a preliminary duel between fuelcell and diesel *buses*, we can expect fuelcell *locomotives* to compete with diesel locomotives for the right to pull North American trains.* Once again this will be a head-to-head competition between technologies, not a competition between oil and hydrogen. Sooner or

---

* I've said North American, although to greater or lesser degrees the competitions will be global. In Europe, with much higher population densities and higher train frequencies, many trunk lines have been economically 'electrified.' In contrast, except for parts of the US eastern seaboard and, for a short time, the Milwaukee Road rail system through the US West, the longer distances and lower train frequencies that characterize most North American railways have precluded the capital cost of direct 'electrification.' Therefore, North America is the most likely venue for early adopter fuelcell locomotives.

later diesel locomotives will lose. Fuelcell locomotives will win. The commuter sheds of great cities will likely host the first skirmishes. Stay tuned.

One day in 1947, Dad started our family's new black Plymouth. Following a brief squabble between my brother and me over who would get to sit up front, we all piled in. Mother joined us too—in the back seat for this adventure (front seat for church). She always enjoyed family adventures, but this time her purpose was to inject a measure of responsibility because we were off to chase locomotives.

At the time we lived in Belleville, a southern Ontario town on the shores of the Bay of Quinte, a meandering, bucolic bay on the north shore of Lake Ontario. The Canadian National Railway line passes through the north side of Belleville, a town about a third of the way from Toronto to Montreal. The Toronto-Montreal run was too long for steam locomotives to complete without taking on coal and water. And in the late 1940s, watering and coaling was done about a mile east of the Belleville station. Locomotives needed water more frequently than coal, so more 'water spouts' were located along the line than 'coal chutes.' At Belleville, the locomotives got both.

Eastbound trains pulled slowly out from the station to halt where the locomotives were 'spotted' under the spouts. Then, coaled and watered, the great locomotives began their run for Montreal. Like monstrous black bulls angrily pawing dirt before charging, these giants from the Coal and Iron Age would, intermittently, lose traction to spin their enormous driving wheels on the track. With each driver revolution, staccato jabs of steam burst from the cylinders* while puffs of smoke shot from the stack to startle the countryside—especially when the frequency jumped as the drivers lost traction to skitter-spin above the track. Gradually the train gathered speed for Montreal. It was grand. And bait for the Plymouth.

---

* In later years, I learned that most of the cylinder's exhaust steam was directed up the stack through a nozzle in the smoke-box, thereby providing forced draft for the firebox.

A road runs east from Belleville along the north side of the railway. Rail grades must be gradual. But in those days, country roads followed the rolling landscape.

The race was on. Yet it wasn't really a race. It was Dad showing his sons the majesty of those proud steam locomotives—and how they worked. We started slowly. It seemed the train would never reach speeds to challenge the Plymouth. Five miles per hour, then ten. The road climbed above the locomotive and we looked down smugly. The engineer waved from his cab. Train and automobile gathered speed in tandem. Now the road dropped beneath the tracks and we looked up to the locomotive's underbelly, the counterweights on the driver wheels turning faster and faster. Now the road was at track level, jets of steam and smoke jabbing at higher and higher frequencies. The Plymouth's speedometer touched 50 miles per hour. . .65. . .the locomotive's flashing silver-painted connecting rods rushed up and down, back and forth, faster and faster, united with the stack's sharp bursts. We were witnessing perhaps the most enthralling sight of the steam era—right at that era's apogee. Finally, Mother gently suggested, 'I think that's fast enough, Gilbert.' Sadly, Dad gave up the chase, as the irrepressible train effortlessly slipped ahead of the Plymouth, gradually at first. . .then faster and faster until it disappeared around a bend, seeking Montreal.

When chasing locomotives, we kids were enveloped within the grandness of this coal-and-iron technology, at its very peak, just before it slipped away into history. We could see, right in front of us, how those lovely locomotives *worked*. We saw the driver rods connecting the drivers to the pistons, the connecting rods joining the drivers to other drivers behind, and the counterweights on all driver wheels counter-balancing the rods so the locomotive wouldn't hippity-hop down the track with each driver revolution. Today I realize the connecting rods were also building links between a dad and his sons.

Wish I could talk to him now.

Today there may be other dads taking their kids to watch locomotives pull out of Belleville. But I doubt it. Not because today's parents don't want to show their kids interesting things, nor because TV traps the youngsters indoors. Rather, I suspect it has something to do with the fact that steam locomotives were designed inside-out and today's diesel locomotives are designed inside-in. You cannot see the important workings of a diesel locomotive as you watch it pull out from a station. Their important bits are hidden beneath their skin. In contrast, the steam locomotive's workings were right before you.

Evolution from inside-out to inside-in is a pattern that seems to characterize how technologies mature. Ships once propelled by oars and sails came to be pushed by steam turbines, then diesel engines and now, sometimes, by nuclear power plants. Merchants who once weighed rice and mangoes with balancing sticks—the Chinese called them *chung*—now use electronic scales. Hydraulic power, originally harvested by splashing buckets hung from water wheels is harvested today by humming turbines buried within concrete generating stations. Yesterday, when farmers wanted to improve their produce they would sometimes graft the branches of one plant onto the trunk of another. Now we implant genes.

Inside-out to inside-in evolution is one aspect of a more general trend for technologies to mature from transparency to opaqueness. Therein lie the seeds of misfortune, because opaqueness often builds mistrust and fertilizes technophobia. The more opaque, the more mistrust. Typically this runs counter to the reality that inside-in designs are usually *better* designs. So our mistrust is often aimed at those technologies that can most improve our lives and the health of our planet.

Steam locomotives were characterized by low efficiencies, poor service and soot-coated downtowns. Having to stop once for coal and several times

for water during a 300-mile run is but one symptom of poor service.* Diesel
locomotives, the first step toward inside-in designs, are more efficient, cause
less environmental damage and provide better service.

In the coming decades, when railroads move from diesel to fuelcell
locomotives, they'll be even more inside-in, more opaque to understanding.
Yet, because they will intrude less on the environment, operate at greater
efficiencies and provide better services, fuelcell locomotives will be *better*.

It's evident we can never claim *all* inside-in technologies are better than the
inside-out technologies they displace. I enjoy recreational sailing, preferring
the inside-out sails of my sailboat, to the inside-in engines of a powerboat. In
part, my preference comes from *seeing*, moment by moment, what's making
Beyond the Stars move, visualizing the wind flowing over the sails providing lift,
watching the behavior of small strings of wool attached to the sails that show
me whether the air flow is laminar or turbulent. Nevertheless, it's sobering to
remember that living things are nearly always designed inside-in. Nature has
her reasons.

Soon today's nascent Hydrogen Age will enter its launch phase. The history
of locomotives has given us some of the patterns that will repeat:

- The future will be determined much more by competition between service
  technologies than between fuels.
- Initially, the competition will be between *hydrogen-fueled* and (what we
  now call) conventional-fueled technologies.
- Later, most competition will be among technologies all of which are
  hydrogen-fueled.
- Fuelcells and electrolyzers—the Hydrogen Age core technologies—
  will more closely mimic Nature than the technologies they replace.

---

* During the mid-20th century, hamlets were often dismissively called 'jerk water' towns. The term
  was born when locomotives needed frequent stops for re-watering. Particularly in the western
  USA, the trains would stop at places (not otherwise meriting a stop) where the fireman would
  climb onto the deck of the locomotive's tender and jerk down a rope, opening the spout from a
  watering tower. I learned the phrase from my mother, who as a little girl grew up in the western
  states of Montana and Washington.

Nevertheless, their inside-in designs may breed a new wave of public mistrust.

Yet we must get started. And in starting, we should anticipate the cultural obstacles we'll find along the way. Many, perhaps most, of these obstacles will be emotional, reinforced by our usual resistance to change.

Today, many argue for a foundation of ethics when we select from among the plethora of technology—allowed choices—from public transit options, to genetic foods, to stem cell research and, of course, nuclear power. I agree. My difficulty is to be confident about what is ethical.

It might be easier to have confidence in what is unethical. I consider it unethical to plead for, or against, one or other technological pathway *without* having paid the entry fee of critical thought. Critical thought does *not* require that we be specialists, just that we make a reasonable effort to inquire, learn and think. And then, sometimes, have the courage to substitute new understanding for old beliefs. Even then, sometimes we might be wrong. But we must make the effort. To me, it's unethical to be intellectually lazy—or an intellectual coward.

Part Eleven

SERVICE TECHNOLOGIES

38. FUELCELLS: CHIP OF THE FUTURE

In which we explore analogies between fuelcells and microelectronic chips, then compare fuelcells with storage batteries and internal combustion engines, and finally place fuelcells within several hundred years of energy system history.

Just as microelectronic chips changed the shape of information systems, fuelcells will change the shape of energy systems. An overstatement? No. But it might be too technocentric because the significance won't be merely technical, it'll be *cultural*. When microelectronic chips pushed aside vacuum tubes, the first result was to improve things we already had, such as radios, TV and big, fat slow computers. Later, chips became the enabling technology for entirely new services—services delivered by mobile phones, heart pacers, camcorders, global-positioning systems, laptop computers and so on. These are technologies that have reshaped our civilization and its culture. Chips are here to stay. Fuelcells will be their partner.

As we entered the 21st century, a gushing media crowned fuelcells 'an exciting new energy source.' That's nonsense. We know fuelcells aren't an energy source. They're an energy conversion technology, usually embedded within a service technology such as a bus or flashlight. Still what is their promise and how do they work?

Way back in the 1840s, barrister Sir William Robert Grove realized that if electrolysis could split water into hydrogen and oxygen, it was likely that the opposite might also work. So he thought that, by using a reciprocal technology, it might be possible to unite hydrogen and oxygen, and make electricity. That technology is now called a fuelcell. I've heard that Grove ran his first experiments in his bathtub, probably to his wife's chagrin. A barrister in a bathtub, building the future long before people dreamed of a Hydrogen Age.

A fuelcell takes a chemical fuel such as hydrogen, and converts a substantial fraction of that fuel's exergy into electricity. Like any exergy conversion technology, there is always some exergy loss during conversion, but in fuelcells the losses are usually much less than in today's chemical fuel-to-electricity technologies (like coal-fired generating stations).

Higher efficiencies are important. Yet beyond efficiency, and because we now stand on the threshold of the Hydrogen Age, it's remarkable that Nature seems to have decreed that the ideal currency for fuelcells is hydrogen.* That is exciting, almost spooky, because once again Nature seems to be looking out for us. What if the ideal currency for fuelcells were coal?

Claiming an analogy between the cultural impact of microelectronics chips and fuelcells is just a start. It's also useful to learn from technological and systemic analogies, which we'll do next. Later in this chapter, we'll compare this new fuelcell technology with the batteries and automobile engines we've known since we were ankle-biters.†

Now we'll consider technological analogies. One of the most important is miniaturization. Vacuum tubes couldn't be miniaturized. But microelectronic chips could, which gave birth to so many of today's electronic products.

Internal combustion engines are also constrained by a lower limit on miniaturization, probably to sizes not much smaller than a model airplane engine. It's not the physical size of moving parts that limits miniaturization. Rather, the fundamental limit on any internal combustion engine is the requirement for large temperature differences *within* the device. The need for high temperature differences is *the* requirement for high efficiencies. So heat engines must be designed with dimensions—and materials—that can separate, sustain and tolerate significant temperature differentials.

In contrast, there is no fundamental lower limit to fuelcell miniaturization. One way to view the prospects for fuelcell miniaturization is to recognize

---

* A few technical reviewers got stuck on this sentence, saying that there are several fuelcell designs that use fuels different than hydrogen—or at least pure hydrogen. They're right. But I hadn't said hydrogen was the only fuel, just that Nature seems to have decreed that it's the best fuel. I stick by my opinion.

† For readers who want to dig further into how fuelcells work, you can turn to the website of the Canadian Hydrogen Association (H_2.ca). The CHA has posted 'Inside Fuelcells,' originally written as a chapter for this book but dropped because it deserved three hot peppers and, it seemed to me, two hot peppers should be the maximum.[58]

that mitochondria, the basic energy conversion organelle of metabolism, are Nature's fuelcells. Typically, the size of a single mitochondrion is one μm* in diameter, and between two and eight μm in length—much the same as a biological virus. Indeed, as fuelcell development proceeds, I expect we'll learn much from how Nature assembled her mitochondria.† This is one more reason I'm persuaded that interdisciplinary teams made up of life scientists, physical scientists and engineers will develop the next wave of energy technologies.

The next analogy is that both fuelcells and microelectronic chips are *component* technologies. They must be embedded within other technologies to become useful. Few of us go out to buy a microelectronic chip. Rather we buy video cameras and laptop computers stuffed full of chips. It's the same for fuelcells. Fuelcell talk was abuzz as we entered the 21st century, but naked fuelcells aren't much use.

Now let's look for analogies in the patterns of how fuelcells will shape services. The impact microelectronic chips had on society came in two phases. First they improved technologies we already had, second they allowed technologies we hadn't yet dreamed of. The adoption of fuelcells will also have two phases. When fuelcells displace internal combustion engines, they'll be used to improve today's transportation technologies, like the power plants in automobiles, locomotives and ships. Fuelcells will also be used to replace batteries in laptop computers, mobile phones, flashlights and camcorders. But later, fuelcells will fit inside a second generation of technologies that will provide services we can't even imagine today. So just as chips before them first improved existing services and then spawned wholly new services, fuelcells will ultimately restructure not

---

* One μm equals 10^{-6} meters, which is one ten-thousandth of a centimeter.

† A fuelcell oxidizes the exergy in a fuel, thereby converting it to exergy in free electrons. Mitochondria oxidize the exergy in food, converting it to electronic exergy that, in turn, is used to drive the synthesis of adenosine triphosphate (ATP) that stores exergy within living systems—like us. Today, an important difference between Nature's mitochondria and most fuelcells is their pH. The ionic conducting media in most fuelcells is a (reasonably strong) acid or base. In mitochondria, the pH is closer to neutral. As *Smelling Land* went to press, some researchers in my Institute for Integrated Energy Systems were hoping to develop fuelcells using enzymes and operating at pH close to neutral.

just the energy system but also civilization—perhaps even more profoundly, because fuelcells will take us a giant step toward cleaning the place up.

One area where the similarities between fuelcells and chips *won't* hold is when we compare market-penetration rates. If we expect that the rate at which chips came to dominate information systems will be repeated as fuelcells come to power transportation systems, we are in for disappointment. Why?

Chip technologies rapidly captured information systems because information technologies could simply be stuck on the end of existing electricity infrastructures. Engineers had to build new things, but these things didn't require new infrastructures or a new currency. Electricity was already here. To change from a vacuum tube radio to a modern radio, we merely unplugged the old and plugged in the new. Later, we unplugged old desktop computers and plugged in new ones—until we unplugged those new desktops and plugged in even newer laptops.

In contrast, fuelcell vehicles will need new refueling infrastructures and a new currency, hydrogen.* In time, the whole energy system chain, from sources through currencies to services, will undergo what will seem an almost natural transformation. Fuelcell vehicles will first be deployed in urban *fleets*—such as buses, taxis or commuter-shed locomotives, which can all be supported by central refueling. Counter to turn-of-the-century hype, I expect fuelcell private automobiles will arrive long after fleets and flashlights (except, perhaps, for those who want to be early adopters). That's because central refueling won't work for private owners who expect the freedom to drive anywhere. Indeed, if we are to believe today's automobile advertisements, the freedom to drive through swamps, splashing mud this way and that, and to the wildest, most inaccessible locations atop craggy mountains.

All the while, fuelcells and microelectronic chips will be teaming up, working together in the Spratt family of hydricity technologies.

---

* During the early stages of the transition to the Hydrogen Age, electrolyzers might be placed on the end of the electricity distribution systems. Ultimately hydrogen networks will replace electricity networks, which we'll discuss in Chapter 42 'A Round Trip to the Deep Future.'

In order to appreciate a new, unfamiliar technology, it's useful to compare the unfamiliar with well-known technologies that have been with us since childhood. So we'll now compare fuelcells with internal combustion engines and storage batteries. We'll start with a compressed summary of the differences:

- Like an internal combustion engine, the currency for a fuelcell is carried in a separate tank. In contrast, a battery's fuel is its guts.

- Like a storage battery, a fuelcell delivers power as DC electricity. In contrast, an engine delivers power along a rotating shaft.

To develop the comparisons further, I'll highlight a single application (powertrains for surface vehicles) and a single fuelcell type (a proton exchange membrane [PEM] fuelcell).* Since fuelcells share many features, if we understand how fuelcells perform in power trains, we'll be a long way toward anticipating how they'll perform in anything—from pushing locomotives to providing electricity and water on Mars.

In Table 38.1, I've set out the characteristics of three different transportation powertrains—those that employ internal combustion engines (both diesel and spark ignition), batteries and fuelcells.

The top row identifies those currencies that feed each powertrain. Fossil-derived liquid fuels are used in internal combustion engines. Electricity is the currency that charges batteries. And hydrogen is the required fuel for PEM fuelcells. I've only identified the inputs that are *paid for* but internal combustion engines and fuelcells also require oxygen which they draw from air†.

Row 2 gives the primary product delivered by the power plant, which will be either electricity or work delivered along a rotating shaft. Both electricity and work have an energy grade of unity, which means both can be *directly* converted into propulsion.

Row 3 identifies the waste products, which can sometimes be by-products. In crewed space vehicles, the output water from fuelcells provides the astronauts'

---

* PEM fuelcells are sometimes referred to as solid polymer electrolyte (SPE) fuelcells.

† In some special cases, like fuelcells in a space vehicle, oxygen is used rather than air. This oxygen must also be paid for.

drinking water, an unlikely use in automobiles but, perchance, a useful wrinkle aboard a fuelcell-powered cruise ship, or sailboat.

Row 4 identifies the role of temperature. As we've noted, internal combustion engines require large internal temperature differences and, therefore, high temperatures. In contrast, most transportation fuelcells operate at much cooler temperatures. The same for batteries.

Row 5 sets out the relation between power and exergy. Batteries are tasked with two functions—storing exergy *and* performing the chemical-to-electrical exergy conversion that delivers power. This is a much greater constraint than many people realize. It means we cannot independently optimize both vehicle range and instantaneous power. Moreover, it contrasts sharply with both fuelcell and internal combustion engine powertrains, where range is determined by the fuel tank's capacity and power is determined by the engine's or fuelcell's rating. Therefore, fuelcell and heat-engine vehicles inherently allow design flexibilities that battery-powered vehicles don't. Power is the top priority for a drag racer, range hardly matters. But range takes priority if you're driving across long stretches of the American West or the Canadian North.

Row	Feature	Characteristics of Powertrains		
		Internal Combustion Engine	Conventional Storage Battery	Fuelcell (using PEM type for illustration)
1	**Inputs (paid for)**	Fossil-derived fuels	DC electricity	Hydrogen
2	**Primary product**	Rotating-shaft work	DC electricity	DC electricity
3	**Waste products** *Sometimes by-products*	Medium-grade heat, CO_2, H_2O, unburned hydrocarbons. Trace: CO, soot.	Very low-grade heat	Low-grade heat, pure water (Sometimes the H_2O is a by-product)
4	*Fundamental* **temperature requirements**	High *internal* temperature differences are *essential* for efficient operation.	Almost no temperature requirements. *Cool operation feasible.*	Some temperature constraint on reaction rates. Cool to warm operation feasible.
5	**Relationship between power (kW) and exergy (kW*h).** *Exergy gives range.* **Power gives thrust**	Power and exergy de-coupled. *(Range is set by the size of fuel tank, power by engine rating. So power and range can be independently optimized.)*	Power and exergy intimately coupled. *(Batteries inherently combine exergy storage and exergy conversion. So power and range can't be independently optimized.)*	Power and exergy decoupled. *(Range is set by size of fuel tank, power by fuelcell rating. So power and range can be independently optimized.)*
6	**'Full' vs. 'empty' performance**	Vehicle performance improves as fuel is consumed because weight decreases.	Vehicle performance is severely degraded when less than 'half full' *(half charge).*	Vehicle performance improves as fuel is consumed because weight decreases.
7	**Effect of repeated refueling cycles**	No effect on performance.	Performance degrades after repeated charge-discharge cycles.	Little effect on performance.
8	**Reactants**	Diesel or gasoline carried on board. Oxygen taken from air.	Battery contains both reactants (e.g., lead and sulfuric acid).	Hydrogen carried on board. Oxygen taken from air.
9	**Additional powertrain requirements**	Fuel tanks. Mechanical transmission.	Electric motor	Fuel tanks. Electric motor.
10	**Power plant change over a 'refueling' cycle**	Almost none, except for 'wear and tear' over *many* duty cycles	Over each duty cycle, battery eats its own guts to deliver exergy.	Almost none, except catalyst degradation over many duty cycles.

Table 38.1 Comparison of powertrains using internal combustion engines, storage batteries and fuelcells.

Row 6 speaks to how vehicle performance changes as the fuel (or charge) drops from full to empty. The performance of fuelcell and internal combustion engine vehicles slightly improves as fuel is consumed, because the vehicle becomes lighter. In aircraft, the difference between 'wet' and 'dry' performance is considerable.* In contrast, the performance of a battery-powered vehicle deteriorates, often precipitously, as the charge approaches empty.

Row 7 is linked to what we observed in row 6, but goes further, because *both* the life and performance of most batteries is severely degraded by repeated deep cycling below 50% charge.

Row 8 identifies the big advantage of air-breathing power plants—the vehicle only needs to carry one of the two reactants on board. The other reactant, oxygen, is taken from the surrounding air. Conversely, in the case of batteries, *both* reactants must be carried within the battery itself. This is a major limitation on batteries, which makes them the heaviest of the exergy conversion devices. It's one of several reasons why battery vehicles will never win as private, free-range automobiles.

Row 9 lists the most important peripheral technologies required for each powertrain. Internal combustion engines deliver their power along rotating shafts. So, in addition to a fuel tank, they need a mechanical transmission to deliver torque to wheels. Both fuelcells and batteries deliver their power as DC electricity and, therefore, require electric motors and controls to convert the electricity to wheel torque, and finally to *thrust* where the rubber meets the road.

Finally, row 10 reminds us that, while neither a fuelcell nor a heat engine undergoes a physical change as it delivers power, batteries eat their innards. This is analogous to a starving animal that gets energy by consuming its own flesh. The animal gets weaker toward its end. So do batteries.

---

* Next time you're on a long, tedious transoceanic flight, you might observe that at some point (usually more than half way through the journey) the aircraft will climb to the next higher altitude available for the flight direction. That's because, as fuel is consumed, the lighter aircraft can climb to higher altitudes where drag is reduced.

Some won't like my observation that electric cars, powered by batteries alone, will never win as private, free-range automobiles. These people will say that I'm not accounting for battery improvements in the future. I'd first respond that battery manufacturing is a multibillion dollar industry that has had more than a hundred years to improve their technologies, and that batteries are now near their development peak. But that is a sideline to the most critical issue.

The issue—or more correctly, the *issues*—are in the fundamentals set out in rows 5, 6, 7 and 8 of Table 38.1. Batteries must carry *both* reactants, so it will be impossible to match the weight reductions easily achieved by either internal combustion engines or fuelcells. In addition, batteries must serve *two* functions (exergy storage and power delivery) so these functions can't be independently optimized. What's more, because battery performance decreases as the charge falls below half-full, we can't use the battery's full capacity. Thus we either need a larger battery, or additional batteries, to extract the same amount of exergy that was stored in the original, fully-charged, battery—which is the amount declared on the label.

While dismissing battery-only powertrains, I don't reject batteries when used as power-components *within* powertrains. Batteries are an essential component in hybrid vehicles, as well as in the so-called 'plug-in' hybrids. The descriptor 'hybrid' means that two or more devices share power delivery—today, these two are internal combustion engines and batteries.

'It was on the Green of Glasgow. I had gone for a walk on a Sabbath afternoon. I was thinking upon the engine.' So spoke James Watt in 1765, reflecting on his idea to condense steam outside, rather than inside, the piston-cylinders that did the steam's work of lifting and pushing*. Steam engines had lived before, but only as bulky, inefficient monsters. Watt's innovation allowed steam engines to launch the Industrial Revolution.

Work is what our energy system really provides. Today it's the work of mining, of flying airplanes, pulling trains and pushing ships—and, via electricity, it's the

---

* To condense steam is to cool it until it becomes a liquid.

work of running computers, TV sets and subway trains. Before steam engines, *every drop* of civilization's work was done by wind, falling water or muscle.

About a century after *condensation* moved from inside to outside steam engines, a new kind of engine moved *combustion* from outside to inside. French engineer Etienne Lenoir built the first practical internal combustion engine in about 1860. Then Nickolaus Otto built a successful four-stroke spark-ignition engine in 1876. Less than ten years later, Gottlieb Daimler and Karl Benz began putting internal combustion engines into automobiles. So about a century after steam engines began pumping water and lifting coal from mines to feed fixed-route trains, internal combustion engines came along to power automobiles.

The invention of the internal combustion engine demanded a new fuel. Solid chunks of coal, then the normal diet of external combustion steam engines, would catch in the craw of internal combustion engines. Internal combustion engines need liquid or gaseous fuels. So the competition between *external* combustion and *internal* combustion engines was launched—and would determine the outcome of the competition between solid and liquid fuels for transportation.

Internal combustion engines also liberalized how and where heat engines could be used. Clunky steam engines could pull trains along fixed routes, push ships across oceans and sit at mine-mouths lifting coal to feed their transportation cousins.* But internal combustion engines widened applications still further, because they allowed *free-range* transportation—first automobiles and trucks, later airplanes.†

Fuelcells are now ready to enter the stage.

---

* The year 1812 saw the first practical locomotive invented by John Blenkinsop hauling coal in Yorkshire, and the first commercial steamship, the *Comet*, operating on the Clyde and built by Scottish engineer, Henry Bell. Curiously, 1812 also saw Baltimore wind-driven clipper ships introduced by US shipbuilders; these wonderful, elegant designs became by far the fastest and (in my view) the most elegant commercial ships plying the oceans. In the language of *Smelling Land* the clippers may be considered 'old technologies fighting back.' They mounted a wonderful fight for almost a century. In the end they lost commercially but never aesthetically.

† Some early automobiles used oil-burning steam engines. Do you remember hearing of the 'Stanley Steamer'?

Fuelcells do not *combust* fuels. Instead the fuel's molecules are split into ions and electrons on catalysts spread over the surface of the fuelcell's anode. In the case of PEM fuelcells, the hydrogen ions then migrate to another part of the fuelcell (the cathode) where they combine with bits of oxygen, while the electrons move off to travel through an external circuit where they power light bulbs, computers or electric motors. Fuelcells produce DC electricity, by-product water, and waste heat. But unlike heat engines (steam engines, internal combustion engines and nuclear power plants, where the heart of the process is to convert heat to work), a fuelcell's heat is only a waste or by-product. Heat is not required to make the process work. Rather, as we've said, fuelcells mimic the mitochondria that produce the work of running, growing and thinking inside us and bullfrogs and caterpillars.

So from the appearance of humans until less than 300 years ago, our ancestors could only use wind, water and muscle to do work.

Then internal-condensation steam engines were conceived. From then on, any fuel that could be burned to supply heat could be used to provide work. Soon after, our ancestors moved from internal condensation to external condensation and then, a little later, from external combustion to internal combustion. Now, with fuelcells, we're about to move from internal combustion to zero combustion. And civilization will have taken another step toward designing its technologies as Nature designs hers.

From the vantage of a hundred years from now, I believe fuelcells will be considered *the* technology that reshaped our energy system and culture perhaps more than any other. Wish I could be here to see it.

39. HYDROGEN ON BOARD

*In which we review the ways hydrogen can be stored on board vehicles—
to ponder what makes sense, what doesn't, what will work for a while
and what for much longer.*

'Whales and Whiskey Barrels' explained why, when hoping to catalyze systemic evolution, removing a barrier is usually better than introducing an attractor.

Today the greatest barrier to H_2-fueled vehicles is the difficulty of on-board storage. Yet compared to efforts to develop attractor technologies like fuelcells, on-board storage technologies have received comparatively little attention. Surprising. Because anyone who breaks the storage barrier will be on a fast track to wealth—and to getting hydrogen vehicles on the road.

It's unlikely that a breakthrough storage technology will come from an entirely new idea—from something now unforeseen. Rather I expect advances will result from doing what we do now, but doing it better, using new tricks that improve old technologies or exploit system synergies.

We all know it isn't practical to use H_2 as a vehicle fuel when the hydrogen is at standard temperatures and pressures (STP)* because the hydrogen's volume would be immense. To appreciate how immense, if a car carried enough hydrogen at STP to give it any significant range, the buoyancy of the fuel would float the whole contraption into the sky—just as it carries weather balloons today.

Any on-board storage strategy has two parts.

First, the hydrogen's thermodynamic state (its pressure and/or temperature) must be changed to a state that can be more easily carried on board. Compressed hydrogen and liquefied hydrogen are the two most promising states for H_2 storage. To make high-pressure gaseous hydrogen, the H_2 must be squeezed (and cooled during the squeezing or the hydrogen will get too hot). To change

---

* Temperatures about 22°C (72°F) and normal atmospheric pressures.

to liquid hydrogen, the H_2 must be deeply refrigerated until it condenses into a liquid. Another change of state is to adsorb hydrogen on metal hydrides.*

The second part of an on-board storage strategy is to choose appropriate tankage. Tanks must be designed to accommodate the thermodynamic state chosen for the H_2, and the vehicle in which the hydrogen will be carried.

The technologies for both high pressure and cryofuel tankage are well understood and continue to improve. The cost of H_2 tanks will be somewhat greater than the cost of today's conventional fuel tanks, but this will not be a constraint on H_2 vehicles. And in spite of popular articles that sometimes tell us the weight of H_2 tanks is a major barrier to using hydrogen, weight is seldom a barrier—and certainly need not be a barrier. The weight of the tanks did not keep Apollo 11 from flying to the Moon—indeed, the low weight of both the tanks and the propellants LH_2 and liquid oxygen allowed the journey.

The major exception to lightweight containment is when the hydrogen is adsorbed on metal hydrides. In this case, the weight of the hydride is much greater than the weight of hydrogen it carries—often twenty to 30 times greater. Although there are no longer credible proposals for metal hydride H_2 storage in transportation vehicles, it could work for things like fuelcell-powered forklift trucks—where the metal hydride's weight would balance the weight of whatever is sitting on the forks.

Let's return to the main players to look at the principles that govern the design of both cryogenic liquid hydrogen and high-pressure tanks.

- *Tanks for liquid hydrogen:* The dominant design principle governing any cryogen tank is to minimize heat leakage *into* the tank. Radiation and

---

* Some folks have proposed carrying hydrogen by binding it chemically to other elements such as methanol (CH_3OH). The reason for promoting methanol is the industry's desire to stay with fuels that have properties similar to those the industry is used to—fuels that are liquids at standard temperatures and pressures. When H_2 is required, say, for a fuelcell power plant, the methanol must be decomposed to release hydrogen. This reformation process requires the addition of water, and releases carbon dioxide. So we must carry more than the methanol which, for the same energy content, is already about six times as heavy as hydrogen. We must also carry the reformation equipment and the extra water. This will *certainly not* be a 'lightweight' way to carry hydrogen on board. How many cars, trucks or locomotives have you seen with refineries on board?

conduction are the two pathways for heat leakage. To minimize radiative heat transfer, it's common practice to use multiple layers of reflective material, each separated by a near vacuum. To minimize conductive heat transfer, the material for structural components is reduced as much as possible—because material provides heat conduction pathways. Cryogenic tanks used for LH_2 will operate at pressures only slightly above ambient, just enough to ensure air cannot leak into the tank which could result in explosive mixtures. At these low pressures, little material is required for structural integrity. All this leads to LH_2 tanks which are inherently lightweight.

- *Tanks for high-pressure hydrogen:* In contrast to LH_2 tanks, high-pressure tanks must be strong—and strength requires material. Advanced materials and construction—such as carbon-filament wound composites—can achieve very high tank pressures with surprisingly low weight tanks. Nevertheless, high-pressure tanks will always be heavier than LH_2 tanks.

While hydrogen tankage isn't a problem, there is a big need to improve the efficiencies of H_2 compression and, especially, of liquefaction. Today the cost of liquefaction can add as much as 50% to the cost of the H_2. It may seem strange, but the promise for improving today's liquefaction efficiencies is large, and exists precisely because today's efficiencies are so poor. Back in Chapter 24 'Exergy Takes Us beyond the Lamppost,' we explained why low exergy efficiencies indicate significant development opportunities. Giant opportunities lie in finding improved liquefaction processes. Magnetocaloric refrigeration may be one approach, although it's still a long way from proven.

I expect compressed and liquefied hydrogen will always be the main players for on-board H_2 storage. Initially, compressed hydrogen will probably win more markets. But in the longer term I predict liquefied hydrogen will dominate. The reason for LH_2's destiny as the king of fuels lies in on-board synergies. To sense the nature of these synergies—albeit not to outline the details—we will now review the advantages of carrying two forms of exergy in a single fuel.

Today's material currencies—such as diesel or steam—carry exergy as either chemical exergy or thermomechanical exergy, but not both. Tomorrow, some currencies will contain both. To appreciate the advantages of carrying both, we must review two ideas drawn from Chapter 23 'It's Exergy!'

Chemical exergy exists when the material of the currency is different from the material of the environment, or, even if it is the same material, if it exists in a different concentration. (Gasoline contains exergy because gasoline is not part of the environment.) Thermomechanical exergy exists when the currency's temperature or pressure is different than the environment. (High-pressure steam contains exergy because both its temperature and its pressure are different than the environmental pressure and temperature.)

Today's conventional fuels, such as gasoline and coal, only contain chemical exergy. In contrast, steam contains only thermomechanical exergy. But liquid hydrogen contains both. About 89% of LH_2's total exergy is chemical because H_2 is not an environmental constituent,* and 11% is thermomechanical exergy because at 20K (-423°F) the cryogen is not in thermal equilibrium with the environment.

Having both chemical and thermomechanical exergy within a single currency offers rich opportunities for technological synergies. Yet most designers have not accepted the offer. The exception are those engineers who, some three decades ago, developed designs for a future generation of LH_2 aircraft. These aircraft will use the dual exergy aspects of LH_2 in ways unique to that application. I expect engineers designing buses and cars will eventually catch up to yesterday's aircraft engineers because, one way or another, the dual exergy advantages of LH_2 await every application.

More than a decade ago, to exploit the dual exergy property of LH_2 my students and I began looking for ways to harvest the thermomechanical exergy of cold in LH_2. We decided one possibility might be to take heat from the environment and deliver it to a heat engine, which would then dump its waste heat into the liquid LH_2. We're not used to thinking about heat engines working this way. We usually think heat engines work by taking heat from a source much hotter than

* You might say that hydrogen is surely part of the environment because all the water in the environment contains lots of hydrogen. But that hydrogen is bound with other elements like oxygen and carbon, it's not free hydrogen.

the environment and dumping their waste heat into the cooler environment. Yet the fundamental principle of heat-engine operation only requires that the heat input be at a higher temperature than the reservoir which accepts the waste heat. With our scheme the heat source is the environment. The heat sink is the cryogen LH_2 fuel leaving the tank. We called these processes Cryogen Exergy Recovery Systems (CERS).

Before it can be fed to a fuelcell, the cryogen LH_2 must be warmed. Something must do the warming, and that something usually requires an energy input. But with CERS, not only do we not need exergy to warm the LH_2, but we get work from the warming process. Seems like magic! But if we routinely think in terms of exergy, it isn't magic at all.

Later we woke up to the obvious: We could draw the input heat from the fuelcell's waste water. This would both increase the temperature differential across the heat engine and, thereby, improve its efficiency. And there was another advantage: The fuelcell's waste water is a liquid, and heat exchange to or from liquids is always more efficient and easier than to or from a gas, like air. We called our refined scheme Enhanced Cryogen Exergy Recovery Systems (ECERS).

We also examined using the exergy of cold to distill pure oxygen from air. The O_2 could then be used, for example, to 'spike' fuelcell output while climbing hills or taking off from a stoplight. There will be many more tricks to exploit the thermomechanical exergy of cold.* They await the attention of bright, creative engineers, many of whom are not yet born—and all of whom will understand the exergy optic.

Compressed hydrogen, of course, is also a dual exergy fuel. The pressure differential between the fuel and its environment gives compressed hydrogen its thermomechanical exergy. So the thermomechanical exergy of compressed H_2 can deliver some benefits, but not as many as liquid hydrogen's thermomechanical exergy. That's because, compared with LH_2, the ratio of

---

* Some of these applications can be found in 'Recovery of Thermomechanical Exergy from Cryofuels.'[59] The theoretic performance of an ECERS heat engine is wonderful. But the requirement for exotic materials dampened our enthusiasm—and reminded me of the wisdom, 'There are no technical problems, just material problems.'

thermomechanical to chemical exergy in compressed hydrogen is lower—and that ratio decreases as the tank empties. In contrast, as an LH_2 tank empties the ratio of thermomechanical to chemical exergy stays constant: As the slogan for Maxwell House coffee once boasted, it will be 'good to the last drop.'

I might not have included these thoughts on ECERS, except for this message: My students and I would never have even begun the hunt for opportunities buried in dual exergy fuels had we not, continuously, been thinking in terms of exergy rather than energy.

The key ideas of this chapter are these:

- While on-board hydrogen storage presents technological and economic challenges, neither is of a magnitude to prevent getting on with Hydrogen Age technologies. We must recognize that hydrogen storage and transport are routine in many applications, and have been for many years. It's just that with hydrogen, storage is a larger component of the cost than when conventional fossil fuels are used.

- Today, most engineers and investors think only of coping with the technical and cost difficulties, rather than looking to the many opportunities that dual exergy fuels open up.

- It's likely that future generations will look back to wonder how we coped with the rather simple-minded storage technologies we use today, none of which exploits the dual exergy synergies that Hydrogen Age folks will consider routine.

The next chapter, which is all about airplanes, illustrates some of the ways we can exploit dual exergy synergies.

40. CONTRAILS AGAINST AN AZURE SKY 🌶

In which the sight of contrails has us thinking about hydrogen-fueled airplanes, a technology that wonderfully demonstrates the multi-layered synergies that will characterize Hydrogen Age systems.

Between 1972 and 1984, Daniel Brewer led a team at Lockheed to develop LH_2-fueled aircraft designs and to compare their performance and operating costs with two other alternative-fueled aircraft. The three competing fuels were synjet (synthetic Jet A made from coal), methane (made from natural gas) and hydrogen (which can be made from anything, but in those times coal was thought the most likely source). As you might expect from the date of the studies, these aircraft designs were motivated by the 'we're running out of oil' mantra—and directed to both civilian and defense applications. Of the three candidates, hydrogen ran away with the prize: *Far better* performance; *much greater* payloads; and sweeteners like lower maintenance costs, less environmental intrusion (both noise and air quality) and greater safety.

In the despondency of those times, people believed oil depletion was imminent—and, worse, that we were without any sensible alternative. So as one consequence, in 1979, an International Symposium on Hydrogen in Air Transportation was held in Stuttgart, Germany. Its purpose was to examine the feasibility of a multinational Liquid-Hydrogen Experimental Aircraft Program (LEAP). It was proposed that a consortium of countries would retrofit a small fleet of Lockheed L1011 jets for hydrogen and fly them between key airports in the United States, Saudi Arabia, England, Germany, Canada and France. Willis Hawkins, then president of Lockheed, and Daniel Brewer had seeded the idea. Carl-Jochen Winter, head of the German Aerospace Establishment in Stuttgart, hosted the meeting.

Unfortunately the LEAP program drifted into dormancy as memories of the oil embargo receded. The Russians and Germans nevertheless continued some development with their cryoplane program, albeit at a modest level. And, for a while, the Americans worked on an LH_2-fueled National Aerospace Plane, and a commercial version which President Reagan called a new 'Orient Express.'

Before the turn of the century however, most work had petered out.* Too bad, because when there has been so much good thinking on a technology that will someday be essential, long periods without thinking can cause an irrecoverable loss of technological memory.

Yet the world was lucky. After his retirement, Dan Brewer wrote *Hydrogen Aircraft Technology*[60] in the mornings and played golf in the afternoons. His book is seminal. I've drawn heavily on it (and clarifying telephone conversations with Dan) for this chapter. While no book can fully capture the thoughts that stimulated the brains of those who someday will be gone, Brewer's book will be a wonderful legacy, a platform for engineers when they once again turn their eyes to designing aircraft that can *really* fly.

For me, perhaps the most memorable impact of the LEAP symposium was a talk by Cesare Marchetti that took me beyond viewing hydrogen as a nice fuel to realizing hydrogen was civilization's destiny.

After returning from Stuttgart, my son Doug and I went for a walk along the northwest shore of Lake Ontario. It was now five years after we'd sailed the Irish Sea to learn the story of the cabin boy smelling land. I hoped to use my walk with Doug to explain how hydrogen could deliver a brighter 21st century and to immunize him against the perception of the day that he'd be growing up in a world that was running out of energy.

We sat on the sun-soaked limestone slabs, gazing southeast over the immense lake where contrails stitched a cloudless sky. We talked of how, within his lifetime, he would look up to see contrails from hydrogen-fueled aircraft. More importantly, we spoke of how these H_2 airplanes would be one of many hydrogen-fueled technologies that would remove people's fear of the future—at least the fear of oil depletion.

Yet by contrails *alone* it will be difficult to distinguish aircraft fueled by hydrogen from those fueled by Jet A. That's because, whether burning hydrogen

* A history of hydrogen aircraft development is well covered in Peter Hoffmann's *Tomorrow's Energy: Hydrogen, Fuel Cells, and the Prospects for a Cleaner Planet.*[61]

or jet fuel, a jetliner fills the hole it drills through the sky with a plume of water vapor—and it's the water vapor that gives birth to visible contrails.

To better understand how the water vapor yields contrails, it helps to remember that this vapor is a *gas* and is, therefore, invisible. A common name for water vapor is steam, which leads to another kitchen experiment. When we watch steam shooting from a vigorously boiling kettle, we're not seeing the steam at all. Rather we're seeing *liquid* water droplets that have condensed out from the steam. That's why the steam 'appears' only after it gets some distance from the spout—the vapor needs time to cool before some of it will condense into visible water droplets. Similarly, the steam in a jetliner's contrail needs a few fuselage lengths before ice crystals form and the contrail appears.

For both hydrogen and Jet A aircraft, the water vapor is cooled by the cold upper atmosphere after leaving the tailpipe, so that by the time it's some distance astern the water will have begun to precipitate out as ice crystals. It's the ice crystals that make the contrails visible. Ice crystals roiling in the aircraft's wake. Ice-crystal brush strokes against azure skies.

The difference between the H_2 and Jet A contrails is the other stuff they contain. In addition to water, a Jet A fueled aircraft leaves behind a *lot* of CO_2 (which is invisible), mixed with a little soot and other unpleasantries (most of which, at that altitude, are also invisible). In contrast, the plume from a hydrogen-fueled aircraft is almost pure water vapor.* This is a profoundly important difference.

Other differences between the aircraft are also important. Take range, for example. Hydrogen-fueled aircraft will be able to fly much farther than today's Jet A-fueled aircraft. There are many reasons for this, but the most straightforward is because the exergy density (exergy per unit mass) of H_2 is almost three times greater than that of Jet A.† Said another way, H_2 has about three times as much

---

* Hydrogen-fueled aircraft plumes can contain other trace constituents. For example, nitrous oxides (NO_x) can be present in the plumes from both hydrogen-and Jet A-fueled engines. Still, in the case of H_2 engines, we can effectively eliminate nitrous oxides by burning slightly lean mixtures—easily accomplished with hydrogen's wide flammability limits.

† Both Jet A and JP-4 are within the kerosene family and have almost identical properties. Jet A is the designation given to commercial aviation kerosene. JP-4 is a wide-cut fuel used by the US Air Force.

oomph as an equal weight of Jet A. So with the same fuel weight, a hydrogen-fueled airplane can fly about three times further than a conventionally fueled aircraft.

For some duty-cycles, increased range isn't important. If we want to fly from New York to Boston, or even to London, there is little need for an aircraft capable of flying, non-stop, around the world. Yet the advantages of hydrogen go much beyond range. They include the ability to design lighter, cleaner, quieter, safer, more efficient airplanes that can carry heavier payloads.

Next time you take a trans-oceanic flight, open up the in-flight magazine and look at the aircraft's specifications.* You might be surprised. The weight of people and cargo is less than the fuel weight at takeoff. During a flight from Frankfurt to Vancouver several years ago, I used data from the in-flight magazine to run a few simple numbers. The great-circle distance was some 4,400 nautical miles, or about 8,150 kilometers. The rate of fuel consumption was a little more than ten kilograms per kilometer—or about 82,580 kilograms for the journey. If every seat were full, the airplane would carry 416 passengers. But my flight was less than half full—let's say it was carrying 208 passengers. I assumed the average passenger weighed about 75 kilos or about 165 pounds. With these numbers and a few tweaks of elementary school arithmetic I found that, on my flight, the fuel used to get the passengers to Vancouver was a little more than five times the weight of the passengers.

Having decided this arithmetic was fun, I went on to think about CO_2. Taking a simplified chemical formula for Jet A to be $CH_{2.2}$, it was easy to calculate the weight of the carbon burned during the flight and, in turn, the weight of the CO_2 combustion products. The total CO_2 was approximately 255,891 kilograms, or about 226 metric tonnes. Dividing this by 208 passengers means that the environmental cost of getting each passenger from Frankfurt to Vancouver was that some 1.23 tonnes of CO_2 dumped into the high atmosphere (where its

---

* These days it seems everything is being dumbed down. As part of this trend, many airlines don't offer this information anymore, concentrating instead on movies, jewelry, perfumes and celebrities. But you can still ask the flight attendant to ask the folks on the flight deck. Pilots are usually pleased to respond to anyone interested enough to ask and, I've found, sometimes invite the curious up to the cockpit, although an invitation to look out the front window is less likely after 9/11.

residence time is about 400 years). Since I weigh more than your average bloke, I suppose my CO_2 guilt quotient might have been as much as 1.4 tonnes. Today, of course, planes are normally much more than half full. On an equivalent flight today, each passenger's guilt quotient might only be about a single tonne of CO_2. Does that make you feel better?

These entertaining calculations based on in-flight magazine data simply remind us that, when an airplane lifts off from Frankfurt on its way to North America, it is mostly Jet A that is being lifted into the sky, not people and their luggage. Can you imagine the economic benefits if the fuel were LH_2 and the saved weight were allocated to payload? More important, can you imagine the environmental benefits?

Let's now learn more about these LH_2 aircraft that will, sooner or later, come up over the time horizon. Among the many hydrogen-fueled service technologies, I'm giving special attention to LH_2 aircraft because they so clearly demonstrate a key attribute of our hydrogen future—technological synergies. But I'll admit: It's also because thinking about them is so much fun.

Some of hydrogen's most interesting benefits and technological synergies are illustrated by the *way* the fuel will be carried aboard. Hydrogen can be stored in many forms, but in aircraft, it will always be carried as LH_2. Don't worry about having to carry heavy refrigeration machinery. Liquefaction will be done on the ground. The fuel tanks will be large thermos bottles designed for very low heat leakage, which eliminates the need for on board refrigeration. This is the same as in today's space programs, where the cryogens—LH_2 and liquefied oxygen (LOX)—are carried aboard but not refrigerated aboard.

I'll start with the unique benefits of LH_2 as an aircraft fuel. In Chapter 39 'Hydrogen on Board,' we saw that, unlike conventional liquid fuels, a cryofuel contains both *chemical* and *thermomechanical* exergy. In the case of liquid hydrogen, about 89% of the total exergy is chemical and 11% is thermomechanical, which in the LH_2 case is exergy of cold rather than of pressure. In high-altitude commercial aircraft applications, the exergy of cold can be used to cool the aircraft's exterior surfaces, suppressing the breakdown of laminar flow over the aerodynamic surfaces, and thereby reducing turbulence and its associated drag. So in addition to the other benefits, the airplane will also be slipperier.

Aerodynamic surface-cooling can only be used at altitudes above 9,000 meters (about 30,000 feet), where the combination of low water vapor and low temperatures eliminates any danger of aircraft icing caused by the cooled aircraft skin. At Mach 0.85—typical of today's long-distance passenger aircraft—the cooling capacity of LH_2 will be sufficient to cool all the wings, the tail, engine nacelles and about 20% of the forward segment of the fuselage. Lockheed's preliminary calculations showed that, for a 400-passenger, 5,500 nautical-mile range aircraft, surface cooling will reduce fuel use by about 30%, and direct operating costs by about 20%.[60] An intermediate recycling refrigerant will transport the exergy of cold from the LH_2 to aerodynamic surfaces.

Time out for perspective: Intuitively, most of us will expect it's the fuel's chemical exergy that pushes the airplane—thinking any benefits from the dual exergy nature of the fuel will be merely minor extras, hardly worth the cost of the additional hardware required to exploit this exergy of cold. We should think again. The exergy of cold accounts for only 11% of the fuel's total exergy, but it's projected to reduce total fuel consumption by about 30%. So it will give almost three times as much bang-per-buck (as we might say in North America), than does the chemical exergy that powers the engines.

As speed increases, first to supersonic (Mach 1.0 to Mach 6.0) and then to hypersonic (above Mach 6.0), hydrogen changes from being just a better fuel to being the *only* fuel. For high-supersonic and low-hypersonic, the exergy of cold is needed to defend structural integrity against frictional heating. We don't know whether hypersonic aircraft will ever become commercially practical. My guess is they won't. But hypersonic military aircraft might.

Because the overall performance of LH_2 aircraft will be much superior to today's aircraft, it may be surprising that the LH_2-fueled engines that push tomorrow's long-haul, wide-body, commercial aircraft through the sky will, except for a few details, be much the same as today's jet engines. I said 'surprising' because we've come to anticipate (correctly) that hydrogen-fueled surface transportation will be dominated by fuelcell power plants, totally unlike the heat engines of today's automobiles, trains, trucks, busses and ships.* So in the future, the hydrogen-fueled ground transportation that carries people to the airport will be powered by fuelcells. But after arriving at the airport, these people will climb on board hydrogen-fueled aircraft that still use heat engines.†

Although these H_2-fueled aircraft will use heat engines for propulsion, they will use fuelcells for their auxiliary power units, which provide electricity for navigation, instrumentation, cabin lighting, movies and warming meals. This will be entirely analogous to the role fuelcells play in today's crewed space flights. Hydrogen and oxygen fuel the propulsion rockets, which are heat engines. But H_2 and O_2 also feed fuelcells that supply both on board electricity and by-product drinking water for the astronauts.

Some special-purpose aircraft might use fuelcell propulsion. One example could be smaller propeller-driven aircraft, where fuelcells would feed electricity to electric motors that turn the propellers. Another application could be for reconnaissance aircraft that require extended time-on-station for surveillance. If civilization still hasn't moved beyond thinking it needs defense technologies, fuelcell propulsion will offer stealth benefits because, unlike jet engines, a fuelcell operates at low temperatures leaving almost no heat signature, which those heat-seeking missiles like to sniff out and chase. Helicopters might also exploit fuelcells.

Let's consider how cryo-hydrogen will influence the performance of the heat engines that power long-range commercial aircraft:

- The thermomechanical exergy of cold will be used for turbine cooling, reducing the need for bypass air and making the engines more efficient. (Unlike cooling aerodynamic surfaces, where an intermediate refrigerant will be used, the LH_2 will be used directly for turbine cooling.)

---

* Most transportation heat engines are internal combustion engines. However, the nuclear power plants that push some of today's aircraft carriers and submarines are also heat engines.

† Ground transportation could, of course, be via electric subways and trains as many airport connector systems are today, especially in Europe. But if they are free-range vehicles like taxis, they will be H_2 fuelcells.

- The low thermal radiation (emissivity) from hydrogen-air flames reduces engine-component temperatures, subjecting the engines to less stressful operating conditions.*

- By burning hydrogen, the engines won't build up carbon deposits.†

- The combination of lower operating stresses and zero carbon buildup means engine life will be increased and maintenance cost decreased, both by about 25%.

Let's now compare commercial aircraft powered by today's conventional Jet A and tomorrow's LH_2:

- Because LH_2 is so much lighter, the gross weight for comparable missions is lower, enabling smaller engines that, in turn, reduce fuel consumption and give quieter operation.

- Although LH_2 is much lighter than Jet A per unit exergy, it is about four times as volumic (when cryotanks are included). Moreover, a basic principle in designing low-heat-leakage LH_2 tanks is to have low surface-to-volume ratios. This is only one reason the fuel will not be carried in the wings (where it is today) but rather in the fuselage. Wings will become more efficient with shorter spans and reduced surface areas. Fuselages will become longer.‡ (At airports, it's the wingspan that pushes loading gates apart; longer fuselage lengths can be more easily accommodated.)

- One pair of LH_2 tanks will probably be located between the cockpit and the passenger section, and a second pair between the passenger section and

---

* The emissivity of H_2-O_2 flames is discussed in chapter 33 'But Doesn't Hydrogen Explode?'

† You may have experienced the annoying phenomenon of 'dieseling' in your spark-ignition car. It's when your engine refuses to stop after you turn off the ignition and continues to burp along with an uncontrollable case of hiccups. Dieseling is caused by carbon buildup that creates hot spots within the cylinder. The cure is to fill up with some higher-octane gasoline and go for a vigorous drive to burn off the carbon deposits.

‡ Or fatter, as in the European cryofuel plane designs, which are modified airbuses with LH_2 cryotanks carried in the top half of a fatter fuselage.

the tail.* In times when we're concerned with hijackers gaining access to the flight deck, having the access between the cabin and flight deck blocked by large, cryogenic fuel tanks will be a security advantage—although, back when the Lockheed designs were developed, preventing hijacking was a low priority.

- Because LH_2 aircraft will be lighter, they will fly higher where the air is thinner and the drag lower. Higher altitudes will take LH_2 airplanes yet another step to still higher efficiencies. During long flights a few years ago, the flight-deck crew used to tell us that for the first hours we'd be flying at about 33,000 feet and then, later, we would climb up to about 36,000 feet. The reason is simple: As the flight proceeds the airplane becomes lighter so it can fly higher and become slipperier. Today, flight-deck crews seldom announce the increase in altitude any more. Probably because little screens silently inform passengers on the altitude and outside temperature, along with a chart of their route.

- Because LH_2 aircraft will be lighter, less fuel will be required for climbing, which is a major component of fuel expenditure.

- Takeoff speed determines the runway length needed. For comparable missions, LH_2 aircraft takeoff weight will be lower, allowing them to fly from shorter runways. Moreover, there will be less wear and tear on landing gear, tires and brakes—major contributors to aircraft maintenance costs.

- On balance, LH_2 is a much safer fuel than Jet A. I said 'on balance' because a fuel's safety is a truly multifaceted property with any one facet usually making the fuel safer in some circumstances and more dangerous in others. In Chapter 33 'But Doesn't Hydrogen Explode?' we discussed the reasons why, by most criteria, hydrogen is safer than today's conventional fuels. But for this chapter, it's worth adding that, in the event of a survivable aircraft accident, LH_2 tanks located in the fuselage will have superior structural

---

* Today's US FAA regulations require a separate tank (usually with cross-feeding) for each engine. I've assumed aircraft employing four engines and therefore requiring four separate tanks—although the drawings, being an elevation view, look as if there might be only a single forward tank and another aft.

integrity and, compared to today's wing tanks, be less likely to be sheared off. This reduces, but doesn't eliminate, the prospect of fuel spillage and fires. Moreover, because the thermal radiation from a hydrogen fire is about one-tenth that from a hydrocarbon fire, passengers who survive the initial accident are less likely to be killed by radiative heat from the fireball.*

- When comparing LH_2 with Jet A aircraft, in both cases the contrails we see are ice crystals. But in comparable travel, LH_2 contrails will contain up to double the water. At these altitudes, water vapor has a long residence time and acts as a greenhouse gas. That said, flying Jet A aircraft with their CO_2 emissions would be much worse. Nevertheless, if we contemplate large fleets of LH_2 aircraft flying at extremely high altitudes, the water vapor issue will need a careful look.

Considering the greenhouse gas properties of water vapor emitted at high altitudes where its residence time is extended, perhaps the only solution is to revert to dirigibles for transcontinental and transoceanic flights. There is no doubt modern dirigibles can be as safe as or safer than today's aircraft. They will fly at much lower altitudes, be much more exergy efficient and, with more room to move about, a whole lot more pleasant for passengers. The trip will be longer. But in such pleasant surroundings that may be a treat, unless you're rushing to close a deal. Then you should stay home and use advanced video conferencing.

We'll now look at two drawings and two tables with data drawn from Brewer's *Hydrogen Aircraft Technology*. Figure 40.1 shows a 400 passenger, 5,500 nautical mile-range, Mach 0.85, commercial LH_2 jetliner. In Table 40.1, that same airplane is compared with an equivalent Jet A aircraft.

---

* An excellent discussion comparing LH_2-fueled aircraft and conventionally fueled aircraft from a safety point of view is given in Chapter 8 of *Hydrogen Aircraft Technology*.[60]

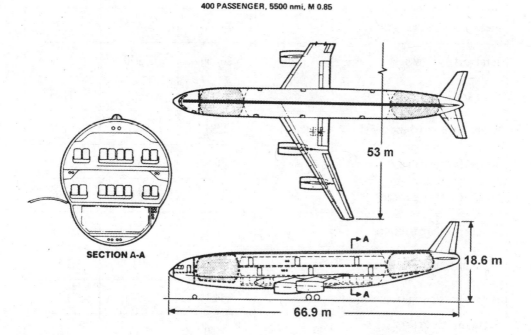

400 PASSENGER, 5500 nmi, M 0.85

SECTION A-A

53 m

18.6 m

66.9 m

Figure 40.1 400-passenger, 5,500 nautical mile-range, Mach 0.85, commercial LH$_2$ Airliner.

	Units	Jet A	LH$_2$	Ratio (LH$_2$/Jet A)
Takeoff gross weight	kg	237,730	177,700	0.749
Operating empty weight	kg	110,000	109,000	0.991
Block fuel mass (weight)	kg	75,070	23,990	0.230
Total fuel mass (weight)	kg	86,550	27,940	0.323
Wing Area	m^2	388.9	312.4	0.803
Ratio: Lift/Drag (cruise speed)	unitless	17.91	16.07	0.897
Weight fraction: fuel	%	36.5	15.2	0.416
Weight fraction: payload	%	16.8	23.6	1.40
Weight fraction: structure	%	26.0	32.4	1.25
Weight fraction: propulsion	%	6.4	9.1	1.42
Weight fraction: misc. equipment	%	14.3	19.7	1.38
Energy usage: kJ per nmi-seat	kJ/ seat-nmi	1460	1307	0.89

Table 40.1 Comparison of Jet A with LH$_2$-fueled, Mach 0.85 Airliners.
Passengers: 400; Range: 10,192 km (5,500 nautical miles).

Because you can read and ponder the numbers yourself, I'll make only a few comments. An empty (dry) LH$_2$ aircraft weighs about the same as a dry Jet A plane, but a fueled LH$_2$ plane weighs about 37% less than a fueled Jet A plane. Because LH$_2$ weighs about a third of Jet A per-unit exergy, we might have expected that the LH$_2$ required for equivalent missions would be about one-third the weight. But, as you can see from Table 40.1, additional synergies reduce the LH$_2$ requirements still further.

Payload is what pays. And Table 40.1 shows that the LH_2 aircraft payload goes up by more than one third. Energy use per passenger-nautical mile is reduced by 12% and the takeoff distance is reduced by about 22%.

Turning to more exotic aircraft, Figure 40.2 shows a 232 passenger, Mach 2.7, supersonic LH_2 design, with a range of 4,200 nautical miles. In Table 40.2 that same airplane is compared with an equivalent Jet A aircraft.

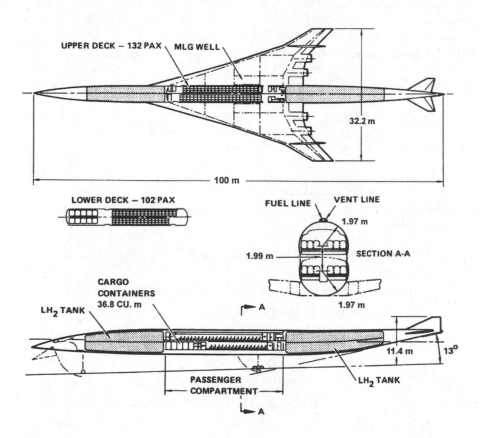

Figure 40.2 A   232-passenger, 4,200-nautical miles, Mach-2.7, supersonic LH_2 Airliner.

	Units	Jet A	LH$_2$	Ratio (LH$_2$/Jet A)
Takeoff gross weight	kg	345,700	179,100	0.518
Operating empty weight	kg	144,000	111,200	0.772
Block fuel mass (weight)	kg	150,000	38,730	0.258
Total fuel mass (weight)	kg	179,500	45,670	0.254
Wing Area	m^2	1,031	739	0.710
Ratio: Lift/Drag (cruise speed)	unitless	8.65	7.42	0.856
Weight fraction: fuel	%	52.0	25.5	0.490
Weight fraction: payload	%	6.4	12.4	1.94
Weight fraction: structure	%	25.9	36.5	1.41
Weight fraction: propulsion	%	9.9	15.1	1.53
Weight fraction: misc. equipment	%	5.8	10.5	1.81
Energy usage: kJ per nmi-seat	kJ/ seat-nmi	6,530	4,730	0.724

Table 40.2 Comparison of LH$_2$-versus Jet A-fueled, Mach-2.7 aircraft.
Passengers: 232; Range: 7,783km (4,200 nautical miles).

The advantages of LH$_2$ become even sharper with Mach-2.7 aircraft. This time the dry LH$_2$ airplane, rather than being about the same weight as its dry Jet A counterpart, is about 30% lighter. But what's truly remarkable, because it's so important, is that the LH$_2$ aircraft takeoff weight is almost half that of a comparable Jet A aircraft, and the payload is almost twice, while the energy use per passenger is reduced by a whopping 38%.

Both designs were developed by Lockheed in 1974 and 1975. The numbers will be even better when we get back to designing and *building* LH$_2$ airplanes, probably around 2030–2040. Then bright, enthusiastic engineers will be pouring over Brewer's book and scouring old Lockheed and European cryoplane files.

Former Lockheed president Willis Hawkins had a wonderful ability to use simple language to say important things. More than two decades ago he told me: 'The more difficult it is to get something to fly, the more it needs hydrogen.'

Pop wisdom likes to tell us that aerodynamic analysis proves bees can't fly. I'm not sure about bees, but I do know that helicopters can give bees a close run for things that shouldn't fly. But they do although suffering high fuel consumption and comparatively short range. Now let's think about how hydrogen can improve helicopter performance.

Consider ship-based helicopters that fly search-and-rescue (SAR) missions. Typically, the fuel for such a mission is apportioned into four phases—getting out, time-on-station, getting back and, lastly, fuel for some margin of error. If a SAR helicopter were powered by an equivalent weight of LH_2 rather than a conventional fuel, we know the exergy content of the fuel would almost triple. So its total available air time would almost triple—without accounting for other synergies that would increase flight duration even further. So for a SAR mission we could more than triple either range or time-on-station—or split the difference.* Thinking back to Willis' wisdom: Flying to the Moon was another difficult task. Jet A could not have got the Apollo mission to the Moon. But LH_2 and its partner, LOX, did.

If LH_2 airplanes are clearly superior, why aren't they filling our skies? It can't be the cost of hydrogen, which we now realize will cost little more than Jet A, and probably less when produced in quantity. The answer lies in the infrastructure barriers, which come from two sides—established conventional-fuel infrastructures, and the absence of LH_2 infrastructures. If you are taken

---

* The cruise ship *Prinsendam* caught on fire well off the north Pacific coast of North America. The passengers (mainly seniors) were able to get into lifeboats and life rafts in the unusually calm seas. The incident was so far offshore that, as the SAR chopper pilots said during a debriefing, 'there wasn't enough fuel left to wet-sponge the tanks by the time we got to shore.' By the good fortune of calm weather and high courage, there was no loss of life.

aback by the first of these barriers, recall that the space program was greatly helped because no infrastructures needed to be razed to fly to the Moon. The path to the Moon was empty, no buildings had to be dismantled, no land had to be appropriated, and no industries or jobs were put in jeopardy.

Established infrastructures are often a much greater barrier to progress than absent infrastructures.

Second, beyond benefits such as efficiency, range and safety, LH_2 airplanes will bring healthier urban and global environments. The global benefit is obvious because CO_2 emissions will be gone. But can you also imagine the cleanliness of airports when LH_2 is the staple aircraft fuel? Next time your cab wends through the ramps of an urban airport—and the tangy bite of partially burned Jet A jabs your nose and coats the roof of your mouth—take solace in the fact that when H_2-fueled aircraft dominate the airports of the world, your olfactory senses will be quiet. If you want to give them a stir, you'll need to walk by the sea, through a pine forest, or into a bakery.

Third, let's look back over the history of human flight. On December 17, 1903, at Kitty Hawk, North Carolina, the Wright brothers were the first to fly a heavier-than-air aircraft—basically bailing wire, bicycle wheels and floppy canvas—for flights measured in seconds and yards.* Sixty-seven years later, Pan Am inaugurated regular trans-Atlantic jumbo-jet flights using Boeing 747s. But since then not much has happened. Avionics have improved and stealth technologies have been developed for military applications. But no major design breakthroughs have occurred in commercial aircraft since 1969.

'Well,' you might ask, 'what about the Concorde?' Although the Concorde was a source of Anglo-French pride, it never became a commercial success, never dented mainstream commercial aviation. Today these marvelous-looking birds are distributed among the world's aviation museums.

As I look back over this history, it seems we had an absolutely extraordinary six decades of aircraft development followed by more than 30 years of almost none. When will the next leap in commercial aviation arrive—a leap like the move from propellers to jets, or from four passengers to 400, or to pressurized cabins?

---

* The flight actually took place not at Kitty Hawk but at nearby Kill Devil Hills.

The answer is: Probably not until LH_2 begins to claim aviation's future. Then efficiencies, payloads and range will take giant leaps toward *better*. Then the technological lessons from the Concorde and space programs will be applied to a new generation of supersonic planes that will be profitable and perhaps able to fly over populated areas rather than just oceans. It will be then that clever designers exploit synergies in ways no one, today, can imagine.

It may take another 30 years until LH_2 jetliners will be stitching the sky. Six decades from Kitty Hawk to jumbos. Followed by six decades of incremental improvements until, before mid-21st century, LH_2-fueled jumbos lift into the sky.

The future brings in the new but also selects from the past. With that thought, I find it fascinating that 21st century LH_2 aircraft will be pushed by steam engines. Steam engines? Yes, because the only product of burning hydrogen is steam. So it will be pure steam that rushes through the turbines of future aircraft. But *these* steam engines will be a whole lot better than those dreamed up by James Watt when, on a Sabbath afternoon, he walked the Green of Glasgow 'thinking upon the engine.'

Part Twelve

INTERMEZZO

41. NINETEENTH CENTURY PRESCIENCE

In which, having thought about a Hydrogen Age more than a century in the future, we marvel at a prescient novelist who wrote more than a century in the past.

I enjoy science fiction, abhor science fantasy. To my mind, Arthur C. Clarke produced some of the very best science fiction. We can learn much about the possibilities the future might hold, and some intriguing physics, from his writings. Many other writers have also left us with wonderful ideas, especially when we consider the times in which they were writing. Take the French author Jules Verne.

Verne published his last book, *Mysterious Island*, in 1874.[62] The story is set during the American Civil War when five northerners escape from a Confederate camp by balloon. They end up, marooned, on a mysterious island some 7,000 miles away. Verne's cast of characters included: Pencroft, a sailor; engineer Cyrus Harding with his servant Neb; and a reporter, Gideon Spillett. Spillett raised concerns for a future when coal runs out. (Jules Verne was writing before oil and natural gas were used as fuels.) Verne has Spillett, the reporter, leading off the discussion with the observation:

'Without coal there would be no machinery, and without machinery there would be no railways, no steamers, no manufactories, nothing of that which is indispensable to modern civilization.'

'But what will they find,' asked Pencroft. 'Can you guess, captain?'

'Nearly, my friend.'

'And what will they burn instead of coal?'

'Water,' replied Harding.

'Water!' cried Pencroft. 'Water as a fuel for steamers and engines! Water to heat water!'

`Yes, but water decomposed into its primitive elements,' replied Cyrus Harding, 'and decomposed doubtless, by electricity, which will then have become a powerful and manageable force, for all great discoveries, by some inexplicable laws, appear to agree and become completed at the same time. Yes, my friends, I believe that water will one day be employed as fuel, that hydrogen and oxygen which constitute it, used singly or together, will furnish an inexhaustible source of heat and light, of an intensity of which coal is not capable. Some day the coal rooms of steamers and the tenders of locomotives will, instead of coal, be stored with these two condensed gases, which will burn in the furnaces with enormous calorific power. There is, therefore, nothing to fear. As long as Earth is inhabited it will supply the wants of its inhabitants, and there will be no want of either light or heat as long as the productions of the vegetable, mineral or animal kingdoms do not fail us. I believe, then, that when the deposits of coal are exhausted, we shall heat and warm ourselves with water. Water will be the coal of the future.'

'I should like to see that,' observed the sailor.

'You were born too soon, Pencroft.'

From the vantage of the early 21st century, Jules Verne's ideas may seem a touch naive. Yet what startles is how much he got right.

Part Thirteen

ON WHERE WE ARE
AND WHERE WE SHOULD GO
FROM HERE

42. ARROWS ON THE WALL

In which memories of arrows on a wall leave me wondering whether we've done enough of the wrong things yet.

Is every prediction equally probable, merely a roll of unloaded dice? Or are some projections more likely to materialize than others?

When we look ahead several decades, population demographics can help us anticipate the number of high schools we'll need, the national cost of pensions and the likely crime rates. Crime rates? Mayors and governors, prime ministers and presidents like to assume (or assign) blame for decreasing (or increasing) crime. But crime rates are often more tied to demographics than to political action. Everyday crime—as distinct from commercial crime—is mostly perpetrated by young men between 18 and 26 years of age. So within decadal time frames, demographics can provide reasonable estimates of crime trends. They can also help us anticipate the national cost of caring for the elderly several decades into the future.

Predictions are more credible when based on quantifiable data that is intelligently analyzed using good logic. Many years in advance of a planned voyage, sailors can use tide tables to predict the best time to leave Buzzards Bay for Cuttyhunk, or Penzance for Le Havre, or Vancouver for Seattle. My electronic charts give tide predictions through to the year 2100. (Unfortunately, before the 22nd century arrives, tide tables will likely be recalculated to account for changed ocean currents and sea levels induced by climate disruption.)

The easiest approach to prediction is simply to extrapolate today's trends into the future—by taking the rate at which life spans are increasing, for example, and assuming this rate will continue. Extrapolation can be useful in the short term, but is frequently misleading for the longer term. Most popular futurists use simple extrapolation, probably because it's easy. I pay attention to extrapolated projections if the time frames are something less than, say, 30 years when based on demographics, eight years when based on political trends, three years for information technologies, and two months for the latest in stuffed animals.

Then there are predictions on whether next year's hemlines will be higher, lower or about the same, and if designers will favor black or chrysanthemum. These predictions are little more than guesses, albeit often self-serving and self-fulfilling marketing tactics. So there are things that we can predict with confidence and things that we cannot, things that matter and things that don't. And we should strive to know the difference.

The need for a Hydrogen Age has nothing to do with prediction. It is simply the consequence of rational analysis of irrefutable facts. Prediction comes along when we ask: When, and by what path, will we get there—and will it be soon enough? Those are the questions.

I first met Ed Schmidt—engineer, and counsel to corporations, think-tanks and NASA—in 1983. We met over breakfast in one of those noisy, Manhattan sidewalk restaurants—this one a short walk from his apartment near the 57th Street bridge. We spoke of many things. Of Ed's remarkable, yet understated, experiences in the US Navy during the War in the Pacific,* of our thoughts on the Hydrogen Age, and of my determination to get on with it. Our conversation had consumed most of the forenoon when Ed suggested we continue our talk later in the day.

So that's how I came to ride a vintage elevator to Ed's apartment. What a pad! Not glamorous, as you might expect in a Manhattan penthouse, but *unexpected*. A faint odor of natural gas swam through the rooms—the result, I suspect, of rather tired burners atop his stove. A pair of vice grips replaced the broken handle on a kettle that was heating water for tea. Charts of Earth and its Moon covered the living room walls. Not that each chart had its own wall space. Each wall had *layers* of charts. Mementos abounded. In the center of the room sat a huge globe—the oceans transparent so viewers could see through to the other side of the world. A photo tacked in one corner showed the Moon's Schmidt

---

* Several years later I asked Ed what action he'd been in. He named several Pacific encounters.
I asked 'Where were you during the Midway battle?' Response: 'I was there.' 'Well,' I asked, knowing Midway was the key naval battle of the war in the Pacific, 'why didn't you tell me you were at Midway?' 'Because, you'd asked what action I was in. At Midway I was aboard a picket destroyer. We saw the action but we weren't in it.'

Crater, named in Ed's honor. Schedules for both the Metropolitan and Vienna State opera companies hung from the kitchen door. To the south, glass doors opened onto a deck with a magnificent view over lower Manhattan, the World Trade Center still standing proud.

Yet I mislead. Not all the living room walls were covered. The east wall had been painted glossy white. On it, dry-erase felt pens had scribbled out equations, diagrams, observations. In the middle of this montage, Ed had erased a space to write, 'Welcome David Scott' with a little maple-leaf superscript. To the left he'd written the word 'today.' Starting from 'today' a crooked, blue line wandered up, down and around, but gradually meandered to the right where it ended with an arrowhead. Ed had labeled this squiggly line 'The real world—the world of mistakes and greed, of ignorance, trials and errors.' Also starting from 'today' he'd drawn a second, beautifully straight arrow, in red, toward the same region on the right. He'd labeled it, 'The world as it should be.' To the right of the arrows he'd written, 'The world as it will be in the fullness of time.'

During the next several years, we met many times. The last was in Sweden when we were both part of an integrated energy system consortium visiting the Siljan region where, millions of years earlier, a large meteor had smashed into Earth. This was where the Swedes were drilling for that abiogenic gas we spoke of in Chapter 25 'A Whiff of Earth from Afar.' Unknown to me or his other friends, Ed, shot through with cancer and surviving on pain pills, had but a few weeks to live. We went for an evening stroll by the Siljan Lake to again speak of many things. Looking west over the lake as the sky reddened I remember thinking aloud, 'I'm coming to realize that some places we visit we may never visit again.' Ed, always a man of few words, answered, 'yes.' I had no idea how my words must have stung. Our conversation returned to the future of energy systems when—echoing what he'd drawn on his white wall almost a decade earlier—Ed turned to me and said, 'David, you must remember that the world always does all the wrong things before the only thing left is the right thing.'

Soon after, Ed was on his cloud. I often wonder if he looks down to watch how things are unfolding. I would like to believe he thinks that, by now, we've just about done our quota of wrong things. But I'm afraid he'd say, 'Seems we need a little longer.'

43. A Journey to the Deep Future

In which we leap into the deep future to explore a fully developed Hydrogen Age, and then return with our findings to identify strategies for today.

Sometimes, for some things, we can predict the deep future better than tomorrow. I can't be sure if I'll catch a cold next month, but I'm certain I won't a century from now. When planning energy system development, looking further out disperses the fog of immediacy, allows us to leap the tangles of today's conventional wisdom and wishdom. So in this chapter, we'll imagine ourselves in a fully developed Hydrogen Age. What we'll find will be an excellent platform for determining what we should do today.

As we've observed, about the future we never know everything but always know something. When applied to energy systems, that something is useful because it includes the shape of infrastructures. Energy system infrastructures typically take more than half a century to build and often much longer to displace. The infrastructures we build today will channel development into the deep future, so understanding where we're going is vital.

For perspective, let's try to imagine living in the early 20th century. In those days people might have anticipated new technologies that would evolve from already existing technologies. It would have been possible to envisage a future of diesel locomotives rather than steam locomotives, or even families having two horseless carriages. Diesel locomotives and automobiles were in the first rank of new 20th century technologies. But it would have been next to impossible to imagine, let alone forecast, the second rank of new 20th century technologies—like global positioning systems and the Internet.

Today, from our vantage at the dawn of the 21st century, we can guess some trends for the first-rank of hydrogen-fueled products, because they will evolve from today's products. This means we can foresee things such as fuelcell-powered mobile phones, laptop computers, flashlights and automobiles. But we can't predict the unimaginables, those second-rank hydricity products that will begin arriving during the mid-to late-21st century. In this chapter we'll think about those things we can predict with considerable certainty. I'll try to list them in sequence from big-picture, big-impact issues to smaller but intriguing bits and pieces.

Today, the infrastructures for continental exergy delivery are high-voltage AC-electricity networks, oil and gas pipelines, and railroads for coal. For oceanic delivery we use ships for both oil and coal and, in some cases, for liquefied natural gas (LNG). So we'll now consider how exergy will be delivered during the Hydrogen Age:

- Pipelines carrying hydrogen will be the staple mode of continental exergy distribution. Shipping gaseous hydrogen by pipeline transmits exergy more efficiently than today's high-voltage AC-electricity, and even more efficiently than high-voltage DC-electricity.*

- Gaseous H_2 pipelines will be wonderful short-term exergy storage reservoirs.

- Large dedicated storage facilities—either underground caverns or above-ground tanks—will accommodate seasonal supply-demand swings and ensure national supply security.

- H_2 pipelines will be immune to severe weather that, today, frequently knocks out electricity supplied by overhead lines. Power outages caused by severe weather will be events of the past.

- Earthquakes that sever pipelines may be the only remaining natural hazard that could bring supply discontinuities.

- Today we need sufficient electricity-generation capacity to supply expected peak demands over extended time-frames. This is capacity far in excess of the average demand. In contrast, when we've reached a fully developed Hydrogen Age, and because H_2 is storable, the hydrogen production capacity needs to be only slightly above the average hydrogen demand. This will result in a significant reduction in the capital cost of generation capacity.

---

* For long distances, DC electricity has fewer losses than AC electricity.

- LH_2 tankers will provide trans-oceanic hydrogen delivery until, perhaps, someone lays down trans-oceanic pipelines (for gaseous hydrogen) in a very deep future. These LH_2 ships will have evolved from today's LNG tankers.

- Underwater, intermediate-length H_2 pipelines will be used in settings like the Mediterranean to carry, say, solar-produced hydrogen from North Africa to southern Europe.

- In urban settings, smaller pipelines will deliver gaseous hydrogen directly to homes, offices and factories—just as natural gas is delivered today.

- Hydrogen will be liquefied at high-capacity plants and distributed, by road or rail tankers, to refueling stations serving LH_2-fueled cars, trucks and so on.

- Long distance LH_2 pipelines are unlikely. First, the capital cost of a cryo-pipeline is much greater than that of a gas pipeline. Second, because LH_2 is an incompressible fluid, a cryo-pipeline does not have the inherent ability to serve as an exergy sponge. Still, there could be niche circumstances when LH_2 cryo-pipelines are used. One might be short-distance runs where an LH_2 cryo-pipeline could contain an embedded coaxial superconducting electrical cable, surrounded by flowing liquid hydrogen.

- If there are significant advances in small-scale LH_2 liquefaction technologies, liquefaction might be done at service stations, or even inside homes.

- The electricity required for electrical appliances, such as computers and microwave ovens, will normally be produced on-site by fuelcells that deliver DC electricity.

- Some of today's electrical appliances—stoves and hot water heaters, for example—may find hydrogen a preferred exergy currency.

- AC electricity will gradually vanish, because its prime value in today's world is for stepping voltages up and down as the electricity moves from generation, to trunk distribution, to domestic loads. Trunk electricity distribution won't be needed because, for longer distances, pipelines for gaseous hydrogen will be *de rigueur*.

- If superconducting materials continue to advance, some electricity could be stored in superconducting capacitors.

- Interplanetary exergy transport won't be needed for several hundred years. But when it is, why wouldn't the delivery currency be electromagnetic radiation? After all, that's what Nature uses.

A geopolitical benefit of the Hydrogen Age flows from its liberation from any particular exergy source. This means that each nation—certainly all larger nations—will have the opportunity to be source independent. Paradoxically, the ability to be independent will reduce the need to be independent. Looking back to late 20th and early 21st centuries, the lack of such independence, particularly in oil, was a cause of fear in many nations, especially the United States. This fear—or strategic need, if you like—was at the root of much international tension.

As noted before, sustainable sources can be grouped as either reliables or whimsicals—the latter, which include wind and Sunlight, are those that deliver exergy when they choose, but not necessarily when we need it. While both reliables and whimsicals will contribute in a fully developed Hydrogen Age, reliables will always have the largest market share. Among reliables, nuclear fission will be pre-eminent during the first half-century of the Hydrogen Age. Thereafter, it gets more difficult to be sure. By the year 2150, controlled nuclear fusion may have become feasible and may have started pushing aside fission. Now some observations:

- *Nuclear fission I:* During the Hydrogen Age, we'll exploit nuclear fission's ability to swing rapidly between electricity and hydrogen production.

- *Nuclear fission II:* Advanced high-temperature fission reactors will be dedicated to hydrogen production. The high-temperature heat from these reactors is particularly valuable for the efficient production of hydrogen using thermochemical processes. But thermochemical processes must run continuously, which is why these reactors will be dedicated to H_2 production.

- *Nuclear fusion:* The extraordinary challenges and great difficulties we've had developing fusion power reactors may prevent their commercialization for more than a century. If fusion ultimately becomes feasible, the output will be dedicated to thermochemical hydrogen production.

- *Hydraulic power I:* Before the mid-21st century, industrialized nations will have developed all their available large-scale hydraulic capacity—and hydraulic power will have increasingly encroached upon other renewable needs, including fish and irrigation. Some hydraulic power installations will have been decommissioned for environmental reasons.

- *Hydraulic power II:* In less-developed nations, new hydraulic capacity will have been introduced in this century. But by the time the Hydrogen Age is at full throttle, hydraulic resources throughout the world will have reached maximum capacity unless (tragically) today's poverty-stricken nations have not moved up the economic ladder.

- *Natural gas I:* The primary component of natural gas, methane (CH_4), will be used to manufacture hydrogen, using advanced steam-methane reformation (SMR) processes. The heat for SMR will be provided by nuclear power, not natural gas. The demand for sustainability will require sequestering the CO_2 waste product. This will need nearby sequestering sites—leak-proof over time frames of several thousand years. Therefore, SMR H_2 is unlikely to make a significant contribution to the H_2 requirements.

- *Natural gas II:* Methane will disappear from electricity generation, because (having been supplanted by H_2 pipelines) the need to feed electricity networks will be reduced, and because it is much more difficult to sequester CO_2 from air-breathing gas turbines than from SMR processes.*

- *Coal and oil:* Both coal and oil will be illegal or irrelevant as hydricity sources, although they will retain high-value roles for such things as medicines, materials and lubricants—or, perhaps as feedstocks for artificial foods. This last might startle. But I expect clever people will learn how to make excellent artificial salmon, beef and tomatoes from oil and coal.

---

* If a turbine is air-breathing (rather than oxygen-breathing) then, *before* the exhaust CO_2 can be sequestered, it must be separated from the large quantities of nitrogen carried by the incoming combustion air.

Moreover, after initial resistance, people will come to appreciate the reliability and uniformity of coal-based tomato soup.

- *Whimsicals:* The intermittent and unpredictable production schedules, that restrict the practicality of whimsical sources today, will be eased by pipelines and stationary storage facilities. Nevertheless, reliables will always provide by far the most of civilization's exergy needs.

We know we can't predict the second rank of H_2-fueled consumer products—the unimaginables. But we can say something about the first rank because they will evolve from today's consumer products. Of these I'll concentrate on transportation vehicles because transportation will always be a major, probably *the* major, exergy consumer:

- *Free-range surface vehicles:* Hydrogen will fuel all free-range private and commercial vehicles.
- *Fixed-route vehicles:* Hydrogen-powered fuelcell locomotives will dominate continental railroads—except, perhaps, in the high population regions of Europe, Japan and the east coast of the United States where established electrification may hold their markets.
- *Fixed-route urban vehicles:* Most fixed-route urban transportation, like subways, will probably use electricity. However, commuter rail will be an attractive setting for H_2-fueled locomotives.
- *Aircraft:* All aircraft will be fueled by LH_2.
- *Ocean commercial shipping:* Fuelcells will be the prevailing power plants for propulsion.
- *Ground-based military vehicles:* Fuelcells will be the power plants for most military surface vehicles.
- *Ocean-based military, surveillance and research vehicles:* Most surface ships will use fuelcell propulsion—although some, like aircraft carriers, may stay with nuclear power plants. All submarines will employ fuelcells, fueled by LH_2 and LOX (liquefied oxygen). For military applications, fuelcells with their low acoustic and thermal signals will offer important stealth advantages.

- *Submersibles:* Fuelcells will power untethered submersibles exploring the ocean floor. But when submersible duty-cycles are both short-range and short-duration, electric storage batteries might win.

- *Forklift trucks:* These vehicles will use H_2 fuelcells. Because, as we said earlier, forklifts need weight to balance the load on their forks, this is one application where hydrogen storage in metal hydrides could win—if costs become acceptable.

- *Mining equipment:* Some types of mining equipment will use hydrogen fuelcells. Others will use electric motors.

- *Skating-rink maintenance:* Cleaning and flooding indoor ice skating rinks will use H_2-fueled fuelcells. These machines will be early adopters because, surely, indoor sports facilities need clean air.

- *Short-range, smaller vehicles:* Small vehicles like golf carts and motorized wheelchairs might use fuelcell propulsion, probably storing their hydrogen in high-pressure tanks.

In this chapter we've enjoyed a kind of fly-by circling a planet Earth that's already enjoying a Hydrogen Age. Now, having had the privilege of glimpsing what could be, we'll glide down to a soft landing on today's world.

44. IF I COULD WAVE A WAND

In which, returning from the deep future, we're motivated to set down strategies for today.

It's fun to contemplate a brighter deep future. But we live today. So let's take stock of our present status to distinguish what's desperately needed from what would be nice—and from what would be foolish. Most of *Smelling Land* has been about patterns and principles, such as how to judge the severity of environmental intrusion by comparing recent levels of the intrudant with historical levels. But eventually we must apply these principles to the question: What should we do today?

So I'll pretend I can wave a magic wand. Since I'm North American, my wand-waving will be shaped more by my North American experiences than if I'd grown up in, say, Germany, China or Bangladesh. Still, I expect that my wand-waving can be adapted to most nations, with a little tweak here and a little reemphasis there.

Let's begin with a cultural issue—but one that shapes our response to technical issues.

I consider our most critical need is a news media with a reasonable measure of science and engineering literacy, and a professional culture imbued with the importance of such literacy.

On balance, our Western media do a good job. Indeed, without them we wouldn't have a Western culture—and we wouldn't have any understanding of other cultures. But there are failures, and most of those come from brazen reporting without adequate knowledge. This happens in all areas. But when the reporting is about technical and scientific issues—and, especially, when explaining how different science and engineering options relate to each other, these limitations have become critical. I suspect the reason some interviewers claim the public won't understand something if it's explained thoroughly, is to justify putting little effort into understanding it thoroughly themselves.

Today, governments are making decisions, or avoiding decisions, critical to long-term planetary health. And, of course, to avoid a decision is to make one. So we must demand that our news-dispensers better understand our choices and give more informed information. We need tough investigative reporters to be as thorough when digging into energy system buffoonery as when digging into political shenanigans. We need a next generation of Bob Woodwards and Carl Bernsteins to write *All Our Energy Nonsense*.

We should at least expect the media to explain what they're talking about. For example, the phrases 'enriched uranium' and 'depleted uranium' populate our news. Yet I expect fewer than one in a thousand people understand what depleted uranium is.* If people don't know what something is, how can they have informed opinions on what to do about it? Depleted uranium can be explained in a fraction of the time now devoted to 'who has it, who doesn't want others to have it, and what we should or shouldn't do with it.'

So remembering we're speaking of things we'd wish for today, here come the first waves of my magic wand:

- Journalism curricula should include courses such as 'engineering and science for journalists.' Exuberant, eclectic-interest engineering and science professors (there are many) will love the challenge. Although we can't expect this science-exposure to give future journalists the answers, we should expect this introduction will help them to ask the right questions— and to have some ability to judge the quality of the responses. Our present engineering school curricula require liberal arts courses. For engineering students, these courses enrich their own lives. For journalism students, engineering-science courses will not just enrich there own lives—they might save civilization's life.

- Consistent with my recommendation for journalism schools, I see no reason why, as a part of their degree requirements, liberal arts students— especially education majors—shouldn't have about an equivalent exposure to engineering and science, as engineering and science students have

---

* If you know, great! If you are unsure, try Appendix B-4.

to liberal arts. At the dawn of the 21st century, when our lives are so dominated by technologies and the great majority of our public decisions are science-and technology-based, to do otherwise means the educational experience can't possibly be considered 'liberal.'

In this wand waving, I placed media and public understanding ahead of legislation. That's because it's the media's reporting that determines the public's understanding and therefore its perception of things. And it's public perception that largely sets what the legislators legislate—which is what we'll consider next:

- Of the legislation that can attack CO_2 emissions, a carbon tax is the most straightforward. It's easy to measure compliance and it's easy for the public understand. Most importantly, it is the most direct shot at the problem. The sole danger in carbon taxes is that the legislation will attract vigorous, well-funded attacks from groups wedded to business as usual—especially those groups without the mental agility to see the new opportunities that carbon-tax legislation will lay out before them.

- Carbon taxes could begin modestly, say by setting an initial target to bring the price of North American fuels at-the-pump up to those found in Europe, Japan and other advanced countries. The tax should then increase on a year-by-year basis. This would provide a robust platform for innovation.

- Any carbon tax must apply at *each step* along the full system chain, from the service to the source. The tax can't be just on emissions from the service technologies, like the transportation sector, that ultimately consume the fuels. This way, the tax on harvesting heavy oils will drive the oil industry toward harvesting and refining technologies that will minimize CO_2 emissions.

- Implementing carbon taxes will be politically difficult. Still, they will provide a level playing field. And it's important to have a level playing field as *the* default legislative test, because special interests will work to twist legislation details this way and that.*

- Closely related to carbon-emission taxes are carbon-reduction incentives. Generally, I like incentives better than taxes, but we should be aware of the potential for abuse. First, carbon-reduction incentives will be more difficult to quantify than carbon-emissions. Second, they must be credited on the real gains delivered rather than on promotional jingles like those that currently populate the nightly news.

- Another closely related scheme is the 'cap-and-trade' proposals. An economist told me: 'With a carbon tax you know the price but not the results. With cap-and-trade you know the result but not the price.' Some policy professionals prefer of cap-and-trade because, they argue, it will give results more quickly. But to me, cap-and-trade will be a sandbox for lawyers and lobbyists, while a carbon tax will be a catalyst for engineers and product developers. I'll go with the latter.†

Leaving carbon taxes and carbon-reduction incentives, the next rank of legislation should be governed by the following principles:

- Governments must see their role to be more one of developing infrastructures rather than as sponsoring individual technologies.‡ (An international program to establish LH$_2$ refueling facilities at the world's major airports is a wonderful example. Participating nations will be positioned to host the

---

* On February 19, 2008, the provincial government of British Columbia in Canada announced the most significant carbon taxes in any North American jurisdiction, with a year-by-year escalating schedule. Newspaper reports were overwhelmingly positive.

† As we've noted elsewhere, the engineering leadership taken by German and Japanese automobile manufacturers—in building more efficient cars than North American companies—can be attributed, in large measure, to their home markets having fuel prices several times greater than the North American fuel prices. This was a price-of-fuel incentive, not an inter-corporation cap-and-trade incentive.

‡ A startlingly effective example from the past was the building of the Erie Barge Canal across upstate New York. It opened the eyes of east coast Americans to the promise of the interior. And it triggered cascading infrastructure development. For example, the need to engineer the canal was the reason Rensselaer Polytechnic Institute was founded, in 1824, as the nation's first technological university.

first generation of LH$_2$ passenger aircraft. Aircraft manufacturers, knowing the timetable for a network of airport LH$_2$ refueling stations, will have the security necessary to invest in LH$_2$ aircraft development.)

- Local and national governments should encourage the installation of central H$_2$-refueling depots for urban fleets such as buses, taxis and couriers.

- Governments should avoid excessive picking of what they consider will be new technology winners. Instead, they should outline what is needed, then ensure level playing fields for innovators and investors without telling them how to innovate or invest.

- Although uncommon these days, prizes—especially large ones—are often better than contracts. Prizes do not trap governments into escalating costs should the technology developers underbid or ill-conceive their proposals. Taking a highly successful example from earlier times, a prize was the bait that led to the discovery of how to determine longitude at sea when, during the frenzied global exploration of the 1700s, nothing was more important to nations flexing their empire-building muscles.*

- Governments should encourage the development of technologies that will fill energy system gaps. Governments should define the *objectives*, not distinct technologies to fill the objectives. Prizes could play a role. These niches must be selected to provide high systemic leverage—using procedures that employ the barrier and attractor optics introduced in Chapter 7 'Whales and Whiskey Barrels.'

- Governments should temper their enthusiasm for taking the advice of stakeholders when formulating policy—especially when identifying which niches most need filling. Often stakeholders are established industries that find it difficult to look beyond their own markets and core-competencies

---

* In 1714, some seven years after Admiral Shovel had driven his ships aground on the Scilly Isles (which the cabin boy is reputed to have been smelling), the British government offered a prize of £20,000 for the discovery of how to fix longitude. Ultimately, by order of King George III, it was awarded to John Harrison after the king became weary of petty jealousies (by members of the Board of Longitude) that frustrated the fair dispensation of the prize money.

to identify the most important niches for advancing the system rather than advancing their own business. Sometimes the stakeholders are researchers, often respected, who are good at their own research but pay little attention to how their research fits a systemic need. So the challenge for governments is to assemble teams of systemically knowledgeable, intellectually honest individuals, and give them an explicit mandate to identify barriers to systemic evolution and to direct funding to reducing those barriers.*

- Whenever there is an increase in fossil fuel prices, American and Canadian governments must resist public and corporate pleading for fuel-price subsidies. The public and automobile industry will be better served if North American fuel prices more closely track those in Europe and Japan. With North American fuel at the pump prices historically less than half those in Europe and Japan, it's no surprise that the Europeans and Japanese developed better engineered, more efficient automobiles while North American companies concentrated on chrome, fins and, later, SUVs.

- Individual legislators should resist promoting projects simply because they will be performed in, or serve, their home constituencies. (This is an idealistic hope, but still very important to hold as a target.)

Because governments today are often involved in major business developments, they should insist that:

- Whenever new power-generation stations are proposed, whether renewable or nuclear, the potential for hydricity co-production should be examined.

- Environmental assessments should be required before new installations can be licensed and built. And, an *additional* assessment should address the flip-side question: What will be the environmental impact if the proposed new power generation is *not* built? This is particularly relevant to nuclear power plant construction if, for example, the alternative is building new, or retaining old, coal-fired stations.

---

* Under contract to the Canadian government, I once prepared a report, *Systemic Considerations in Selecting Hydrogen R&D Priorities*, Catalogue No. M91-7/384-1996E, 1996. The report was well received but, to my knowledge, its counsel has never been followed.

- Whenever new nuclear power stations are proposed, we should look for ways to match the exergy-grade of discharge heat to appropriate industrial uses. District heating is obvious. But it's often possible to be more imaginative—for instance, building attractive and clean industrial parks around generating stations. The best way to design such parks is to match the material and exergy-grade requirements of satellite industries, from plastics molding to greenhouse agriculture—even fish farms. Few renewable sources provide waste heat, but any that do should consider directing this heat to other services.

- Whenever new natural gas pipelines are proposed, techno-economic analyses should determine the marginal cost of assuring that the system can be adapted for hydrogen delivery. Unless the marginal cost is prohibitive, new pipelines should be 'hydrogen-ready.'

- Long haul freight should be shipped by rail rather than road. 'Intermodal transport' would off-and on-load freight to and from road transport used for short-haul delivery or pickup only. Freight transport by road requires up to ten times as much fossil fuel (and emits up to ten times as much CO_2) as shipping the same tonnage of freight by rail.

- As part of their mandate, national security and defense departments should accelerate the deployment of hydrogen-fueled technologies whenever appropriate. A few examples, both large and small, include:

  - *Search and Rescue:* H_2-fueled fuelcell helicopters that will greatly extend the range and/or time-on-station during SAR missions.

  - *Remote sites:* H_2-fuelcells that will replace diesel-powered generators thereby improving power plant efficiency and, most importantly, requiring a much lighter fuel be airlifted to inaccessible locations.

  - *Forest fire fighting:* LH_2-fueled helicopters and water-bombers that will carry larger loads of fire-suppressants.

  - *Military:* H_2-O_2 powered fuelcell submarines and submersibles that will enhance stealth and time-on-station.

  - *Military:* Fuelcell-powered surveillance aircraft that will enhance stealth and time-on-station.

- We must accelerate advanced nuclear power development, including high-temperature reactors and fast-breeder reactors.* This development must not be at the price of either reducing, or deprecating, efforts to improve conventional reactor designs, for these will be the workhorses of hydricity production during the early transition to the Hydrogen Age.

- The United States should reverse today's 'draining-away' of institutional memory in the nuclear power field. It seems a strangely skewed policy that retiring nuclear weapons scientists and engineers undergo extensive debriefing, while no such debriefing is required of retirees from the nuclear power industry.

- Another so far neglected opportunity is to use H_2-fueled locomotives for commuter trains. I'll give two illustrations from my North American experience. One opportunity is Toronto's commuter-rail network where, at one end of the line, North America's only urban-setting nuclear power plant resides. Another is the Chicago and Northwestern commuter rail system that serves Chicago and its surrounding communities.

I also foresee many private sector niche opportunities. Two examples are:

- The immense markets to be won by developing efficient, low operating and capital cost, liquefaction technologies.

- A reciprocal process opportunity exists in the development of technologies that can harvest the exergy-of-the-cold from LH_2 as it leaves the onboard tank on its way to the drive-train. This concept was introduced in Chapter 39 'Hydrogen on Board.'

In 1980 I attended my first World Hydrogen Energy Conference. It was also my first visit to Japan. I set about to absorb, to breathe in as much as I could of Japanese culture. I visited the historic sites in Kyoto and Nara, walked alone

---

* The 'fast' in fast breeder refers to fast neutrons, not fast operation. In today's conventional reactors, the neutrons which all fly 'fast' away from a fissioned uranium-238 atom are slowed, by what's called a moderator. Slower neutrons increase the probability that, upon striking another uranium-235 atom, the collision will result in another fission event. 'Fast' reactors are designed to not require moderators.

through ordinary streets, stopping, listening, watching. I stumbled upon a bookstore where I found an English translation of Yukio Mishima's *The Temple of Dawn*,[62] in which I read:

> He had not abandoned his idea, the one he had stressed long, long ago when talking with the nineteen-year-old Kiyoaki: The will to engage oneself in history is the essence of human purpose.

I was stunned. The privilege! To be alive during these most crucial of times. To have the chance to engage ourselves in history when:

> *The needs are critical*
> *The fundamental ideas simple*
> *The lack of understanding terrifying*
> *The dithering scary*
> *The promise brilliant.*

45. IF SHOCKED AWAKE, HOW FAST COULD WE GET TO THE HYDROGEN AGE?

In which we find we can get to the Hydrogen Age within two decades.

People often ask me how long it will be before they can buy a hydrogen-fueled car.* It's an interesting yet, in a vital way, irrelevant question. The critical question is: how fast can we get to the Hydrogen Age? *That's* the issue. That will determine if we have a chance to be lucky, a chance to escape climate upheaval. Hydrogen cars will simply come along as part of the package.

The answer may be surprising because, if all nations joined to go flat out, we could get to the Hydrogen Age within twenty years. That's how long it would take for the twin hydricity currencies, hydrogen and electricity, to power 70- 90% of the world's energy services.

Using historical precedents, this chapter will set out the rationale that led me to the twenty-year conclusion.

Still, instead of getting on with it, we hear people argue that it would be better to spend money on adaptation strategies rather than on new energy systems that can eliminate CO_2 emissions. They claim that if we too quickly replace our fossil-fueled system with a sustainable non-carbon hydricity system, we'd face economic ruin. But I suspect their real fear is the disruption of their own way of doing business. And they neglect the reality that major technological and infrastructure missions almost always spur economic growth.

That said, I endorse some efforts toward adaptation. Indeed, because the early consequences of climate change have started sooner than anticipated, we will need adaptation strategies. We can build seawalls a little higher (to defend against modest sea-level increases), or change agricultural and reforestation

---

* As pointed out in Chapter 2 'Charting the Course: Toward a Cleaner, Richer Hydrogen Age,' we can buy one today, but we can't take into remote locations and expect to refuel.

practices (to defend against regionally warmer, drier, or wetter weather).* However, adapt-rather-than-change proponents seem to have little concept of the range and magnitude of the changes to which future generations will need to adapt.† *These folks simply don't get it.*

If climate topples over its metastable tipping point, a sudden rise in ocean levels is almost certain to be one of the consequences. It will result from a rapid disintegration of the West Antarctic ice shelf and will lead to a rise in ocean levels of five meters or more. Moreover, because there are numerous positive feedback mechanisms, once this phase of ocean rising starts, I know of no plausible reason why it should stop.

In order of population, the world's ten largest cities are: Tokyo, Mexico City, Mumbai (Bombay), São Paulo, New York City, Shanghai, Lagos, Los Angeles, Calcutta and Buenos Aires.‡ Except for Mexico City, all these cities are on the coast of an ocean. So a sudden rise in ocean levels of a magnitude deemed possible by many scientists would destroy nine of the world's ten largest cities. Moreover, many other ocean-exposed cities, London and Venice for example, although not as massive by population, are surely important to civilization's economic and cultural heritage. With sea level increases of five meters, these cities too will be gone.

We should think not only of cities but also of nations and regions—like the low-lying communities of southern Florida, the Netherlands, Bangladesh and the island states of Oceania. To physically visualize a sudden rise in ocean levels is difficult but possible. But the human consequences are impossible to grasp

---

* A straightforward adaptation to changes in temperature or wetness is easy to understand, but it only skims the problem. In my home province of British Columbia, for instance, warming has reduced winter frosts that kill off pine beetles. The result is that our pine beetle infestation has grown to cover an area some 50,000 square miles—about four times the size of Vancouver Island, or 35 times the size of Long Island. This forest plague has now spread to Alberta and 12 western American states. It's depressing to drive through mile after mile of dead and dying forests which pine beetles have destroyed.

† If you'd like a reminder, return to Chapter 14 'Controversy, Conveyors and Consequences.'

‡ This ranking might change a little depending on the reference used and the criteria for selecting metropolitan boundaries. But the message won't.

and climate shockers even more difficult to envisage could be just round the corner.

I can't scrub from my mind a staple of 20th century B-grade movies—a panicked damsel hanging by her fingers from a crumbling ledge high above rock-strewn, thundering rapids. Will the hero arrive in time to save her? Will the last piece of ledge break, or will her fingers weaken to send her tumbling into the chasm?

Let's leave dangling damsels to the movies and turn to the future of civilization. Our cliff-hanging questions are:

- How fast can we reverse CO_2 growth and begin returning atmospheric CO_2 levels toward pre-Industrial Revolution concentrations?
- Will that be fast enough to pull us back from the brink of climate disaster?

There is no definitive answer to these questions; we simply don't know. But when I compare today's carbon dioxide levels with historical levels, it's difficult to avoid the conclusion: We're probably already toast. Yet history is replete with people fighting seemingly insurmountable odds, and sometimes they win. The premise of this book is that we'll try. The optimism of this book is that our trying will be resolute enough and soon enough.

The Necessary Ingredients for Winning. The first requirement is the determined will of nearly all peoples. But when contemplating an appropriate defense against climate disruption, we must face an unfortunate reality. Communities of people seldom respond to threats when the threat builds gradually, no matter how robust the evidence or how serious the peril. Individuals sometimes plan and respond with foresight, communities seldom do. So as tragic as it is, it will likely require the death or displacement of millions before enough people accept what we must do to save billions.

And there is another sting. If a sudden global catastrophe is required to drive home the need for the Hydrogen Age, will that catastrophe leave standing enough infrastructures to build it? We'll need a shocker to wake us up, but let's hope it's a Goldilocks shocker—big enough to get us off our butts, but not so big as to destroy our ability to accomplish what we got off our butts to do.

Second, we need to find ways to expedite approvals for action, especially in western democracies. While public input is important, we must find ways around ever-prolonged public hearings.

Third, we must select strong leaders with clear vision and understanding, and the authority to drive the mission. War management precedents can be models for how to get things done quickly.

Fourth, we must shed fears that don't stack up when analyzed rationally. One is our fear of nuclear power.

Fifth, every nation must participate. But we can't wait for every nation to agree before we get started. For the first decades of the 21st century, I think American leadership will be essential—I hope in concert with Germany, Canada, France, Russia, Japan, China and many other nations. The United States could employ her economic, industrial and technical strength while building upon her information-age and space-age legacies to guide the way to a brighter future. She could establish energy independence In so doing she would restore her place as a beacon of world leadership.

Fast Track Precedents. The Hydrogen Age will require major technology development. We've done these before. Let's start with a technology precedent.

It's Friday evening, October 4, 1957. I'm with my girlfriend in my mother's '51 maroon Plymouth, listening to the radio, looking out over Lake Ontario. I no longer remember my girlfriend's name, but I surely remember the beep . . .beep. . .beep. . . from the car radio. It was Sputnik. Sputnik telling the world: I'm up here! I'm up here first!

Thus began the space race.

On May 25, 1961, less than four years after those now famous beeps, President John F. Kennedy told the world, 'I believe that this nation should commit itself to achieving the goal, before this decade is out, of landing a man on the moon and returning him safely to the Earth.' On July 21, 1969, eight years after Kennedy's challenge, and almost half a year ahead of the target, Neil Armstrong and Buzz Aldrin stepped onto our Moon.

The United States, with some help from other nations, completed this engineering and logistic marvel in double-quick time. What's more, the achievement bequeathed vital by-products. Fuelcells, hydrogen storage, refueling technologies, indeed, almost all the components of our future Hydrogen Age were all born of the space race. It's an astonishing legacy lying in wait.

So we've built complete and integrated hydrogen system infrastructures over very short timelines and for the challenging environment of space. My point is this: the challenge of designing, building and then deploying these early hydrogen systems, jammed as it was within the twelve years between Sputnik and stepping on the Moon, *exceeded* the technological challenges required to take us from today to the Hydrogen Age. The fundamental principles needed for a terrestrial Hydrogen Age were developed to put men on the Moon. The improvements we now need lie in systemic efficiencies and in cost reductions. That's happening. Advanced designs, materials and modern manufacturing are making these technologies cheaper, more efficient and better.

So recognizing that it took less than ten years to get more than half way through the technological development needed for the Hydrogen Age, there is no reason we'd need more than ten years to complete the job. Then steady improvements on into the future will follow.

Now we'll consider a precedent for fast-track infrastructure development.

On December 7, 1941, Pearl Harbor pulled the United States into World War II. On June 6, 1944, with the support of her allies, the United States played the central role delivering men, materiel and logistics to mount the largest military invasion in history.* It was a mere two-and-a-half years, less one day, between Pearl and Normandy.

Within those two-and-a-half years, car manufacturing factories were converted to build self-propelled gun carriers, jeeps, tanks and on and on. Ship builders expanded their output to produce landing craft, destroyers, submarines, aircraft carriers and, of course, many more merchant ships. All

---

* While the Americans landed on two beaches, Omaha and Utah, the British landed on Gold and Sword, the Canadians on Juno.

this required the redeployment of human resources. Men left their jobs to join the military. Women left their kitchens to become Rosie-the-Riveters.

Still, as gargantuan as was the Normandy invasion, it was a small part of the total war effort by many nations. I chose the 30-month period from Pearl Harbor to the invasion simply to provide a sense of speed—in just one of many theaters of the war.* My point is not to argue which of the world's nations committed more of their human, technical and industrial resources—and certainly not which are to be blamed more, or which suffered more. Rather, my point is that the magnitude of the *worldwide* commitment during World War II to science and engineering, to building infrastructures and to training and deploying humans, was *far* greater than the effort and commitment needed to establish a worldwide Hydrogen Age. Indeed, I doubt a worldwide effort to establish the Hydrogen Age will surpass that of the US war effort during the two-and-a-half years between Pearl and Normandy.

And there is a very important difference. The reason for building the Hydrogen Age will not be to conquer peoples and nations. The purpose will be to give all peoples a chance to escape climate catastrophe which, if not deflected, will likely cause much more death and certainly more wealth destruction than did World War II.

We are repeatedly told by the go-slow folks that economic ruin will result from a major effort to displace today's fossil age with a Hydrogen Age.

But the history that followed the fast infrastructure precedent we just reviewed—at least for countries whose infrastructures weren't severely damaged or destroyed, such as the US, Canada and Australia—demonstrated that the post-World War II era was one of economic boom. And if we turn to our technology development precedent, the space race unleashed technological innovations that, to this day, continue to enrich consumer products and the companies that produce them. These patterns will be replicated in the wake

---

* The building of World War II infrastructures began about September 18, 1931, when Japan invaded Manchuria and continued through the summer of 1945, when Germany and Japan surrendered. This was less than fourteen years. Nevertheless, most World War II production occurred within the decade between 1936 and 1945.

of a massive effort to quickly build the Hydrogen Age. Exuberant economic growth and intriguing new consumer products will result.

If we coalesce the infrastructure development timeline with the shorter, overlapping timeline for technology development, the answer to how fast we could get to the Hydrogen Age is: give us two decades and we'll have it done.

That's the take-away message.

Part Fourteen

CODA

46. OUR SLIVER OF TIME

In which we reflect on the three extraordinary milestones in the maturing of our moist blue planet—the dawn of life, the arrival of oxygen and the first Earthlet spores.

Almost five billion years ago, our planet condensed out from cosmic debris. Then, about a billion years later, life began. This was the first of three extraordinary milestones in the epic journey of our planet's evolution. Early life required neither sunlight nor oxygen, and came to be called anaerobic. Oxygen was a poison to this primordial life that had grown up in a world without an oxidizing atmosphere.

The second milestone came a little more than two billion years ago, when aerobic life—deploying blue-green bacteria as mop-up infantry—achieved victory over anaerobic life. In the winning, oxygen, the victor's excrement, came to dominate atmospheric chemistry. This pushed the atmosphere out of chemical equilibrium with the rest of Earth, chasing anaerobic life off into small ghettos—like our intestines.* Aerobic life had waged the first chemical warfare. Like so many disruptive events, this was a two-edged sword. By changing the atmosphere to an oxidizing environment, aerobic life opened the door for newer kinds of life. These new species took oxygen from the atmosphere, used it to mine exergy from material originally produced by photosynthesis and, thereby, won for themselves the prize of mobility. And so, in time, aerobic life came to allow us.

The third milestone in the maturing of our planet occurred on July 20, 1969, when life flew off Earth to visit another celestial body. This was more than 'One small step for man, one giant leap for mankind,' which I think is to trivialize. Rather this was the beginning of the next epoch for our planet. For it marked

---

* Although constituting only a little more than 20% of the atmosphere, oxygen dominates atmospheric chemistry.

not just mankind's, but life's first baby steps to colonize the universe.* A few Earthlet spores had taken flight.

At the time, the Moon landings engaged the imaginations of all people. Today, that engagement is little more than a warm ember. Nonetheless, flying to the Moon bequeathed wonderful legacies—many still too close in time to fully appreciate. Of these many legacies, two are especially relevant to our odyssey toward a cleaner, richer Hydrogen Age:

The view from space. The photographs of the Earthrise appearing over a lunar landscape bestowed an altered, visceral perspective of our home. Although unable to prevent tragedies born of greed, religion, stupidity and power lust, the view from space did change how we think about our planet. Seen from afar, Earth became something we must protect. Soon after, space programs provided something more—they quantified many of the risks Earth faced. If it were not for research satellites, we would not have learned until much later about ozone holes and other intrusions upon the global commons. And we would never have learned these facts with as much precision.

In the fullness of time, this less insular view of Earth may grow into a near universal *Zeitgeist*, which in the early 21st century might ultimately save us from ourselves.

The energy currencies of space. The staple currencies of space flight are electricity and the protonic pair, hydrogen and oxygen. We were lucky that the missions that first deployed these technologies were driven by a grand objective, not by immediate economic payback. Subsidized by national pride, we learned much about systems needed for the coming Hydrogen Age. We learned about fuelcells, about hydrogen production, handling, safety, liquefaction and about hydrogen-compatible materials.

Today, these are precisely the technologies that must take over civilization's terrestrial energy systems—if we are to defend our planet against the dangers those same space programs identified and quantified.

---

* At least, Earth-based life. None of us can be sure extra-terrestrial life hasn't gone before.

Since the Moon landings, our eyes have turned down from the stars to the mud around our feet. Although the prospect of global wars may have diminished after the fracturing of planned economies in eastern Europe and northern Asia, we now have regional conflicts scattered in pockets around the continents—and growing religious fundamentalism is fertilizing nightmarish killing fields. Above all, we have climate catastrophe roaring toward us with many leaders still in denial and none having a clear idea what to do about it.

Still, within the distemper of our times, good things are happening. After the first Moon emissaries showed space travel possible, we're now in a period resembling those when Europeans were waiting to settle the Americas after Columbus and others had shown they were there. Today, imaginative and courageous investors and innovators are improving efficiencies and lowering costs of technologies originally deployed in space. In so doing, they are preparing these technologies to introduce new energy systems on Earth during the next 50-100 years. Not all the bits and pieces are complete, but they will come. And when they do, a wave of new energy systems will begin to roll, pushing back environmental stress, improving public health, seeding economic revitalization, enriching the joys of being alive.

Let's look further into the future. A codicil to the terrestrial spread of hydricity systems will be that these technologies will be available to the next grand epoch of space exploration. It would be impossible to mount a substantial move out into space today using today's expensive hydrogen-electricity systems. So the sequenced steps along the path to the future seem evident. We must first develop hydrogen technologies for deployment on Earth. Then, having lowered costs and raised efficiencies, we will have technological readiness for the colonization of space.

Curiously, this sequenced perspective parallels the systemic evolution of our energy system as outlined in earlier parts of this book. We have merely changed the time frame from decades to centuries, thereby placing our evolving energy system in the context of planetary destiny. That is why I've come to see our 'sliver of time' in three phases:

1. The second half of the 20th century: A time glorified by a cost-be-damned demonstration that we can send emissaries from Earth's biosphere to another celestial body. The energy currencies for these first missions were electricity and the hydrogen-oxygen pair.

2. The first half of the 21st century: A time when performance improvements and cost reductions will allow these hydricity technologies to transform down-to-Earth applications. They will power a wave of wealth-creating industries that will bring improved quality of life and environmental security on Earth.

3. The second half of the 21st century and the early 22nd century: A time when people again turn their eyes to the stars. Using hydrogen technologies, Earthlet spores will set off to fulfill our planet's destiny. Earth will be going to seed.

While reflecting on the three milestones of our planet's first five billion years, let's try to imagine what milestones might punctuate Earth's remaining years. To do that, we must first assume we'll be sufficiently intelligent and non-parochial to deflect climate catastrophe. If we are, as time unfolds more and more people will live outside our moist blue planet.

In prehistoric times, when peoples living on different continents were truly separated by oceans—before ships and airplanes bound us all together with ever-tightening hoops of people and microbes—isolation was a defense against catastrophes that could otherwise have brought about global annihilation. In contrast, the extraordinary connectedness of present-day life has brought flourishing culture and wealth, but these same bonds have greatly increased the dangers of annihilation. The tighter the links, the greater and more brittle our global fragility. We live on a tight little island in space.

If everyone remains confined to Earth then, sooner or later, civilization will do some damn foolish thing to destroy itself. As an optimist, I trust we have favorable odds over the next hundred years. I hope the odds remain with us for the next thousand. Even then, a thousand years is just one-thousandth of the next million. But when the population living outside Earth is large enough to be self-sustaining, the odds on the survival of our species—and other Earth-conceived life—will be greatly improved. Earth, life's womb, will be fiercely protected as the museum of life's beginnings. Earth-conceived life might even be able, one day, to live outside our solar system—and therefore could outlive it. Wouldn't that be something?

Five billion years ago our corner of the universe was taking shape from stardust—the debris of supernova. Our local patch of star detritus was spinning, condensing, becoming our solar system. Most debris was gathering at the center of things, shaping up to be the Sun. However, a bit out from the main event, Earth was also condensing and twirling, much faster than today—a frenzied planetary ballerina. All the while gravity was collapsing the hydrogen-dominated nascent Sun, spinning it faster, increasing its temperature until, flash!—its hydrogen began fusing to helium, igniting a furnace destined to burn for some ten billion years.

Helium, the ash of fusing hydrogen, began filling the Sun's atmosphere. Helium is more opaque to radiation than is hydrogen, so it acts as a solar greenhouse gas that increases the Sun's temperature. This is why the Sun is now about 30% hotter than it was when terrestrial life was conceived. In about five billion years, our fevered Sun will be in its death throes, and it will take Earth along with it in a cosmological suttee. Our favorite star will have swelled into a glowing red giant, boiling off Earth's oceans, crisping Earth's life and then, in the final scene, swallowing Earth whole—after which, bloated, growing sleepy after eating its planets, our Sun will quieten down and shrink into a long afterlife as a fading white dwarf.

Some five billion years ago our planet started, five billion years from now it will end. In more ways than one, it's high noon for planet Earth.

APPENDICES

APPENDIX A:
ABOUT NAMES, NUMBERS AND UNITS

About *words*. Sometimes words can mean different things or, worse, change meaning depending on where you live or what you do. Take the word 'gas.' In North America, it's the diminutive for gasoline. But in elementary school we were taught that a gas is the third state of matter in the sequence: solid, liquid, gas. So gasoline is not a gas at all. It's a liquid. Moreover, natural gas is also called 'gas,' which compounds confusion. In Britain a less confusing word, 'petrol,' describes what North Americans call gasoline. In Germany it's called *benzin*.

Throughout this book I've tried to use words that won't mislead. That's why you won't find the diminutive 'gas' replacing gasoline. And you'll always find the adjective 'natural' when I'm speaking of natural gas.

I've also used the new word 'hydricity,' which stands for the twin currency pair, hydrogen and electricity. 'Hydricity' is a term dreamed up by the founding Board of General Hydrogen when searching for the best way to describe the business they hoped to develop. Later they trademarked the word. Perhaps because I was a member of the founding board when we had a lot of fun thinking up names, Paul Howard, board chair in 2007, gave me permission to use 'hydricity' whenever I thought it appropriate.

Now to *numbers*. While plodding along behind tour guides during visits to engineering achievements like a bridge or the Grand Coulee Dam, I have found most guides like to strike awe by pronouncing large numbers: X thousand tonnes of concrete or Y hundred miles of wiring. Looking about for a place to sit away from the crowd, I wonder: what does all that mean? Who cares? To me it doesn't mean much unless I can compare the guide's numbers with others from a different bridge or dam. Numbers have their place, especially when comparing options that will allow us to decide what makes sense and what doesn't. During most of *Smelling Land*, especially during the early legs of our voyage, I used few numbers because I wanted to emphasize concepts. Numbers came later when we needed to make choices, because, then, numeracy was essential.

I must also say something about the numbers that are not included. This book does not speculate on how much it will cost to buy a fuelcell car in 2027. Why? How would we have predicted, in 1987, the cost of a laptop computer in 2007? My purpose is to show the patterns about which I can be confident, not to throw out numbers that can only be WAGs.

Finally to *units*. Units throw us a heap of grief whenever we talk about energy systems. Popular usage makes things worse.

Americans think in terms of gallons when they purchase fuel for their car. The rest of the world thinks of liters. This leaves us with two problems. The first is that both gallons and liters are units of volume. We expect (or should expect) the flipping meter to be telling us how much energy (or exergy) we're buying. It isn't. It's telling us how much volume, which would translate to energy only if all liquid fuels had the same volumetric energy density. But fuels don't. Take gasohol, a mixture of gasoline and alcohol. The volumetric energy density of any alcohol, like ethanol, is about two-thirds that of gasoline. So depending on the percentage of alcohol, you're buying a lot less energy when you fill-'er-up with gasohol.

Throughout *Smelling Land*, I've tried to use units that make the reading easy, which often means using familiar (to most readers) names and units, whether these are miles, feet, knots, kilometers, gallons or liters. But when it's important to ensure there is no confusion when comparing options, I've always included Universal System International (SI) units, commonly known as metric units.

The biggest disadvantage to the old Imperial System of units is not the difficulty of becoming familiar with temperature measurements in Celsius rather than Fahrenheit, or volume in liters instead of gallons. The biggest problem is that, working in these units retained from bygone eras—when a foot was the length of some dead king's appendage on the bottom of his leg—makes technical calculations much more difficult.

Therefore using the Imperial System leaves us with at least two *practical* impediments. First, it is a significant barrier to international trade and trade standards. Second, and perhaps a more serious disadvantage, North America faces challenges in maintaining technological competitiveness—in part because, per capita, so many more engineers are graduating in countries like Japan, China, Korea, Germany, France and India. There are many reasons for this. But why add to the burden by requiring North American engineers to be fluent in both the awkward Imperial System and SI units?

APPENDIX B:
MORE LEARNING THE ROPES

In Chapter 2, we began learning the ropes, a never-ending task. So this Appendix expands on some topics we introduced during our voyage and then presents a few new ideas.

B-1 More about Electricity

To understand how electricity delivers power, it helps to compare a current of flowing electric charges with a current of flowing water. Water flows over a dam because the gravitational potential above the dam is higher than below. The maximum energy that can be delivered by this falling water is simply the amount of water multiplied by the height it drops. The maximum energy delivered by electricity is the product of the electronic current measured in amps (analogous to the amount of water), multiplied by the electric potential through which the electrons drop measured in volts (analogous to the height of the dam).

The electricity we use is either alternating current (AC) or direct current (DC). In AC electricity the voltage alternates back and forth from positive-to-negative, at frequencies of 60 Hz (cycles per second) in North America—50Hz in Europe. When we speak of 115V AC electricity, we mean the root-mean-square (rms) value of the voltage is 115V. But the peak voltage swings between +170V and -170V. This is in contrast to DC electricity, where the voltage is constant.

With both AC and DC electricity, the current increases to follow load, while the voltage characteristics are almost unaffected. The current in AC electricity swings in the same cyclical pattern as does the voltage, but with a phase shift.

The advantage of AC is that it allows transformers to *easily* change the 'carrier' voltage—say, from a transmission line voltage of 120kV down to 115V for your living room. High-voltage electricity allows for the more efficient transmission of power across the country. Remember, power (in Watts) is the product of the current (in Amps) times the voltage (in Volts). Thus, as the voltage increases, to deliver the same power the current will increase. Now the major cause of transmission energy wastage (losses) is due to the resistance the wire has pushing current through it. The higher the current, the higher the losses. So, for the same power (energy transmission) if we increase the voltage we can decrease the voltage and thereby decrease the losses. That's why long distance transmission lines operate at high voltages. But you wouldn't want transmission voltages of say 120kV (120,000 Volts) in your home. So for safety reasons, before the power arrives at your home, transformers reduce its voltage.

In Chapter 2, we spoke only of electrons as charge carriers. But there are other charge carriers, like ions and cations. Flowing ions and cations are two more ways of creating electricity. What's more, sometimes, as in microelectronic chips, there may be an *absence* of an electron, which is called a *hole*. Flowing holes also constitute electric currents.

Ions are atoms or molecules stripped of one or more of their electrons, and this leaves them with a net positive charge. Just as a lone electron will search for a positive environment, a positive ion will search for a negative one. Flows of ions constitute the electric currents inside batteries and fuelcells. Cations are atoms or molecules with an extra electron, and so they have a net negative charge.

Electrons dropping into lower energy states produce electromagnetic radiation such as microwaves or sunlight. You can think of electromagnetic radiation as a 'current' of photons (particles of electromagnetic energy). I consider electromagnetic radiation to be part of the extended electricity family —a kind of *ersatz* electricity—because microwaves, radio waves and their kith and kin can do many of the same things as everyday electricity. In particular, they can transport energy and information.

B-2 *The Distinctions among Electric, Hydrogen and Hybrid Vehicles*
I'll report what seems to be common usage.

An 'electric vehicle' is usually thought to be a road vehicle that stores its motive energy *solely* in electric storage batteries. Such a vehicle must spend considerable time refueling (having the batteries recharged) compared with the time it takes for refueling conventional liquid-fueled vehicles. The batteries deliver DC electricity to electric motors that turn the wheels. (I'm excluding diesel locomotives as an electric vehicle, although diesel engines generate electricity which is fed to electric motors on the driving wheels. I'm also leaving out electricity 'fueled' metropolitan-subway systems.)

A 'hydrogen vehicle' is one that stores its motive energy solely as hydrogen. The hydrogen may be carried on board in either compressed or liquefied form. Rarely (and in most cases I think unwisely), the hydrogen can be stored in metal hydrides, or produced on board by a reformation process that converts liquid fuels like methanol into hydrogen. The hydrogen might be used to power a fuelcell that produces DC electricity which, in turn, is delivered to electric motors that rotate the driving wheels. In this case, the hydrogen vehicle is also an electric vehicle, except that compared with a battery-driven one, the hydrogen vehicle can be much lighter and have a much longer range. Alternatively, hydrogen can power an internal combustion engine similar to fossil-fueled cars, buses and trucks.

'Hybrid vehicles' carry fuels like gasoline on board. Typically, the fuel is fed to an internal combustion engine that generates electricity. There is also the possibility of employing regenerative braking for recharging the batteries. The electricity can either directly power electric motors that turn the driving wheels, or be stored in batteries to feed electric motors later. The advantages include smaller internal combustion engines, the prospect of optimized powertrain duty cycles and, usually, less pollution. In my view hybrid vehicles are an interesting, important approach to improving vehicular efficiencies—but they are transitional technology.

B-3 Atoms, Elements, Isotopes and Molecules

All materials consist of atoms, the smallest part of any substance that participates in chemical reactions. A group of atoms that behaves identically in chemical processes are the same element. All atoms of the same element have the same number of protons (positively charged sub-atomic particles) in their nuclei. It is the number of protons that determines the element's chemical behavior. Molecules are tightly bound combinations of elements, like hydrogen and oxygen in water.

All elements, from hydrogen, to oxygen, carbon, neptunium and beyond, have several naturally occurring isotopes. Each isotope exhibits the same chemical behavior as other isotopes of the same element—because each isotope has the same number of protons. But an isotope is distinguished from a different isotope of the same element by having, in its nucleus, a different number of neutrons (uncharged sub-atomic particles with the same mass as a proton).

In high school, we learned that hydrogen had three isotopes each with a single proton, but either zero, one or two neutrons. We also learned that carbon (C), whose chemical properties are defined by its six protons, has several isotopes. The two most common have either six or seven neutrons—corresponding to the isotopes carbon-12 (^{12}C) and carbon-13 (^{13}C).

The numbers 12 and 13 indicate the sum of neutrons and protons—which also identifies the isotope's atomic mass (weight). Throughout *Smelling Land*, I have used two conventions to identify isotopes, either the name of the element followed by a dash with the atomic mass (like carbon-12) or by using a superscript before the element's chemical symbol (like ^{12}C). If the particular isotope is unimportant to the discussion, I have used either the element's name (carbon) or its chemical symbol (C).

The nucleus of the most common hydrogen isotope contains zero neutrons and a single proton. It is identified by the symbol 'H.' The nucleus of the much rarer hydrogen isotope, deuterium, has a single neutron in addition to its proton, and is often given the symbols 'D,' 'hydrogen-2,' or sometimes '^{2}H.' Next is the even rarer (man-made) hydrogen isotope with two neutrons that is called tritium (or hydrogen-3). Oxygen has two important isotopes: the common oxygen-16 with eight neutrons, and the much rarer oxygen-18 with ten neutrons. Clearly an atom of D is heavier than an atom of H, and an atom of oxygen-18 is heavier than an atom of oxygen-16.

Remember that burning any chemical fuel, made up of, say, H and C or any other atom, is a chemical reaction. That means *all* the fuel's carbon and hydrogen atoms—no matter which isotope—can combine with oxygen to release energy. For chemical reactions, the isotope doesn't matter.

B-4 Depleted and enriched uranium

Having just discussed isotopes, we should describe the frequently named but rarely explained descriptors, 'depleted' and 'enriched' uranium. Like all elements, uranium, a metal, has several isotopes, but the two naturally occurring isotopes are uranium-235 and uranium-238. Of these, only the much rarer isotope, uranium-235, can be 'fissioned' (split) to release what we call 'nuclear energy.' The uranium we mine from the ground consists, overwhelmingly, of the non-fissionable isotope uranium-238. Indeed, the isotopic ratio of natural uranium is about 99.3% uranium-238 and 0.7% uranium-235.

While some nuclear power plants, in particular Canada's first-generation CANDU systems, use natural uranium as a fuel, most cannot. The fuel for most power reactors is uranium that's been 'enriched' by increasing the uranium-235 content to somewhere around 3%. In contrast, weapons-grade uranium is enriched to well above 90%. That's just one reason a nuclear power plant cannot have a nuclear explosion—power plants use much different fuels than nuclear bombs.

Technically, it's very difficult to separate the uranium isotopes in uranium ore. In contrast, it is comparatively easily to separate chemical elements from molecules, by using chemical or electrochemical reactions. (*Smelling Land* is full of talk about separating the hydrogen and oxygen in water.) But separating isotopes of the same element is always a very much tougher job. That is why today's news is full of talk about centrifuges, which exploit the *extremely* small mass differences between uranium-235 and uranium-238 isotopes in order to separate them using centrifugal-centripetal forces. It is a time-consuming and expensive process.

'Enriched' uranium has a higher content of uranium-235 than occurs naturally in the ore. 'Depleted' uranium is what's left behind after the uranium-235 has been extracted. Therefore, depleted uranium has *much less* than the 0.7% uranium-235 that occurs naturally in the ore and, therefore, almost zero radio-toxicity.

B-5 What about all those '-sphere' words?

'Atmosphere' and 'biosphere' have entered everyday language. Other 'spheres' are less common. All these spheres can be understood from the prefix in the composite word. For convenience I'll list a few—most of which appear in the body of *Smelling Land:*

Atmosphere: All the 'air' that envelopes Earth.

Stratosphere: That portion of the atmosphere at altitudes between about 10-50 kilometers (with a slightly lower bottom boundary at the poles). The stratosphere has higher temperatures at the top (positive thermal gradient upwards), which distinguishes it from the *troposphere*, which lies beneath the stratosphere and has the opposite thermal gradient, becoming colder as altitude increases. As a consequence of the positive thermal gradient, the stratosphere is much more stable than the *troposphere*—which is one reason why long-range commercial aircraft like to fly close to 10,000 meters. The *mesosphere* is above the stratosphere.

Hydrosphere: All the water that envelopes Earth. Overwhelmingly, this water is in the oceans. But the hydrosphere also includes rivers, glaciers, lakes and atmospheric moisture.

Cryosphere: As you might expect from the prefix, the cryosphere is the frozen bits of water on Earth's surface. It includes snow, floating sea and lake ice, glaciers, polar ice caps and, I suppose, permafrost. I think of the cryosphere as a component of the hydrosphere—so as the cryosphere shrinks, the liquid and vapor portions of the hydrosphere grow.

Lithosphere: The prefix 'litho' comes from the Greek word for 'rock.' It is the outermost rocky shell of our planet. It includes what we call Earth's crust and some of the upper layers of the mantle. Its distinguishing feature is its motion rather than its material—for it participates in the movement of tectonic plates. Under the oceans, it can be as thin as 1.6 kilometers and, under the continents, as thick as 150 kilometers.

Biosphere: This is a tricky one. From the broadest geophysiological viewpoint, it includes all parts of Earth's outer shell where life occurs—the land, ocean and rocks—and, of course, life itself. I'm sympathetic to this view, in part because it seems entirely consistent with the Gaia view of planetary life, which I consider a valuable optic. However, my use of the descriptor 'biosphere' in *Smelling Land* is more limited, meaning the totality of living things, not including their habitat. For me this was a tough choice, done more for the convenience of putting things in boxes, rather than from the more intellectually holistic— and probably more legitimate—viewpoint encompassed by the Gaia model of Earth.

B-6 What about all those '-genic' words?

The word *anthropogenic* means 'created by humans.' It is a composite word squeezed together from anthro-(which means 'pertaining to human' as in 'anthropology') and -genic (which means 'source of' as in *Genesis*). When discussing greenhouse gases, we often need this word to distinguish between the atmospheric CO_2 that comes from civilization's emissions and CO_2 from natural sources.

Abiogenic, is a three-part composite word. The prefix 'A-' means 'not;' while 'bio' means 'biological.' We already know about -genic. So 'abiogenic' means it does not come from a biological source. It's used in *Smelling Land* to describe methane not produced by biological processes. There is a lot of abiogenic methane on other planets and the occasional fly-by comet whose main constituent can be frozen methane.

B-7 The meaning of 'albedo'

When radiation (let's say sunlight) strikes the surface of any material, some of the radiative energy is reflected or scattered away, while the remainder is absorbed by the material. Albedo is the *ratio* of the reflected and scattered radiation to the total incoming radiation. An albedo of unity means that *all* incoming radiation is reflected or scattered and none is absorbed by the material. An albedo of zero means all the radiative energy is absorbed. Albedo is sometimes expressed as a percentage: 100% means all the incident radiation is reflected or scattered, while 0% means all is absorbed. Albedo depends on the radiation's frequency, that is, its wavelength.

To help me remember whether a high albedo reflects or absorbs, I recall that the word is Latin for 'white'—which is clearly related to the English word albino (white animals). On our planet freshly fallen snow has an albedo of about 90%, whereas the average albedo of Earth is about 30%. The albedo of dense swampland lies between 9 and 14%. The oceans' can be as low as 3.5%.

Of course the albedo also depends on the radiation's angle of incidence and, most importantly, on the radiation's frequency—which gives rise to blue skies and rainbows.

B-8 The link between entropy growth and exergy destruction

As you worked your way through the chapters on entropy and exergy, sooner or later you will have guessed that entropy production and exergy consumption are linked. They are. The linkage is described in this simple relationship:

> Exergy destruction equals the product of entropy production multiplied by the environmental temperature.

The clear-cut relation between entropy production and exergy destruction can be encapsulated by this equation:

$$\dot{I} = T_e \dot{\Pi}$$

where $\dot{I}$ is the rate of exergy consumption, $\dot{\Pi}$ is the entropy production rate and T_e is the environmental temperature as measured on an absolute scale. Because exergy consumption is a measure of irreversibility, it's been assigned the symbol I.[*] If we want to describe exergy consumption in terms of entropy production, we can tip over the previous equation so it becomes

$$\dot{\Pi} = \dot{I}/T_e$$

This equation shows entropy production is equal to exergy consumption divided by the environmental temperature. The linkage between entropy and exergy is a great tool for finding the root cause of systemic inefficiencies and for designing efficient technologies.

---

[*] The symbol I (with out dot atop) means irreversibility, or an amount of exergy destroyed, while $\dot{I}$ (with dot atop) means a rate of exergy destruction.

B-9 The Value of Planting a Tree

On August 25, 1992, the *Globe and Mail* reported that the National Community Tree Foundation (NCTF) had claimed that by planting just 12 trees, each Canadian would offset the approximately 2.3 tons of carbon each of us releases to the atmosphere annually. To emphasize how engineering training can help them comment on social issues, I asked my students to calculate how far an average car could drive if all the carbon emitted was absorbed by an average tree.

They jumped into the assignment, but realized they needed to check some facts. They discovered, from Statistics Canada data, that each of us is responsible for emitting more than 4.4 tons of carbon annually, almost twice the 2.3 tons claimed by the NCTF. They also needed to resolve such questions as, is the tree a seedling, mature or geriatric? And where does the tree live?

For an average tree, the average car could drive about 37 km (23 miles) per year. If the tree lives on the North American Pacific Coast, where trees are larger than in most of the world, it lets you drive another 190 km (118 miles) per year without feeling guilty.

The students then put this result together with the data from a 1990 *Science* article, which showed that a one-hectare forest of 60-year old Douglas firs and hemlock would absorb about 4.4 tons of carbon each year. Thus, to assuage our carbon-emission guilt, *each* of us would have to plant and tend almost 1,000 of these trees—until they reached their maximum carbon-eating age of some 60 years. Of course, later, when the trees die, they will put back into the atmosphere almost all the CO_2 they absorbed during their life.

The students concluded that planting a tree in your backyard may be a good idea. As it grows, the tree will provide shade and a home for squirrels and birds. But for balancing your greenhouse gas guilt-quotient 12 backyard trees is a non-starter.

Appendix C:
Some Bits and Pieces

C-1: On Fairness

While following the coverage of the December 2007 United Nations Climate Change Conference held in Bali, I was dismayed by how little was said about solutions—except for some sessions on coping strategies. I found little reported in newspapers or other media on the potential role for nuclear power. And certainly I saw nothing on the essential role of hydrogen, which could allow non-carbon sources to enter transportation markets. Indeed, any substantive discussion of technologies seemed conspicuously absent.

This caused me to write a rather bitter email to several of my policy-guru friends, one of whom responded: 'David, you must realize this conference was not about solutions, it was about 'fairness'—fairness on whether developing nations should be expected to set emission reduction targets equivalent to those set for developed nations.'

Over the next few days I thought about fairness. But my thoughts drifted from allocating CO_2 emission targets to thinking about a fairness issue that all of us must face, even with a fast transition to the Hydrogen Age *if* it turns out not to be fast enough.

If it's not fast enough, all of us will be hurtling towards climate upheaval. We can't dismiss this prospect. So it seems that to have a chance of delivering future fairness we must, today, begin to develop international protocols that anticipate the displacement and migration of millions whose homelands might be drowned, or otherwise destroyed. Where will these people go? How will they be integrated within less damaged nations and cultures? Which nations will use their wealth and wisdom to help? Which will use their military as a bulwark to fend off those seeking new homes?

This is why I'd like to see the United Nations establish a date and venue where all nations would come together to address this issue—a kind of daughter of Bali, a granddaughter of Rio. If you think it is difficult to agree on a fair allocation of emission reduction targets, just wait; we haven't seen anything yet.

As challenging as planning for this diaspora would be today, the difficulty will be greatly magnified as the need becomes more certain and the timing closer—when millions of south-sea islanders, Bangladeshis and probably Netherlanders are massing on the borders of countries that have most of their territory well above sea level. I suppose south Floridians may want to keep their feet dry too.

C-2: On Fuelcell Market Strategies

In the spring of 2008, General Motors and Honda announced they would be consumer-testing small fleets of hydrogen-fueled automobiles. They will use fuelcell powerplants. This is impressive, because I think private automobiles are probably the most difficult market for fuelcells to crack.

Chapter 36 'Fuelcells: Chip of the Future,' placed fuelcells within their *systemic* role. It's a role that will mirror the role of microelectronic chips, in that both will always be *imbedded* technologies *within* service technologies. Just as microelectronic chips found their way into a plethora of service technologies, so will fuelcells. But today many people think the prime application for fuelcells is as a powerplant for road transportation, particularly cars. I consider this perception unfortunate.

When a new, higher capital-cost technology is introduced, it is prudent to first target those applications that can operate on a nearly continuous basis. This allows their capital cost disadvantage to be offset by operating advantages—not just cost, but often environmental gentility and better service—over substantial in-service times.* But private automobiles spend most of their lives parked. So if we target road applications, it makes more sense to think of urban fleets of buses, trucks or couriers, which can be on the road many hours each day.

For fuelcell vehicles it would also be advantageous if the vehicle provided large physical platforms for powertrain installations—not just for the fuelcell

---

* Electric utilities refer to this as a high 'capacity factor.'

itself but also for easy physical access to the propulsion system. This is a major reason I suggested commuter-shed rail locomotives in chapter 44.

Both urban road fleets and certainly commuter rail can benefit from large central refueling depots. The private automobile can't.

GM and Honda may protest my thoughts by responding something like: 'You may be right, but we'll gain experience with these consumer-testing fleets and expand public exposure.' My concern is that, after such exposure, it will probably be many years until you'll find fuelcell vehicles running about town and to the cottage. Then people will observe: 'I guess they weren't so promising after all'—and fuelcell automobiles may come to be perceived like the electric-battery powered vehicles now relegated to the dustbin of failed technologies.*

So when reflecting on the promise of early-adopter fuelcells niches, I suggest this: *think big* (like locomotives) or *think small* (like mobile phones and laptop computers), but avoid the mid-size (like private automobiles)—at least until we've learned more from the big and the small.

C-3: On Blue-boxes for Spent-Fuel

These days, most of us fill our 'We recycle' blue boxes with cans, plastic bottles and old newspapers—all of which, we hope, will get remanufactured as new products. Although faintly amused by the sanctimonious, smug 'We recycle' printed across these blue boxes, I consider this cultural trend entirely appropriate.

But if recycling is correct for plastic, aluminum and newsprint, why is it seldom considered for spent nuclear fuel? As we pointed out in Chapter 31 'You've Got to be Carefully Taught—Know Nukes,' the spent-fuel from once-through nuclear powerplants, the standard in North America, still contains almost 100 times as much exergy as was harvested in the first-pass through the reactor. Even today, we could take the spent fuel from American light-water moderated reactors—which contains about 0.9% U-235—and use it to fuel Canadian heavy-water CANDU reactors. Natural uranium, which contains

---

* In the case of all-battery EVs, the technological components didn't fail. Rather the systemics— which had been there from the beginning for those who thought systemically—were simply an insurmountable barrier. Chapter 38 explains these fundamental barriers.

0.7% U-235, is the staple fuel for CANDU. These Canadian reactors could become the first level of recycling for US light-water reactor spent fuel.

If there was ever a recycling opportunity, this is it. But instead of using spent nuclear fuel again and again to extract its full potential exergy, we say go-away-bad-dream and make plans to bury it under a mountain. Stupid!

C-4: On Engineering, Culture and the Abolition of Slavery

Most of us appreciate the tight links between the quality of our lives and technology. We know that over the last 200 years, human life spans have increased from about 38 years to now more than 80. Many contributed to the fight against infection—Florence Nightingale and Dr. Lister, for instance. And there were those who achieved breakthroughs in medical science that told us more about how we tick. Yet for longevity, perhaps engineers gave us the most significant advances—when they took sewage running along open gutters and pushed it underground into sewers. In so doing, they built what today we'd call a 'firewall' between drinking water and sewage.

Another example of the intimate links between quality of life and technology is the tie between the invention of steam engines and the abolition of slavery. It's common to think that abolition resulted from cultural maturing. But what allowed culture to mature?

In chapter 38 'Fuelcells: Chip of the Future,' I recalled James Watt's reflection 'It was on the Green of Glasgow. I had gone for a walk on a Sabbath afternoon. I was thinking upon the engine.' Watt's walk marked the inauguration of steam engines and their impact on civilization.

If the setting is appropriate, I like to remind folks that, before steam engines, every drop of work (lifting, plowing, digging and so on) could only be done by wind, falling water, or muscle.* But after steam engines, heat produced from *anything*—by burning wood, coal, oil, or even fissioning uranium—could be converted into work.

Can we get our heads around the significance?

---

* The muscle needn't belong to humans; often it belonged to an animal.

Steam engines and their progeny freed people from drudgery; fewer people were needed to pick cotton, lift stones, or dig canals and ditches. Many more people could write symphonies and play them, pen books and read them, or design and build highways and automobiles. And many more could research the causes of disease and deliver health care.

Until the steam engine, almost all great civilizations were built upon slavery—from the Egyptians, to the Romans, to the British. America started that way too. But then along came steam engines, followed by their kith and kin, like internal combustion engines.

So after we had technologies that could convert heat to work, what big tasks were there left for slaves to do?*

In America it's often said, 'Lincoln freed the slaves.' Lincoln played an important role. But I think he couldn't have proclaimed emancipation were he not standing on the shoulders of James Watt.

---

* In William Styron's *The Confession of Nat Turner*[64], a historical novel about a Virginia slave uprising that killed 60 whites (and perhaps 200 Blacks as a consequence), Nat Turner overhears a conversation among landowners. One of the white folks raises the question of when slaves could be freed. Another guest says it would be impossible, unless, he chuckles, machines are invented that can rig the horses to the buggies and then take the reins to drive the buggies into town. This is a fictional wise man—but not wise enough to foresee a horseless carriage.

APPRECIATIONS

I'm overwhelmed with appreciation for the many people who have contributed to what appears on the pages of both the First and Enhanced Editions. What strikes me is that so many busy people, many with demanding careers, took innumerable hours to carefully read and critique draft sections. As every author says, 'the errors are mine.' But I must add, without the input from these readers the errors would be many more.

The book-writing process began when Dr. T. Nejat Veziroglu, editor-in-chief of the *International Journal of Hydrogen Energy*, invited me to write a series of articles for the *Journal* that, he wrote in his introduction to the series, would provide a '. . . package, presenting new concepts important to understanding our energy system in general and the hydrogen energy system in particular.' Forty-three such articles were published between 1994 and 2004. They served as the feedstock for this book. I therefore thank Dr. Veziroglu and Elsevier Press for this kick-start. Without this support, I would never have had the perseverance to write *Smelling Land*. In this same context, I also thank my wife, Marianne, and my daughter, Sue, for months of editorial effort and wisdom—which finally pushed both editions out the door. I could not have done it without them.

It's impossible to group, in tight categories, the nature of the contributions made by so many. All technical contributors added literary insights—some quite wonderful—and all literary contributors helped clarify technological issues. Nevertheless, I've placed the names of all these generous people within two, albeit overlapping, groups. It pains me to be unable to say more about individual contributions.

For *comments on science and engineering:* Addison Bain, Ben Ball, Geoff Ballard, Daniel Brewer, James Cannon, Phillip Cockshutt, Peter Coy, Jerry Cuttler, Andrew Dicks, Ned Djilali, Zuomin Dong, Romney Duffey, Bill Escher, Matthew Fairlie, Mike Flaherty, Doug Fletcher, Neil Fricker, Aaron Fyke, Barry Glickman, Gary Gurbin, Wolf Häfele, Martin Hammerli, Ken Hare, Willis Hawkins, John Hayward, Peter Hoffmann, Jerry Hopwood, Michael Ivanco, Jon Jennekens, Murray Love, Cesare Marchetti, Dan Meneley, Sam McGregor,

Ged McLean, Walter Merida, Katrina Messiner, Alistair Miller, Art Mountain, Tokio Ohta, Niel Pearce, Mark Petrie, Keith Prater, Terry Rogers, Hans-Holger Rogner, Marc Rosen, Andrew Rowe, Ed Schmidt, Douglas Scott, Karina Shaw, Ken Shultz, Ernest Siddel, John Stone, Sandy Stuart, Harry Swain, Bez Tabarrok, Maarten van Emden, Kevin Waher, Patricia Wall, Alan Walter, Leon Walters, Andrew Weaver, Jeremy Whitlock, Jim Wilcox, Carl-Jochen Winter, and Sean Wright.

For suggestions on structure and style: Nick Beck, Dawson Brenner, Sherry Brown, Vern Burkhardt, Anne Burkhardt, Carie Costin-Woolford, Katherine Gibson, Richard Handler, Stan Hartfelder, Phil Hughes, Nat and Leslie Klein, Max Mallinick, Georgina Montgomery, Jim Munro, David Price, Peter Reilly-Roe, Ed Schreyer, Carolyn Tabarrok, Tabitha Takeda, Tony Taylor, Annette Toth, Penelope Lee Scott, Peter Scott, Don Swagar, Harry Swain, Ron Venter, Peter Vivian, Sue Walton and Peter Ward.

REFERENCES

1. Lee, T. et al. *Energy Aftermath*. Harvard Business School Press, Cambridge, MA, 1990.

2. Galbraith, John Kenneth. *The Affluent Society*. Houghton Mifflin Company, New York, NY, 1958.

3. Karney, B. W., Ivan, M., and Waher, K.J. 'To Switch, or Not to Switch: A Critical Analysis of Canada's Ban on Incandescent Light Bulbs.' *Proceedings of the IEEE Electrical Power Conference on Renewable and Alternative Energy Sources* (IEEE EPC07), 2007, Quebec, Canada. If you'd like a more widely accessible publication which is in preparation, I suggest you contact Kevin Waher, the graduate student who performed most of the analysis, kevin.waher@gmail.com.

4. Pirsig, R. *Zen and the Art of Motorcycle Maintenance*. William Morrow & Company, New York, NY, 1974.

5. *Newsweek*, Vol. 93, p.100, April 16, 1979.

6. Yergin, Daniel. *The Prize: The Epic Quest for Oil, Money and Power*. Simon and Schuster, New York, NY, 1991.

7. *United States Statutes at Large* (1978, Volume 92, Part 3). The relevant law is titled the 'Powerplant and Industrial Fuels Use Act of 1978.'

8. Joseph, Lawrence E. *Gaia: The Growth of an Idea*. St. Martin's Press.New York, NY, 1990.

9. Clarke, A. C. *The Exploration of Space*. Harper & Brothers, New York, NY, 1951.

10. Brock, E.J. 'Tiny Bubbles Tell All.' *Science*. Vol. 310, November 25, 2005. (This article discusses the correlation between the Vostok and Dome C data.)

11. Petit, J.R. et al. 'Climate and atmospheric history of the past 240,000 years from the Vostok ice core, Antarctica.' *Nature*, 399, 429-436.

12. Siegenthaler, U., et al. 'Stable Carbon Cycle-Climate Relationship During the Late Pleistocene.' *Science*, Vol. 310, 25 November 2005.

13. Petit, J.R. et al. 'Historical Isotropic Temperature Record from the Vostok Ice Core'. Carbon Dioxide Information Analysis Center, 1999, Oak Ridge National Laboratory, U.S, Department of Energy, Oak Ridge, TN.

14. Barnett, T. P. et al. 'Penetration of Human-Induced Warming into the World's Oceans.' *Science*, Vol. 309, 8 July 2005.

15. Hansen, B., et al. 'Already the Day After Tomorrow?' *Science*, Vol. 305, pps. 953-954, 13 August 2004.

16. Tannen, Deborah. *The Argument Culture: Moving from Debate to Dialogue*. Random House, New York, NY, 1998.

17. Ramstorf, Stefan. 'The Thermohaline Ocean Circulation - A Brief Fact Sheet.' www.pik-potsdam.de/~stefan/thc_fact_sheet.html.

18. Hansen, B. and Turrell, W. R. 'Decreasing overflow from the Nordic seas into the Atlantic Ocean through the Faroe Bank channel since 1950.' *Nature*, Vol. 411, pps. 927-930, 21 June 2001.

19. Ocean acidification due to increasing atmospheric carbon dioxide.' *Policy Document 12/05*, The Royal Society. www.royalsoc.ac.uk.

20. Wright, Ronald. *A Short History of Progress*. House of Anansi Press Inc. Toronto, 2004.

21. Häfele, W. et al. *Energy in a Finite World: A Global Systems Analysis*.Ballinger Publishing Company, Cambridge, MA, 1981.

22. Clark W. G. and Munn R. E. (ed.). *Sustainable Development of the Biosphere*. Cambridge U. Press, Cambridge, UK, 1987. (An excellent discussion of technological monocultures and other aspects of institutional surprise written by Harvey Brooks can be found in Chapter 11.)

23. Bruntland, Gro, editor. *Our Common Future: The World Commission on Environment and Development*. Oxford U Press, 1987.

24. Moran, M. J. and Shapiro, H. N. *Fundamentals of Engineering Thermo-dynamics*. John Wiley and Sons, New York, NY, 1988.

25. Yergin, Daniel. *The Prize: The Epic Quest for Oil, Money and Power.*Simon & Schuster, New York, NY, 1991.

26. Clark W. G. and Munn R. E. (ed.). *Sustainable Development of the Biosphere.* Cambridge U. Press, Cambridge, UK, 1987.

27. Geisel, Theodore S. and Audrey S. Geisel (better known as Dr. Seuss). *One Fish Two Fish Red Fish Blue Fish.* New York: Beginner Books, Inc., 1960.

28. Wilson, Edward O. *Consilience: The Unity of Knowledge.* Alfred Knopf, New York, NY, 1998, p. 4.

29. Zinsser, William. *On Writing Well.* Harper Collins, New York, NY, 1994.

30. Grobstein, Clifford. *The Strategy of Life.* W.H. Freeman and Company, New York, NY, 1964.

31. Bejan, A. *Advanced Engineering Thermodynamics.* John Wiley & Sons, New York, NY, 1988.

32. Schrödinger, Erwin. *What is Life?* Cambridge University Press, Cambridge, MA, 1944.

33. Margulis, Lynn, & Sagan, Dorion. *What is Life?* Simon and Schuster, New York, NY, 1995.

34. Lovelock, James. *The Ages of Gaia, A Biography of Our Living Earth.* Norton, New York, NY, 1988.

35. Coveney, Peter, Highfield, Roger. *The Arrow of Time.* Flamingo Press, London,1991.

36. Greene, Brian. *The Fabric of the Cosmos.* Vintage Books, New York, NY, 2005.

37. Rifkin, J. (with Howard, T.) *Entropy: A New World View.* Bantam, New York, NY, 1981.

38. Thomas, Lewis. *Lives of a Cell.* Penguin Books, New York, NY, 1979. (Floating free...' is one of Thomas's magnificent phrases.)

39. *World Energy Assessment and the Challenges of Sustainability.* United Nations Development Programme, 2000.

40. Freese, B. *Coal: A Human History.* Perseus Publishing, Cambridge, MA, 2003.

41. *BP Statistical Review of World Energy.* www.bp.com/.

42. Rogner, Hans-Holger. 'An Assessment of World Hydrocarbon Resources.' *Annual Review Energy & Environment,* 22, pps. 217-62, 1997. (In particular, see p. 259 and Figure 8.)

43. Diamond, J. *Collapse.* Viking Press, New York, NY, 2005.

44. Galbraith, John Kenneth. *Name Dropping.* Houghton-Mifflin Co., New York, NY, 1999.

45. Coll, S. *Ghost Wars: The Secret History of the CIA, Afghanistan, and Bin Laden, from the Soviet Invasion to September 10, 2001.* Penguin Press, New York, NY, 2004).

46. Cohen, Bernard L. 'Test of the Linear-No Threshold, Theory of Radiation Carcinogenesis of Inhaled Radon Decay Products.' *Health Physics,* Vol. 68. No. 2, 1995.

47. Hiserodt, Ed. *Under Exposed: What if Radiation was Actually Good for You?* Laissez Faire Books, Little Rock, AR, 2005.

48. Chen W. L. et al, 'Effects of Cobalt-60 exposure on Health of Taiwan Residents Suggest New Approach Needed in Radiation Protection.' *International Hormesis Society* (formerly Nonlinearity in Biology, Toxicology and Medicine) Dose-Response 5:63-75 2007.

49. Reynolds, Albert B. *Bluebells and Nuclear Energy.* Medical Physics Pub Corp. Madison, WI. 1996

50. Orwell, George. *Animal Farm.* Random House, New York, NY, 1993.

51. Sperling, D. and Cannon, J. S. *The Hydrogen Energy Transition: Moving Toward the Post Petroleum Age in Transportation.* Proceeding of the IX Biennial Asilomar Conference on Transportation and Energy, Institute of Transportation Studies, University of California, Davis, Elsevier Academic Press, Burlington, MA 2004.

52. Bain, Addison. *The Freedom Element, Living With Hydrogen.* Blue Note Books, FL. 2004.

53. Häfele, W. et al. 'Novel Integrated Energy Systems: The Case for Zero Emissions.' Chapter 6, *Sustainable Development of the Biosphere,* Eds. Clark and Munn, Cambridge U. Press, 1987.

54. Scott, D.S., and W. Häfele. 'The Coming Hydrogen Age: Preventing World Climatic Disruption.' *International Journal of Hydrogen Energy,* Vol. 15, No. 10, pp. 727-737, 1990.

55. Scott, David, et al. *Hydrogen, National Mission for Canada,* Report of the Advisory Group on Hydrogen Opportunities. Ministry of Supply and Services, Canada. Cat. No. M27-86/1987E, June 1987.

56. Koplow, D. and Martin A. *Fueling Global Warming: Federal Subsidies to Oil in the United States.* Industrial Economics Inc., Cambridge, MA, 1998.

57. Häfele, W. et al. *Energy in a Finite World: A Global Systems Analysis.*Ballinger Publishing Company, Cambridge, MA, 1981.

58. Scott, D. S. 'Inside Fuelcells.' *International Journal of Hydrogen Energy,* Vol. 29, pps. 1203-1211, 2004.

59. Fyke A. et al, 'Recovery of Thermomechanical Exergy from Cryofuels.' *International Journal of Hydrogen Energy,* Vol. 22, April 1997.

60. Brewer, G. Daniel. *Hydrogen Aircraft Technology.* CRC Press Inc., Boca Raton, FL. 1991.

61. Hoffmann, P. *Tomorrow's Energy: Hydrogen, Fuel Cells, and the Prospects for a Cleaner Planet.* MIT Press, Cambridge, MA, 2001.

62. Mishima, Yukio. *The Temple of Dawn.* Translated from the Japanese by Saunders and Seigle, Alfred A. Knopf 1973. Charles E. Tuttle, 2nd edition 1976. Page 21.

63. Verne, Jules. *The Mysterious Island.* Translated from the French. Paris: Pierre-Jules Hetzel, 1874.

64. Styron, William, *The Confessions of Nat Turner.* Random House, New York, NY, 1966.

INDEX

ABOUT THE AUTHOR

Engineer, scientist and environmentalist, David Sanborn Scott has been active in energy system analysis, design and strategies for more than three decades. He is vice-president (for the Americas) of the International Association for Hydrogen Energy.

After receiving his doctorate in mechanical engineering and astronautical sciences from Northwestern University, he joined the University of Toronto where he subsequently served as chair of Mechanical Engineering and founded the university's Institute for Hydrogen Systems. He later joined the University of Victoria to found the Institute for Integrated Energy Systems. The institute is engaged in developing fuelcells, hydrogen production and storage technologies, and in evaluating feasible ways of integrating renewable sources. Scott chaired Canada's Task Force on Hydrogen Opportunities, which produced *Hydrogen: National Mission for Canada,* a report now internationally recognized for its prescience.

Scott consults for US and Canadian corporations and national laboratories. In 2006, he became the first Canadian awarded the internationally prestigious Jules Verne Award, for 'Outstanding Contributions to Hydrogen Physics, and Hydrogen Energy Sociology and Philosophy.' A year later, he received the inaugural D.Sc. (honorary) from the University of Ontario Institute of Technology.